AF522152

ENCYCLOPEDIA OF ORGANIC CHEMISTRY

ENCYCLOPEDIA OF ORGANIC CHEMISTRY

Vol. 1

By

Sananda Chatterjee

DAE, DCA, CMCNet

Scientific Consultant

Institute for Natural Sciences

Kolkata

(West Bengal)

DISCOVERY PUBLISHING HOUSE PVT. LTD.

NEW DELHI-110 002

Published by:
Tilak Wasan
DISCOVERY PUBLISHING HOUSE PVT. LTD.
4383/4B, Ansari Road, Darya Ganj
New Delhi-110 002 (India)
Phone : +91-11-23279245, 43596064-65
Fax : +91-11-23253475
E-mail : parul.wasan@gmail.com
discoverypublishinghouse@gmail.com
web : www.discoverypublishinggroup.com

***First Edition:* 2012**

ISBN: 978-81-8356-875-3 (Set)

Encyclopedia of Organic Chemistry

Printed at:
Shree Balaji Art Press
Delhi

Preface

Organic Chemistry is that branch of chemistry, which deals with the reactions related to the life form, or one of the source of the life form, carbon (we can say the end of it).

This carbon mainly conjugates with different elements such as nitrogen, hydrogen and oxygen sometimes metals and other non-metals also). Organic chemistry studies mainly the extraction, structure, properties, of these compounds.

In this book chapters deals with the different methods of purification, extraction of compounds, some processes were used and devised by Alchemists.

The extraction preservation of natural dyes are examples of alchemical work, these been discussed in the chapter "dyes indicators and pigments".

Dyes is the mal-pronunciation of the damathi word dyum meaning colour, it was, the ancient Egyptians who use different natural pigments to extract different colours.

In this book the organic compounds has been mainly classified into two broad groups, the aromatic and the aliphatic compounds. The extraction of alcohol is definitely an alchemical process'.

The destructive distillation of wood is one of the finest processes devised by the alchemists for the preparation of methanol is one of the important topics of this book.

Looking ahead in to the modem world, the world of future, the subject comes in vision is the biochemistry, the base of which is organic chemistry, the chapter containing carbohydrates, and the chapter, chemical compounds of biological and biochemical interest is the best element of focus for the students stepping foot in the advanced staircase of biochemistry.

The last chapter, chapter 41, discusses the different organic reactions and their suitably mechanisms, this will help those students who are very much afraid of organic formulas and reactions. This chapter also deals with the chirality one of the interest of Dr. Palash Gangopadhyay (author's friend).

Organic Chemistry is not a mere subject, it is the rhythm of life tuned in the essence of carbon, which looks black but carries the dazzleness of diamond in its heart.

SANANDA CHATTERJEE

Contents

CHAPTER

1

Introduction

Organic chemistry is that part of the branch of chemistry which deals with the compounds of carbon. As early as 1675 Nicolas Lemery classified chemical substances into three main groups. According to origin from the three main group - according to their origin from the three kingdoms of nature – Vegetable, Minerals and Animals. Compounds such as sugar starch, fats, oils, gums, resins, fruits, etc, which are obtained from animals and plants, were called organic compounds and were nomenclated separately from inorganic chemistry or mineral substance like metals, alum, common salt, nitre, etc.

Lavoisier (1743-1794) did study many organic compounds and their compositions and deputed that they are all made up of carbon, oxygen, hydrogen and nitrogen.

In view of complex nature of organic compounds Berzzelius (1779-1884), the great Swedish chemist pointed out that they could be produced only under the influence of a mysterious force of nature, this force was called vital force and the theory is still remembered after his name Berzzelius vital force theory. It was also shown that the formation of such substances were regulated by Laws which were not applicable to the formation of organic or mineral substances and consequently it was believed that organic substance could not be prepared artificially or synthetically in the laboratory. The idea of vital forces remained in prominence for some years to discriminate organic and inorganic chemistry.

Wöhler, a German chemist in 1828 accidentally prepared Urea – a constituent of animal urine and animal waste from the ammonium cyanate. Which truly in turn could be obtained from the two organic substances potassium cyanide and ammonium chloride.

Ammonium cyanate was first formed and under the influence of heat changed into urea due to the rearrangement of the atoms in the molecule.

The belief on the vital force theory was thus shaken off and the barrier between organic and inorganic chemistry was finally swept away in 1845 when Koble synthesized acetic acid from inorganic material. Dumas, Kekule and others also succeeded in preparing a large number of organic compounds in their !aboratories from inorganic series.

Hence, it was thought that there was no such fundamental difference between organic and inorganic compounds and the same laws of chemical combustion govern their properties.

As a matter of convenience however, the term Organic Chemistry was retained. At present, therefore organic chemistry is regarded as the "***chemistry of carbons***" though some compounds such as carbonates, bi carbonates carbon dioxide, carbon monoxide carbon di-sulphide and carbides which also contain carbon are included amongst the inorganic compounds as they are common to the mineral world.

Reactions for treating organic chemistry are a separate branch of chemistry.

Organic chemistry is treated as a distinct and separate branch of chemistry because of following reasons:

1. **Vast number of known organic compounds:** The position of the carbon atom in the periodic table is unique an thus it possess an unusual degree, a tendency to share a pair of electrons with other carbon atoms and so by doing, it is capable of forming long chains and rings of the type shown below:

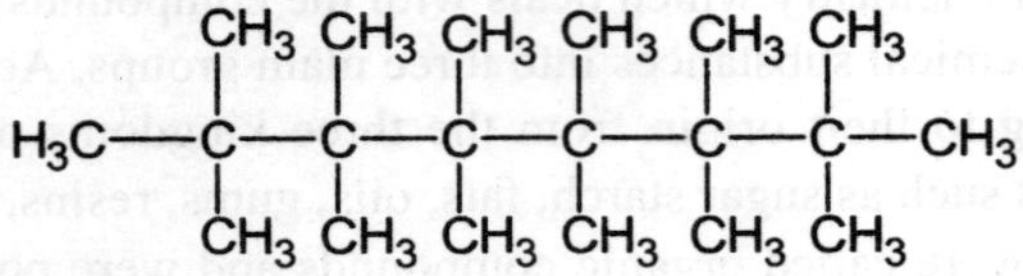

straight chain

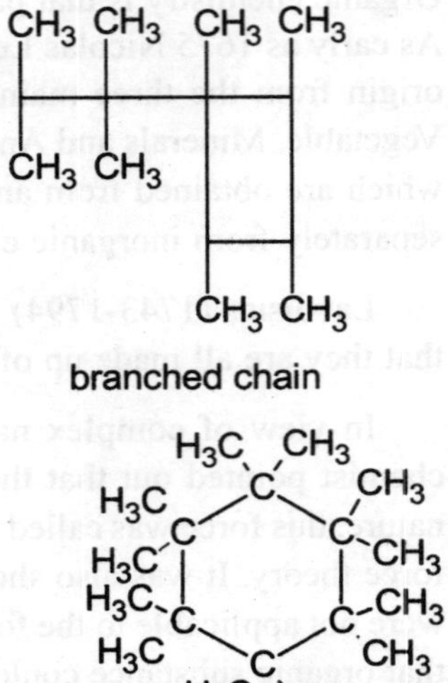

branched chain

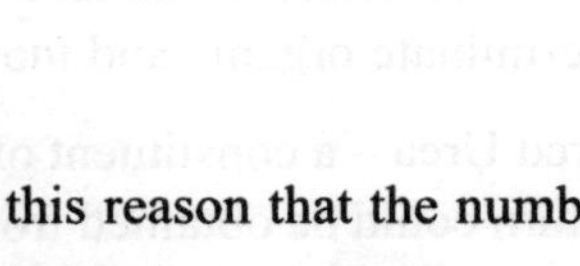

closed chain

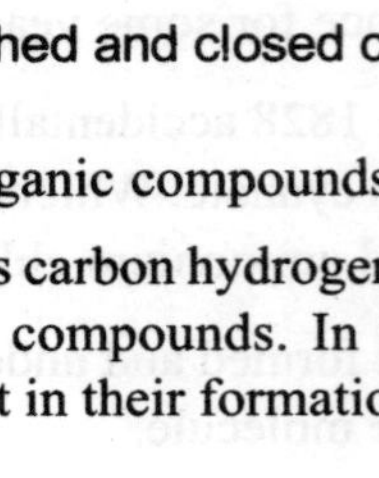

branched and closed chain

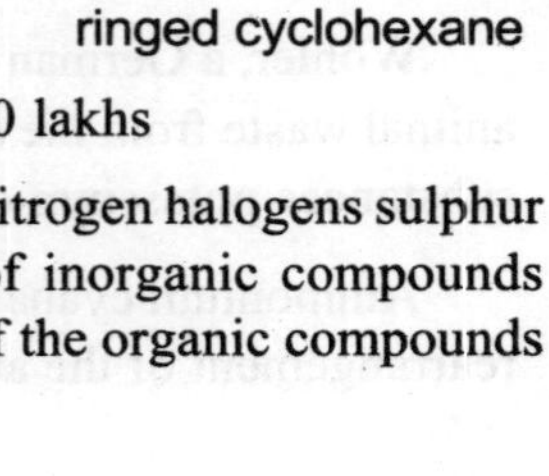

ringed cyclohexane

It is for this reason that the number of organic compounds is over 10 lakhs

2. **Composition**: Only a few elements such as carbon hydrogen oxygen nitrogen halogens sulphur and phosphorous are present in organic compounds. In the case of inorganic compounds almost all elements (over 92) can take part in their formation. Most of the organic compounds very complex.

3. **Homology:** Similar organic compounds with identical characteristics groups have been placed together into various families. These families are called homologous series and form a distinctive feature of organic chemistry, for e.g.

Paraffin	Alcohol
Methane CH_4	Methyl Alcohol CH_3OH
Ethane C_2H_6	Ethyl Alcohol C_2H_5OH
Propane C_3H_8	Propyl Alcohol C_3H_7OH

There is a difference of CH_2 between any two consecutive members in each series.

4. **Isomerism:** (Greek Esos = equal, Meros = parts). This property is met with only in the case of organic compounds. One molecular formula, in organic chemistry can represent two or more different substances, which are called isomers of one another. Thus the formula C_2H_6O stands for two isomers ethyl alcohol C_2H_5OH (liquid) and dimethyl ether CH_3-O-CH_3 (gas). $C_4H_{10}O$ represents as many as seven different compounds.

(a) $H_3C-CH_2-CH_2-CH_2-OH$ Normal Butyl Alcohol

(b) $(H_3C)_3C-OH$ Tertiary Butyl Alcohol

(c) $H_3C-CH_2-CH(OH)-CH_3$ Secondary Butyl Alcohol

(d) $(H_3C)_2CH-CH_2-OH$ Iso Butyl Alcohol

(e) $H_5C_2-O-C_2H_5$ Diethyl Ether

(f) $H_3C-O-C_3H_7$ Methyl Isopropyl ether

(g) $H_3C-O-HC(CH_3)_2$ Methyl Isopropyl ether

This phenomenon is called Isomerism.

5. **Polymerism**: (Greek Polloi = many Meros = parts). This means the union of two or more substances (without elimination of any thing) to form a new compound of the same composition but of different (higher) molecular weight. The reaction is reversible.

Example:

(i) Fomaldehyde $\xrightarrow{\text{evaporation}}$ Para-formaldehyde

n - HCHO ⟶ (HCHO) - (HCHO) - n

(ii) Fomaldehyde $\xrightarrow[\text{temperature}]{\text{Ordinary}}$ Metaformaldehyde

3(HCHO) ⟶ $(CH - CHO)_3$

6. **Monomer:** This is the simplest organic compound ready to get converted into polymers having simplest structures or molecules, e.g. styrene, adyphic acid vinyl chloride hexamethylene diamine

7. **Resonance:** Some molecules contain double bonds so they are quite stable than their conventional structures. This property is quite noticeable among the compounds having intermediate double and single bonds. These bonds re often called conjugated double bonds. For example resistance of benzene having hexatriene valance bond structure, to oxidation and its greater tendency to form substitution products than addition products, cannot be explained by its valance bond structure. The theory of resonance states that neither of the two equivalent structures I and II adequately represent benzene but a resonance hybrid bind all the six carbon atoms equally does so. The unusual stability of benzene is due to the depression of p electrons over a large region, its 36 Kcal/mol and is called the resonance or stabilization energy (calculated form the heats of hydrogenation of the benzene and cyclohexane)

I ⟷ II III

8. **Tautomrism or Mesomerism:** - (Greek Mesos = middle, Meros = parts). This includes the sifting of the hydrogen atom, which takes place in the two different isomers and they both stay in the equilibrium state the two isomers are known as **Keto** and **Enol.** e.g. the vinyl alcohol and acetaldehyde

H—C=C—H (H on first C; O—H on second C)

vinyl alcohol (keto structure)

CH_3—CH=O

acetaldehyde (enol structure)

9. **Chirality:** The concept of "chirality" has been known in chemistry since the 1870s although it would be nearly a hundred years before chemists began using this term. In fact, in the first edition of Eliel's "Stereochemistry of Carbon Compounds" in 1962, the word chiral is not mentioned, although it would be prominent in later editions. In extremely simple terms, chirality is "handedness", - that is, the existence of left/right opposition. For example, your left hand and right hand are mirror images and therefore "chiral". The term 'Chiral' is derived from the Greek name which meaning "hand" and apparently was coined by Lord Kelvin in 1904, in his Baltimore Lectures on *Molecular Dynamics and the Wave Theory of Light* in which he stated ...*"I call any geometrical figure, or group of points, chiral, and say it has chirality, if its image in a plane mirror, ideally realized, cannot be brought to coincide with itself."*

While the concepts of "asymmetry" were developed by J.H. vant Hoff and J.A. Le Bel in 1874 following the resolution by Louis Pastuer of a mixture of tartaric acid salt isomers during the period 1848-1853, in which he picked out the differing crystal types by hand - doing so on the basis of the differing physical appearance of the salt crystals. Pastuer recognized that two of the isomers polarized light differently (one to the left and the other to the right) and that this must be due to an asymmetric grouping of atoms in the optically active molecules. Following Kekule's recognition in 1858 that carbon has a valence of 4, vant Hoff and Le Bel independently recognized that when four different groups are attached to a carbon atom, arrayed at the corners of a tetrahedron, then the arrangements can be in two different forms, as depicted schematically to the right.

As the number of carbons with assymetry (chirality) increase in a molecule the number of possible optical isomers (enantiomers) also increases. With one asymmetric caıbon, 2 isomers . . . with two asymmetric carbons, 4 isomers, with three asymmetric carbons, 8 isomers . . . that is, the number of isomers is 2^n, where n = number of asymmetric atoms.

In the early days, chemists often assigned trivial names to differentiate isomers, and enantiomers generally were specified by d- = dextrorotary and l- = leavorotary based on which direction the molecles polarized light. But Cahn, Ingold and Prelog devised a system based on assigning sequence rules based on decreasing atomic number (and respective rate of substitution for atoms of the same atomic number) for projection formulas that allows the **absolute configuration** assignments of **R** (for rectus, Latin for right) and **S** (for sinister, Latin for left). Very occasionally, DS Viewer provides incorrect assignments (for example, with the enantiomers of gamma-dihydroionone, gamma-damascone & gamma-ionone. Thus, even without knowing the sequence rules, chemist's today can rapidly establish the R/S configuration at each asymmetric atom for a given molecular structure in just a few minutes.

Some other Aspects of Organic Chemistry

I. **Reaction:** Organic compounds are non-polar (non-ionizable) compounds and their reactions are therefore non-ionic in character and produced with measurable velocity. Two compounds may react together under different experimental conditions to produce different products. Thus sulphuric acid and ethyl alcohol may react together under different conditions to give ethyl hydrogen sulphate, ethylene and diethyl sulphate as shown below:

Ethyl Alcohol → Ethyl Hydrogen Sulphate + Water

$$C_2H_5OH \xrightarrow{H_2SO_4} C_2H_5HSO_4 + H_2O$$

Ethyl Hydrogen Sulphate → Ethylene + Sulphuric Acid

$$C_2H_5HSO_4 \longrightarrow C_2H_4 + H_2SO_4$$

Ethyl Hydrogen Sulphate + Ethyl Alcohol → Diethyl Ether + Sulphuric Acid

$$C_2H_5HSO_4 + C_2H_5OH \longrightarrow C_2H_5-O-C_2H_5 + H_2SO_4$$

Ethyl Alcohol → Diethyl Sulphate + Water

$$2C_2H_5OH \xrightarrow{H_2SO_4} (C_2H_5)_2SO_4 + H_2O$$

On the other hand inorganic compounds are polar (ionizable) compounds and their reactions are ionic and take place all most instantly. Thus NaCl (sodium chloride) KCl (potassium chloride) and $BaCl_2$ (barium chloride), all can ionize and form precipitate with silver nitrate. While chloride atoms of CH_3Cl_3 (chloroform) cannot produce the precipitate with silver nitrate.

II. Solubility: Organic compounds are usually soluble in organic solvent, e.g. alcohol benzene ether chloroform etc. are insoluble in water. Inorganic compounds have quite in the opposite way.

Sources of Organic Compounds

(a) *Plants*: For sugars starches, cellulose, oils alkaloids, perfumes, dyes, oxalic acid, citric acid, etc.

(b) *Woods:* The source of the pyroligneous acid (obtained by the destructive distillation of wood) used for methyl alcohol acetone and acetic acid.

(c) *Animals:* For gelatine, proteins.

(d) *Coal Tar*: Obtained from the distillation of coal, for benzene, toluene naphthalene phenol anthracene dyes etc.

(e) *Petroleum*: For petrol, gasoline kerosene oil vaselene, paraffin wax etc.

(f) *Synthesis:* Besides the above sources a large number of organic compounds simple and complex have been prepared synthetically in the lab.

CHAPTER

2

Purification of Organic Compounds

Introduction

Before the properties and structure of an organic compound can be successfully investigated. It is essential to have a complete molecular diagnosis involving the following process:

1. ***Purification:*** This is the process of the substance obtained from the natural sources all the associated substances are completely separated.
2. ***Detection:*** of the elements present in the organic compound.
3. ***Determination:*** of the percentage composition and hence the empirical formula and molecular formula
4. ***The Synthesis of Organic Compound***: This will confirm the structural formula, which summarizes the chemical behaviour of the substances and indicates the structural type to which it belongs.

Purification of Organic Compounds

The method applied for the purification of a given organic compound, which depend upon the nature of the compound and the nature of the impurities. In this case of a solid containing a solid impurity. Processes like filtration, crystallization, fractional distillation and sublimation may be used. For liquids we may employ either distillation or fractional distillation. Vacuum distillation or steam distillation.

The following processes are used for the purification of the organic compounds.

1. **Filtration**: This process is used for the separation of solid mixed in a solvent this generally means a physical separation. An ordinary funnel and a filter paper is employed for this process (see Fig. 2.1). The quick filtration is executed by introduction of a vacuum pump, which increases the rate of filtration. For this purpose a filter flask buchner funnel and funnel and vacuum pump is used (Fig. 2.2).

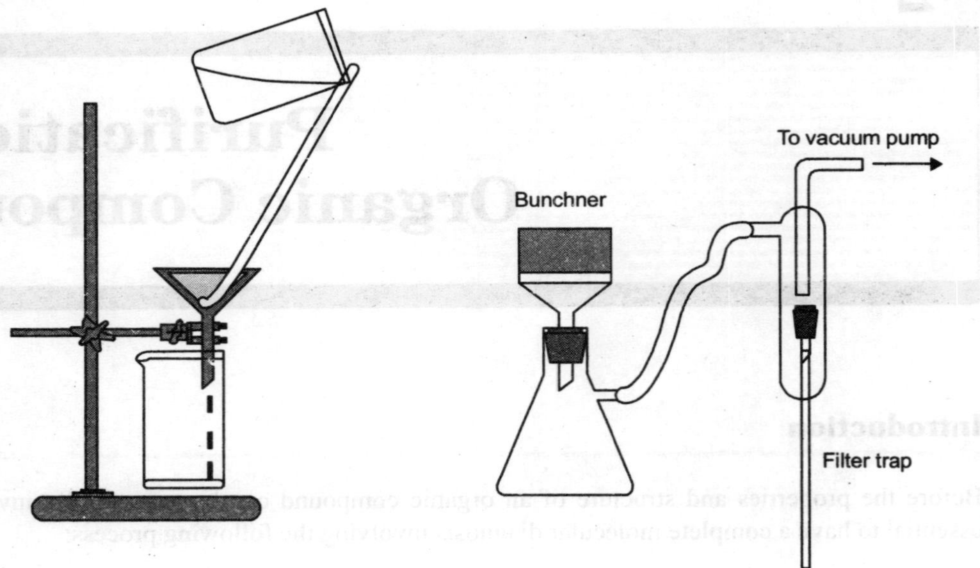

Fig. 2.1. The process of simple filtration

Fig. 2.2. The Vacuum Filtration

If the sample is of low quantity then centrifuge can also be employed here, rotating the substance separates the composition in the tube with the help of an high seed motor (Fig. 2.3).

2. **Crystallization**: If the substance is in the solid state it is the best purified by crystallization form a suitable solvent. The process is repeated until a sharp melting point is obtained.

   A suitable solvent is one which dissolves much more of the substances when hot than cold so that the hot saturated solutions depends a large quantity of the solid on cooling. Some of the important solvents and this boiling points are given in the Table 2.1.

Fig 2.3. The Centrifuge

Solvents like benzene, light petroleum and ether are particularly insoluble, i.e. immiscible in water and are called "dry sdolvents". Most of the organic solvents are highly inflammable and thus need careful manipulation. In order to avoid the loss off a valuable solvent it must always be heated under reflux condenser.

Table 2.1

Water	100°C
Alcohol	78°C
Petroleum ether	40-80°C
Ether	35°C
Benzene	80°C
Methyl alcohol	65°C
Toluene	110°C
Chloroform	61°C
Carbon tetrachloride	78°C
Glacial Acetic acid	118°C
Pyridine	115°C
Carbon Di Sulphide	46°C

The substance to be purified is dissolved in the solvent in a flask. More solid is added and the flask is heated on the heating mantle, if the boiling point is above 100°C then use air condenser. The solvent thus gets saturated with this substance.

During heating a few glass beads or porous porcelain pieces may also added to avoided bumping. If the substance is coloured due to impurities, a little animal charcoal is added to the solution before heating it.

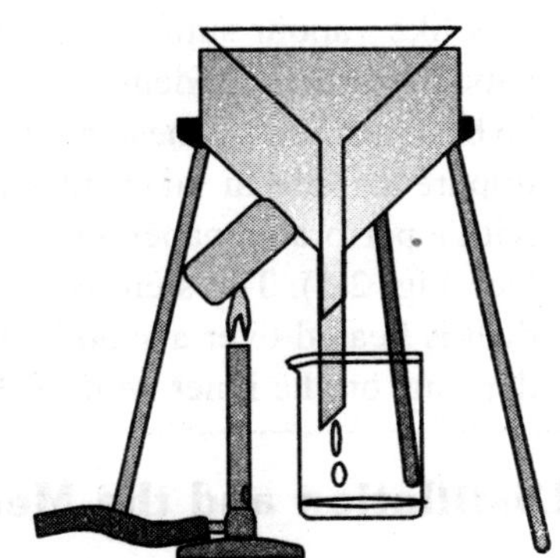

Fig. 2.4. The Sublimation

Animal charcoal is prepared by distillation of bones out of contact with air. It has the property of absorbing the coloured impurities and this is used for purifying organic compounds.

The solution is now quickly filtered while still hot sometimes it is observed that during filtration the substance starts crystallizing. The stem of the funnel thus gets clogged and filtration becomes difficult. In these cases hot water funnel is used. (See Fig. 2.4).

The circulation of the water around the stem of the funnel keeps the crystals dissolved in the solvent and the residues only gets separated from the solvent, the filtrate while cool only crystallizes.

Crystallization may sometimes be facilitated by scratching and rubbing the liquid with glass rod at the sides of the vessel.

It has been observed that when a saturated solution is cooled slowly, well defined, but impure crystals are obtained. On the other hand sudden cooling and vigorous scaling of the solution gives small pure crystals.

After the crystallization is over the crystals are filtered off, washed and dried benzoic acid, iodoform etc. are refined by this process.

3. **Fractional Crystallization**: This is an interesting system for the separation of two or more solids at which have quite different solubilities in the same solvents. Suppose a given sample contains two substance which have different solubilities, one is higher and the other one lower. Now if this solute is allowed to crystallize during filtration the substance having lesser solubility separates out in a comparatively pure form and the other with more solubility remains soluble in the mother liquor and that crystallizes on evaporation.

Separation and Drying of Crystals

The crystals are separated with a small quantity of the mother liquor. They are then dried by pressing then gently between the folds of filter paper or on a piece of unglazed tile or plate, where the liquid is absorbed. Crystals may also be dried in stream of air oven and kept in the desiccator if the substance is volatile or hydroscopic.

Sublimation: There are certain solid substances like benzoic acid camphor, or naphthalene, which directly pass into like vapour suite on heating without melting. These vapours further condense on cooling. Such substances can be best purified by these methods called ***sublimation.*** The impure sample in taken in a porcelain dish and covered with a perforated paper and a funnel is placed on top of it (see Fig. 2.5). The stem is closed with a cotton plug; the dish is heated over a sand bath. The pure solid substance deposits on the inner wall of the funnel near the stem.

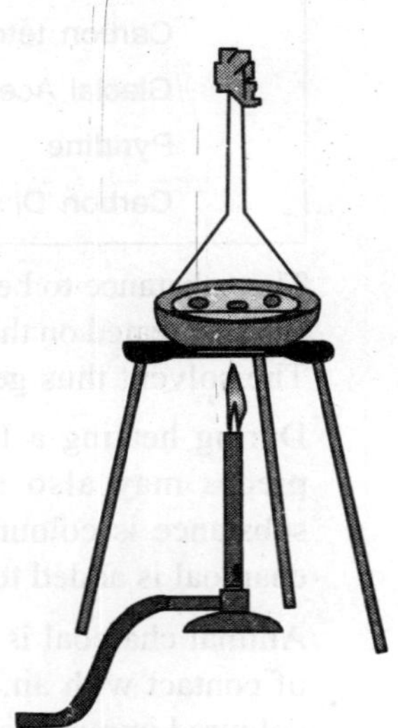

Fig. 2.5. The Sublimation.

Distillation and the Methods

Distillation is a widely used technique in chemical analysis for characterizing materials by establishing an index of purity and for separating selected components from a complete matrix. The technique is even more widely used in preparative chemistry and throughout manufacturing industry as a means of purifying products and chemical intermediates. Distillation operations differ enormously in size and complexity from the semi-micro scale to the 'thousands of tonnes per annum' production operations. For analytical purposes the scale employed is usually bench-level.

Numerous quoted standard specifications refer to distillation ranges as criteria of purity or suitability for use, or as indicators of performance. Published standards for analytical reagents and similar documentation by the American Chemical Society refer to distillation ranges as criteria of purity for appropriate materials.

Distillation is the process that occurs when a liquid sample is volatilized to produce a vapour that is subsequently condensed to a liquid richer in the more volatile components of the original sample. The volatilization process usually involves heating the liquid but it may also be achieved by reducing the pressure or by a combination

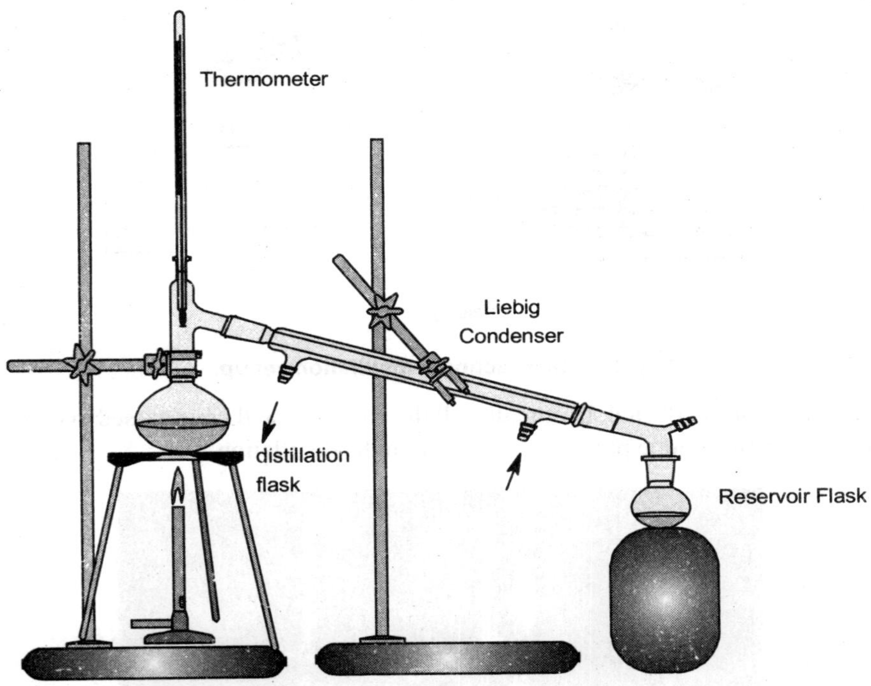

Fig. 2.6. Set up for simple distillation process simple distillation apparatus comprising distillation flask (DF), distillation head (DHI), thermometer m, condenser (C) and receiver (or collection) flask (RF)

This can be demonstrated in a simple laboratory distillation apparatus comprising a flask, distillation head, condenser and sample collector (Fig. 2.6). A thermometer is included in the apparatus as shown to monitor the progress of the operation. In its simplest form this procedure results in a separation into a volatile fraction collected in the receiver flask and a nonvolatile residue in the distillation flask. When a distillation column is incorporated in the equipment (Fig. 2.7), the evaporation and condensation processes occur continuously. This results in a progressive fractionation of the volatiles as they pass up the column. The most volatile components emerge from the top of the column initially and the less volatile components emerge later. By changing the receivers throughout the course of the distillation a

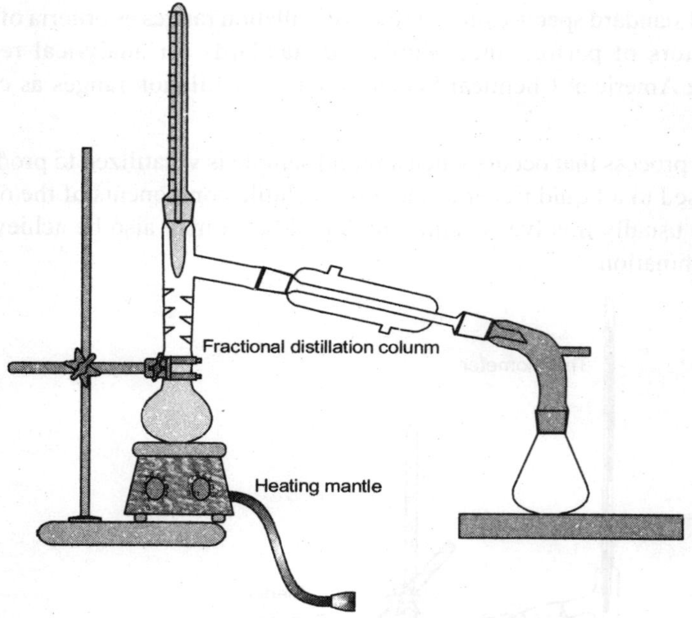

Fig. 2.7. The Fractional distillation set up.

separation or fractionation is effected. Eventually, all the vol-atiles will have passed over into the sample collectors a any in volatile residue present will remain in the distillation flask.

Fig. 2.8. Picture of Fractional distillation set up

Principles

The underlying principles are conveniently illustrated reference to a vapour-liquid equilibrium diagram (Fig. 2.9). The diagram relates to a binary mixture containing components P and Q. The lower curve gives the composition of the liquid boiling at various temperatures whilst the upper curve gives the composition of the vapour equilibrium with the boiling liquid. Points x and y the fore give the boiling points of the individual component P and Q respectively. For example, point A shows that at X degrees the vapour has a composition of approximately 90% p, whilst point B shows that the boiling liquid with which it is in equilibrium, has a composition of approximately 80% P. In a continuous distillation process, such as occurs in a distillation column, liquid of composition C (90% Q. 10% P) vapourizes to vapour of composition D which condenses to liquid of composition E. Subsequently liquid E becomes vapour F and liquid G (composition: 50% Q, 50% P). This continuous process of vaporization and condensation occurs in the distillation column until a volatile fraction leaves the top of the column and is removed from the process by being collected in the collection flask. At the same time the liquid in the distillation flask becomes progressively more concentrated in the in volatile component.

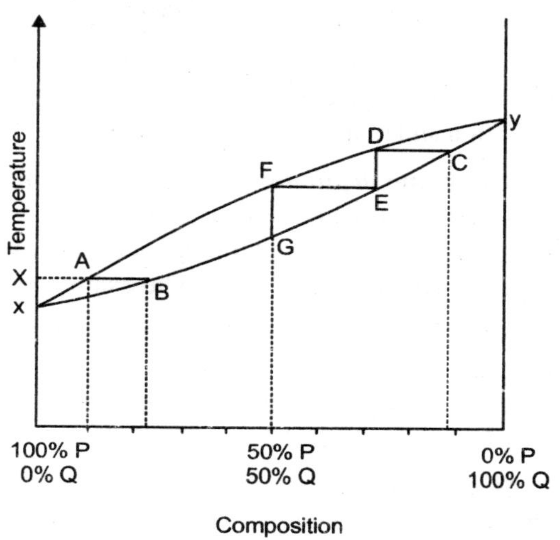

Fig. 2.9. The vapour liquid diagram of a binary mixture P and Q composition

Distillation techniques may be classified into several different types including:

- Distillation at atmospheric pressure
- Distillation under reduced pressure
- Steam distillation
- Fractional Distillation
- Molecular distillation (short-path distillation)
- Azeotropic distillation
- Isopiestic distillation
- Destructive Distillation

Distillation at atmospheric or reduced pressure produces a separation according to the general principles dis-cussed in the introduction.

Steam distillation is a means of distilling that part of a sample that is volatile in steam at a lower temperature than would otherwise be the case. This method is typically used for removing phenols from an aqueous sample. A means of introducing steam into the distillation flask must be provided.

Molecular distillation sometimes termed short-path distillation. Is used principally for compounds normally having high boiling points. In such cases, very low press-ures are needed to achieve the desired low boiling points. The apparatus is constructed such that the condensing surface is located only a short distance from the distilling liquid and the pressure is reduced so that the process is governed to a large extent by the mean free path of the molecules involved. Hence, the terms short-path distillation and molecular distillation.

Azeotropic distillation occurs when a mixture of two materials distils at constant composition. This technique is commonly used to remove water from samples. As an example, toluene may be added to a complex sample containing water, the distillation process results in the toluene-water azeotrope distilling. The distillate can then be examined to determine the water content of the original sample.

Azeotropic solutions are non-ideal solutions. Some display a greater vapor pressure than expected; these are said to exhibit *positive deviation.* Within a certain com-position range such mixtures boil at temperatures higher than the boiling temper-ature of either component; these are *maximum-boiling azeotropes* (see Table 2.3 and Fig. 2.4).

Mixtures that have boiling temperatures much lower than the boiling temperature of either component exhibit *negative deviation;* when such mixtures have a particular composition range, they act as though a third component were present. In Fig. 2.10 and 2.11, the minimum boiling point at Z is a constant boiling point because the vapour is in equilibrium with the liquid and has the same composition as the liquid does. Pure ethanol (bp 78.4°C) cannot be obtained by fractional distillation of aqueous solutions that contain less than 95.57% of ethanol because this is the azeotropic composition; the boiling point of this azeotropic mixture is 0.3° lower than that of pure ethanol.

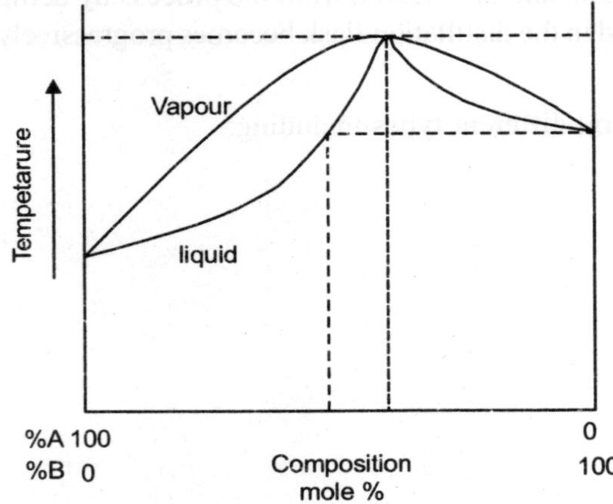

Fig. 2.10. Maximum Boiling Point Azeotrope

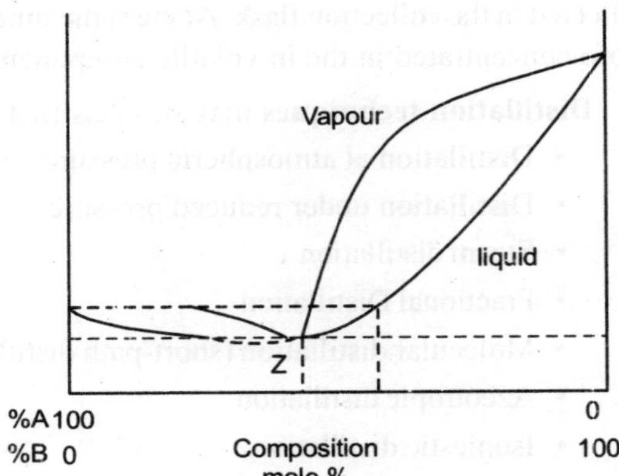

Fig. 2.11. Minimum Boiling Point Azeotrope

Other examples are:

1. A mixture of 32.4% ethyl alcohol and 67.6% benzene (bp 80.1°C) boils at 68.2°C.
2. A ternary azeotrope (bp 64.7°C) is composed of 74.1 % benzene, 18.5% ethyl alcohol, and 7.4% water. (See also Table 4.)

Table 2.3. For The Maximum Boiling Point for the Azeotropic Mixtures

Component A		Component B		bp of Azeotropic mixture, °C	% of B (by mass) in mixture
Substance	bp, °C	Subtance	bp, °C		
Water	100.0	Formic acid	100.8	107.1	77.5
Water	100.0	Hydrofluoric acid	19.4	120.0	37
Water	100.0	Hydrochloric acid	-84.0	108.6	20.22
Water	100.0	Hydrobromic acid	-73	126	47.6
Water	100.0	Hydriodic acid	-35	127	57.0
Water	100.0	Nitric acid	86.0	120.5	68
Water	100.0	Sulfuric acid	10.5 (mp)	338	98.3
Water	100.0	Perchloric acid	110.0	203	71.6
Acetone	56.4	Chloroform	61.2	64.7	80
Acetic acid	118.5	Pyridine	115.5	130.7	65
Chloroform	61.2	Methyl acetate	57.0	64.8	23
Phenol	181.5	Aniline	184.4	186.2	58

Absolute ethyl alcohol can be obtained by distilling azeotropic 95.5% ethyl alcohol with benzene. The water is removed in the volatile azeotrope formed.

The procedure is described above under Distillation of Pure Liquids.

Fractional Distillation. Simple distillation is effective only when separating a volatile liquid from a non-volatile substance or when separating two liquids that differ in boiling point by 50 degrees or more. If the liquids comprising the mixture that is being distilled have boiling points that are closer than 50 degrees to one another, the distillate collected will be richer in the more volatile compound but not to the degree necessary for complete separation of the individual compounds.

The basic idea behind fractional distillation is the same as simple distillation only the process is repeated many a times. If simple distillation was performed on a mixture of liquids with similar volatilities, the resulting distillate would be more concentrated in the more volatile compound than the original mixture but it would still contain a significant amount of the higher boiling compound. If the distillate of this simple distillation was distilled again, the resulting distillate would again be even more concentrated in the lower boiling compound, but still a portion of the distillate would be the higher boiling compound. If this process is repeated several times, a fairly pure distillate will eventually result. This, however, would take a very long time. In fractional distillation, the vapors formed from the boiling mixture rise into the fractionating column where they condense on the column's packing. This condensation is tantamount to a single run of simple distillation; the condensate is more concentrated in the lower boiling compound than the mixture in the distillation flask. As vapours continue to rise through the column, the liquid that has condensed will revapourize. Each time this occurs the resulting vapors are more and more concentrated in the more volatile substances. The length of the fractionating column and the material it is packed with impact the number of times the vapors will recondense before passing into the condenser; the number of times the column will support this is referred to as the number of theoretical plates of the column.

Table 2.4. For The Minimum Boiling Point for the Azeotrope Mixture

Component A		Component B		bp of Azeotropic mixture, °C	% of B (by mass) in mixture
Substance	bp, °C	Subtance	bp, °C		
Water	100.0	Ethyl alochol	78.3	78.15	4.4
Water	100.0	Isopropyl alcohol	82.4	80.4	12.1
Water	100.0	*n*-Propyl alcohol	97.2	87.7	28.3
Water	100.0	*tert*-Butyl alcohol	82.6	79.9	11.8
Water	100.0	Pyridine	115.5	92.6	43.0
Methyl alcohol	64.7	Methyl iodide	44.5	39.0	7.2
Ethyl alcohol	78.3	Ethyl iodide	72.3	63.0	13
Methyl alcohol	64.7	Methyl acetate	57.0	54.0	19
Ethyl alcohol	78.3	Ethyl acetate	77.2	71.8	31
Water	100.0	Butyric acid	163.5	99.4	18.4
Water	100.0	Propionic acid	140.7	100.0	17.7
Benzene	80.2	Cyclohexane	80.8	77.5	55
Ethyl alcohol	78.3	Benzene	80.2	68.2	32.4
Ethyl alcohol	78.3	Toluene	110.6	76.7	68
Methyl alcohol	64.7	Chloroform	61.2	53.5	12.5
Ethyl alcohol	78.3	Chloroform	61.2	59.4	7.0
Ethyl alcohol	78.3	Methyl ethyl ketone	79.6	74.8	40
Methyl alcohol	64.7	Methylal	42.2	41.8	18.2
Acetic acid	118.5	Toluene	110.6	105.4	28

Since the procedures of simple distillation are so similar to those involved in fractional distillation, the apparatus that are used in the procedures are also very similar. The only difference between the equipment used in fractional distillation and that used in simple distillation is that with fractional distillation, a packed fractionating column is attached to the top of the distillation flask and beneath the condenser. This provides the surface area on which rising vapours condense, and subsequently revapourize.

The fractionating column is used to supply a temperature gradient over which the distillation can occur. In an ideal situation, the temperature in the distillation flask would be equal to the boiling point of the mixture of liquids and the temperature at the top of the fractionating column would be equal to the boiling point of the lower boiling compound; all of the lower boiling compound would be distilled away before any of the higher boiling compound. In reality, fractions of the distillate must be collected because as the distillation proceeds, the concentration of the higher boiling compound in the distillate being collected steadily increases. Fractions of the distillate, which are collected over a small temperature range, will be essentially purified; several fractions should be collected as the temperature changes and these portions of the distillate should be distilled again to amplify the purification that has already occurred.

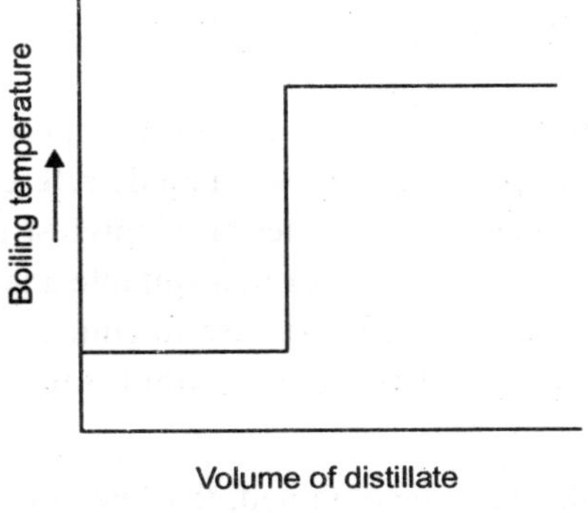

Fig. 2.12

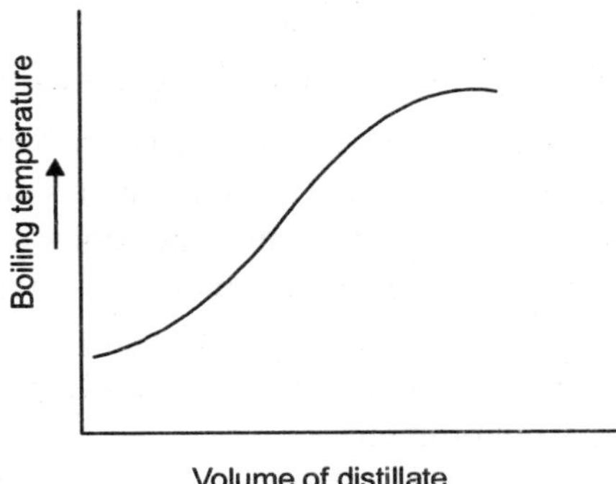

Fig. 2.13

The separation of mixtures by this means is a refinement of ordinary separation by distillation. Thus a series of distillations involving partial vapourization and condensation concentrates the more volatile component in the first fraction of distillate and leaves the less volatile component in the last fraction or in the residual liquid. The vapor leaves the surface of the liquid and passes up through the packing of the column. There it condenses on the cooler surfaces and redistills many times before entering the condenser. Each minute distillation causes a greater concentration of the more volatile liquid in the rising vapor and an enrichment of the residue, which drips down through the column in the less volatile components.

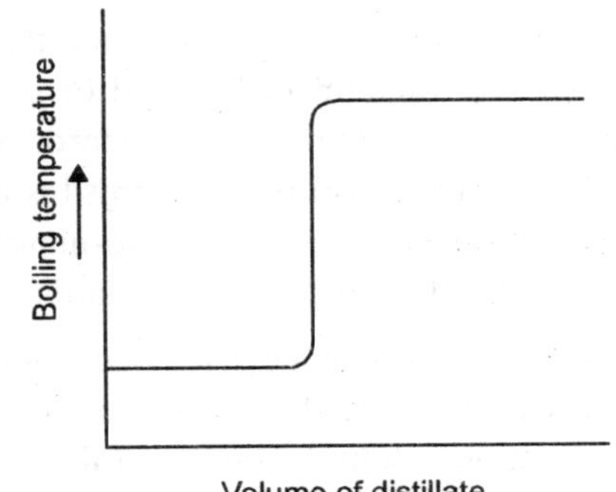

Fig. 2.14

By means of long and efficient distillation columns, two liquids may be completely separated.

1. Select the type of fractionating column to be used, one that offers a large surface contact inside (see Fig. 2.14).

2. Set up the equipment.
3. Open the inlet cooling-water valve.
4. Apply heat.
5. Keep a large volume of liquid condensate continually returning through the column.
6. *Distill slowly* to effect efficient separation

The charcoal left behind is much higher in carbon content than the carbohydrate of the original wood.

Additionally, there are other forms of sample purification and separation that are either a type of distillation or are related to a distillation process:

- Simultaneous distillation/extraction (see application section)
- Dean and Stark distillation (see application section)
- Simulated distillation (gas chromatographic technique)

Analytically, distillation is used for two principal purposes, firstly as a criterion of purity and secondly as a means of preparing a sample for analysis. Many specification tests include reference to a distillation range within the limits of which a stated percentage of the material of interest distils. Alternatively distillation may be used to separate volatiles from a sample prior to a suitable analytical technique being employed on the distillate or on the residue. Standard tests are documented that involve distil-lation as a sample pretreatment method prior to titrimetry, potentiometry and spectrophotometry.

It is of course essential, if meaningful comparative results are to be obtained, that the design and use of the apparatus are standardized for such determinations.

Apparatus

A wide variety of apparatus is available to satisfy the different distillation techniques. The appropriate design of apparatus depends upon the type of distillation to be per-formed, considering, for example, whether a vacuum is required or steam is needed. Descriptions of apparatus are to be found in a number of different texts. Standards referring to the design and use of distillation apparatus have been published by the British Standards Institute and the American Society for Testing and Materials. Simulated distillation, which is a gas chro-matographic technique, is dealt with in a recent review by Robillard. and referred to in several standards.

Apparatus may be discussed in terms of the distillation flask, the distillation column, the condenser and the collecting flask(s). By far the most effort has been expended in the design and operation of the distillation column, which is at the heart of the separation efficiency. The form of the column, its size and the packing used are very influential upon the results that are achievable. A summary of some different types of columns is given in **Table 2.5** and of packings in **Table 2.6.**

Table 2.5

Column type	Description/comments
Dufton	An open tube into which a glass spiral fits closely
Hempel	A simple tube normally filled with a suitable packing (rings/helices) and having a side-arm near the top
Oldershaw	A column with fixed but perforated plates that maintains a fixed amount of liquid on each plate
Podbielniak	A simple tube with a wire packing to provide large contact area between liquid and vapour to effect high efficiencies
Spinning band	A tube fitted with a closely fitting spiral of PTFE or metal gauze that can be rotated at typical speeds of 600 to 3000 rev min-1 as the vapour-liquid equilibrium is maintained in the column.
Vigreux	A tube having pairs of indentations down its length that slope downwards and provide a large and designed surface area to enhance the liquid-vapour equilibrium

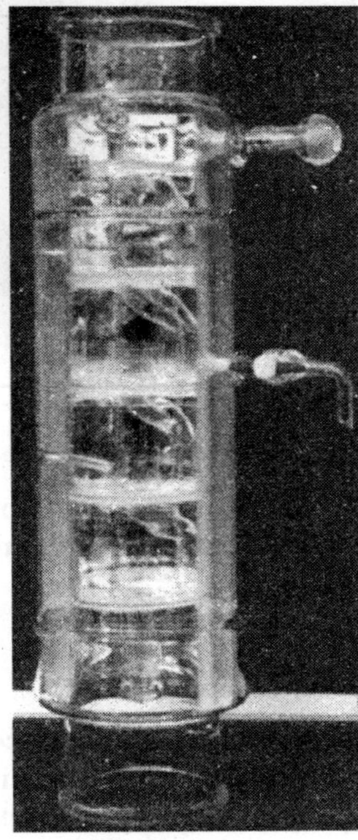

Oldershaw Column used in petroleum Fractionation

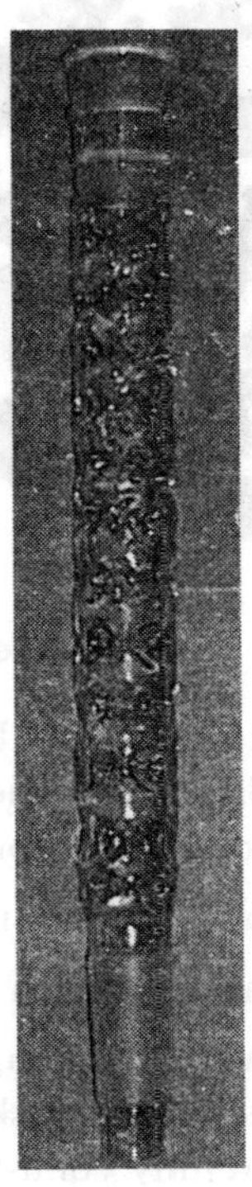

Vigreux column used in organic synthesis

Table 2.6

Packing	Description
Balls	Mostly made of glass. Columns have a tendency to flood easily
Berl Shaddles	These are used in the petroleum fractionation
Helices	Made from metal or glass. although metal may be packed mechanically to produce a more uniform column
Rings	Usually made of glass of an appropriate size for the column but can be made of porcelain, stainless steel, aluminium, copper or nickel. Depending upon design they can be termed Raschig, Lessing or Dixon rings
Wire packings	Produced as .Heli-Grid' and 'Heli-Pak' packings especially for use with Podbielniak columns

berl shaddles

raschig rings

Once apparatus has been chosen carefully to compare with previously used apparatus or to conform to standards, the operation of the equipment must be considered. The following factors are among the most important to be controlled:

- The heating of the distillation flask must be carefully controlled.
- The distillation column must be operated so that it does not become flooded.
- The reflux ratio, that is the ratio of material returning via reflux to the distillation column or the distillation flask compared to the amount presented to the condenser in unit time must be carefully controlled. That higher the reflux ratio, the purer the material collected from the distillation. Reflux ratios are controlled it simple distillation apparatus by adjustment of the heating rate and by maintaining stable thermal condition throughout the apparatus.

Applications

Documentation of analytical applications of distillation is widely dispersed. However, there are numerous references to distillation as a means of characterizing materials an as means of sample pretreatment in the lists of BSI standards, the ASTM methods documentation, the analytic methods of the Institute of Petroleum and those of other worldwide standards organizations.

Distillation is used widely to determine the moisture or water content of a variety of samples from petroleum products to cereal feeds. As the azeotropic distillate condenses, the water separate from the immiscible organic and can be estimated direction a specially graduated collection arm.

Some methods for the determination of water quality involve distillation, for example the determination of 'phenol index', nitrate content or ammonium content.

The determination of nitrogen by the Kjeldahl method involves a preliminary distillation of the sample. The methods for the determination of ammonical and of nitrogen in ammonium nitrate, urea, sulphuric acid al fertilizers for industrial purposes involve a preliminary distillation followed by titrimetry

Methods for the determination of available fluorine involve distillation prior to a potentiometric or spectrometric method.

The determination of distillation range is a method establishing the purity of materials. Specific standard methods are available, for example, for methanol, ethylene glycol and propylene glycol. Many unpublished methods are used for products and intermediates to validate purity standards and to establish the suitability of materials for subsequent use.

As trace analysis of residual compounds in consumables has become more important, methods of extracting these compounds have been developed. A method known as simultaneous distillation extraction developed from the original work of Likens and Nickerson has been particularly popular and effective for extracting the volatiles from foods and plant materials, and the herbicide and pesticide residues in agricultural products. The method involves steam distilling the compound of interest from an aqueous suspension of the crude sample while the condensed steam is continuously extracted with an immiscible organic solvent refluxing within the apparatus. The design of the apparatus allows the volatiles that are extracted from the condensed water to be flushed into the flask containing the organic solvent. After a previously determined time of extraction, the apparatus may be disassembled and the organic solvent removed by evaporation from the now concentrated extract. Further analytical techniques can be used to identify and quantify the components of the residue according to the particular requirements.

The applications of distillation in analysis are wide-spread, with the technique being used to characterize materials and as a means of preparing samples prior to analysis. Standard apparatus and methods are described for many specific applications.

Efficiency of Fractionating Columns

We measure the efficiency of a fractionating column in terms of the number of theoretical plates that it contains. A column with one theoretical plate is one in which the initial distillate has a vapour composition that is at equilibrium with the original solution. It is impossible to operate fractionating columns at equilibrium.

Bubble-cap fractionating columns (Fig. 2.15) have definite numbers of trays of plates and are fitted with either bubble caps or sieve perforations - or modifications of these two - to enable the achievement of intimate vapour liquid disperser. However, because fractionating columns cannot be operated at equilibrium (practically), the number of theoretical plates is always lower than the number of actual plates, depending upon the rate of distillation, reflux ratio, and other factors.

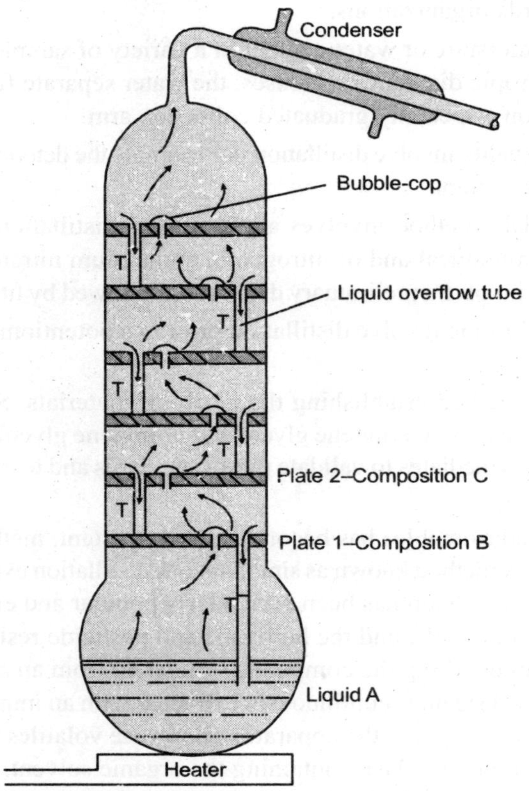

Fig. 2.15. The bubble cap distillation system.

CAUTION If you heat the pot *too* vigorously and remove the condensed vapour *too* quickly, the whole column will heat up uniformly and there will be no fractionation. The fractionating column will become flooded by the returning condensate.

Total-Reflux Partial-Takeoff Distilling Heads

Exercise good judgement in the control of the amount of heat applied, and, for the effective fractionation, use a total-reflux-partial-takeoff distilling head as shown in Fig. 2.16. With the stopcock 'S' completely closed, all condensed vapours are returned to the distilling column, a total reflux condition. With the stopcock partially opened, the number of drops of condensate falling from the condenser which returns to the fractionating column can be adjusted. The ratio of the number of drops of distillate allowed to pass through stopcock 'S' into the receiver to the number of drops of reflux is called the *reflux ratio*. With an efficient column, reflow ratios as high as 100 to 1 can be used to effectively separate compounds that have very close boiling points.

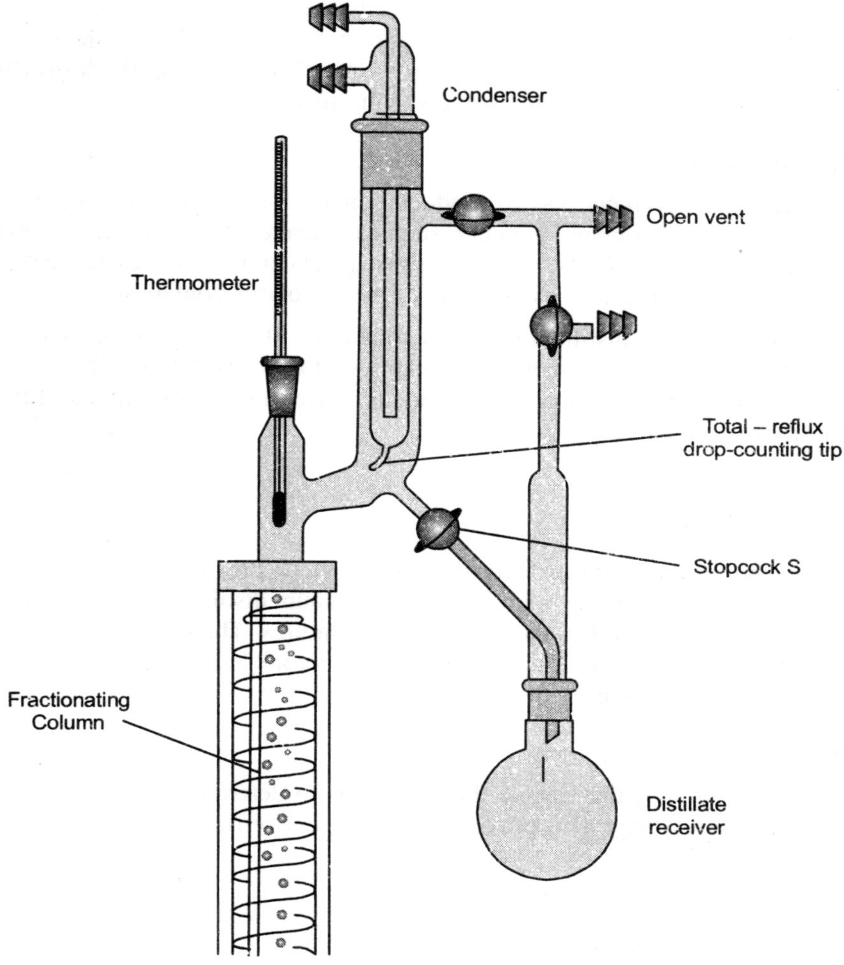

Fig. 2.16. The Total-Reflux-Partial-Takeoff Distilling Heads

Destructive Distillation

Destructive distillation is a process in which wood is heated to form charcoal and methanol. Some wood shavings are placed into a test tube and heated with a Bunsen burner. A small nozzle has been placed on the tube so that any gases coming out of the heated mixture will pass through the nozzle. As the wood is heated, it becomes darker, beginning to char. The wood is decomposing, and one of the decomposition products is methanol or methyl alcohol. Other products also vapourize from the wood and collect on the wall of the test tube, forming a brown-coloured deposit. Destructive distillation was used in the past for generating methyl alcohol, which is used today in windshield washer solvent and many other applications.

A beaker is placed over the end of the nozzle on the tube so that the vapors coming out of the tube will condense on this beaker. Brown liquid is forming from the vapours that condense. A match can be used to ignite the methyl alcohol vapor coming out of the nozzle on the tube. By destructive distillation of wood it is possible to produce the combustible vapour, methyl or wood.

Isopiestic Distillation

Isopiestic distillation is a convenient way of producing metal-free aqueous samples of volatile acids. The 'crude' acid is placed in an open container, such as a beaker, in a desiccator containing also an open beaker of pure water. The acid vapourizes and subsequent condensation in the pure water produces an aqueous sample of the volatile acid without any of the in volatile contaminants such as metals.

The alternative terms 'flash' distillation and 'fractional' distillation are sometimes used to describe some of the above procedures carried out in a particular way. Flash distillation effects a crude separation into volatiles and residue, whilst fractional distillation produces a series of 'cuts' of different volatility (or boiling point) ranges.

Vacuum Distillation

Principle

Many substances cannot be distilled satisfactorily at atmospheric pressure because they are sensitive to heat and decompose before the boiling point is reached. Vacuum distillation, distillation under reduced pressure, makes it possible to distill in much lower temperatures. The boiling point of the material is affected by the pressure in the system. The lower the pressure, the lower the boiling point; the high the pressure, the higher the boiling point.

Nomographs

Nomographs are special graphs that enable the technician to determine more accurately the boiling points at different pressures; they also provide a method converting boiling points to the pressure desired.

Procedure for using the nomograph (see Fig. 2.17):

1. Select the desired boiling point at the reduced pressure.
2. Use a transparent plastic ruler and connect that boiling point (column A) with the given corresponding pressure (column C). It will intersect column B at a definite point. Record that point.
3. Using the point obtained from column B in step 2, select the new pressure desired on column C. Align the plastic ruler with these two points and read corresponding temperature for the boiling point at the new pressure where the ruler intersects column A.

Example

A liquid boils at 200°C at 10 torr pressure. What would be (1) the normal boiling point at 760 torr and (2) the boiling point at 1 torr pressure?

Solution

Given: B.P. = 200°C (column A) at 10 torr (column C).

1. Intersection point in column B is 350°C, the boiling point at *760* torr pressure (corrected).
2. The line connecting the points 350°C on column B, and 1 torr pressure on -column C intersects column A at 150°C. Thus 150°C is the boiling point at 1 torr pressure.

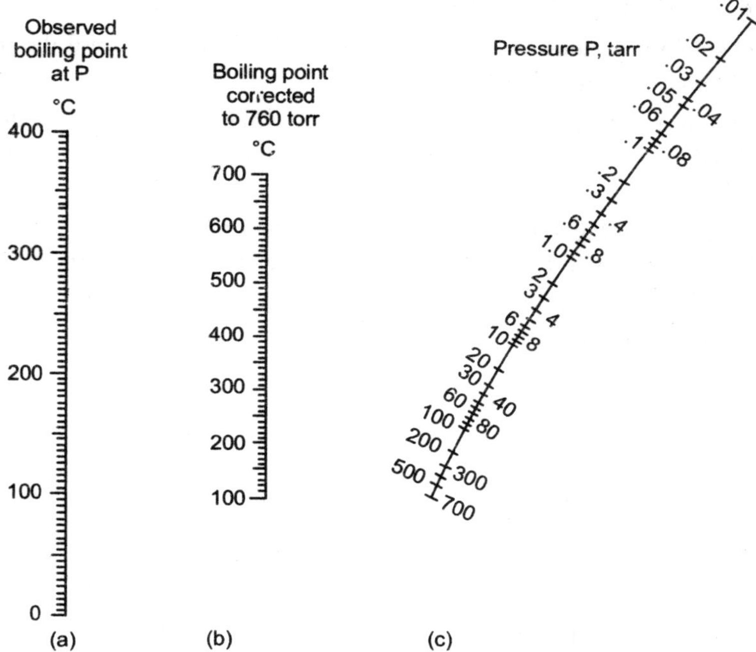

Fig. 2.17: The vacuum distillation nomographs.

General Requirements

The experimental setup is as shown in Fig. 2.18.

1. *A source of vacuum.* Efficient water pumps, aspirators, will theoretically reduce the pressure in the system to the vapor pressure of the water passing through the pump. In practice, the pressure is usually about 10 mm higher. *(a)* Oil mechanical vacuum pumps. *(b)* Use rubber pressure tubing. *(c)* The entire distillation system should be airtight, free from leaks. *(d)* Lubricate all joints and connections.
2. Safety trap to protect manometer and vacuum source from overflow-liquid contamination The trap must be correctly connected. Vapours

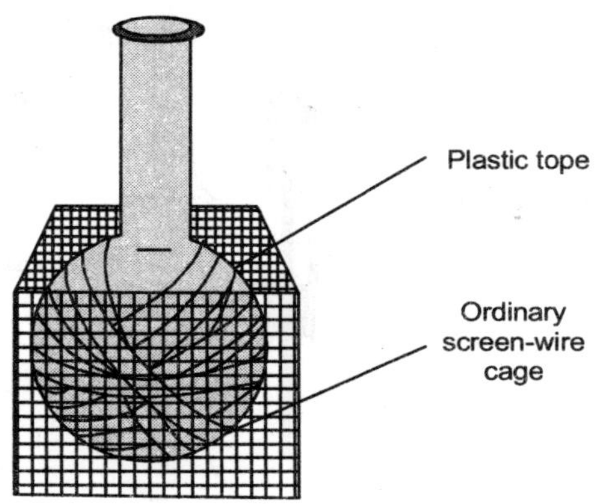

Fig. 2.18. The safety precaution in the vacuum distillation for the large flasks

condense on the sides of the trap and fall to the bottom if they are not solidified. Dry ice will freeze your skin; do not handle with bare fingers or hands.

(*a*) Crush dry ice in a cloth towel with a hammer.

(*b*) Use a scoop to fill the Dewar flask after the trap has been inserted.

(*c*) Add solvent, acetone, or isopropanol in small increments until the trap is filled and the liquid level is near the top.

3. Pressure gauge (manometer). Exercise great care when allowing air into the evacuated system. It must be done slowly to avoid breakage when the mercury column rises to the top of the closed tube.
4. Manostat (pressure regulator). To maintain constant pressure in the system, it automatically opens and closes needle valves, permitting air to enter or keeping the system airtight because of vacuum variations.
5. Capillary air inlet (Fig. 2.20).

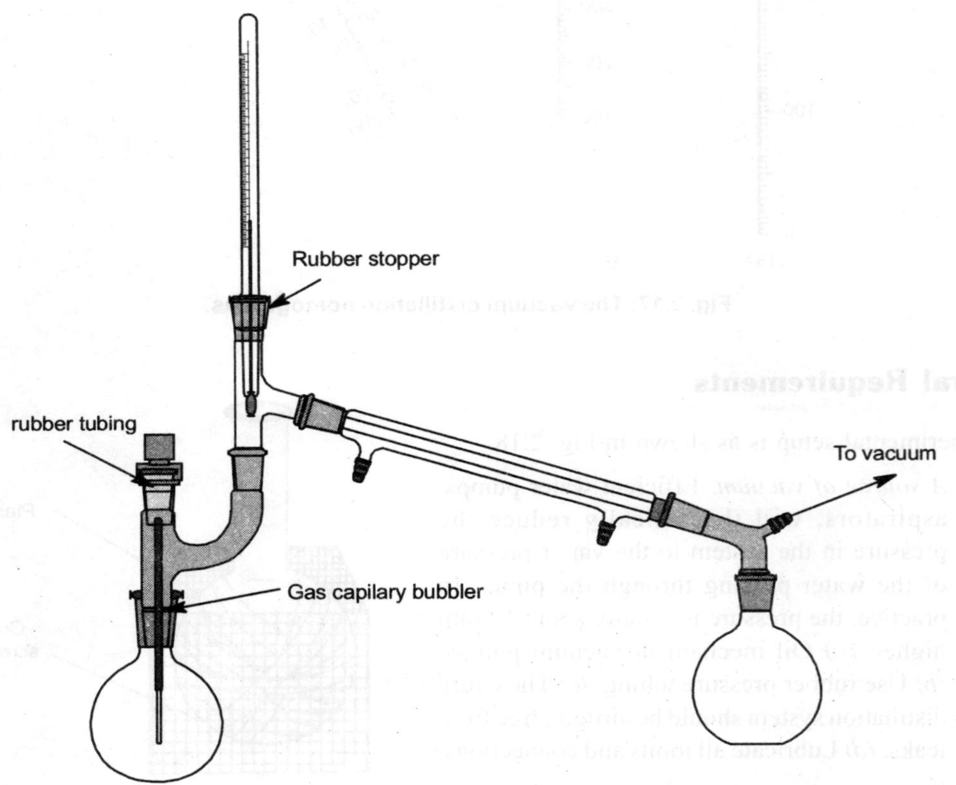

Fig. 2.19. The set up for the Vaccum Distillation in Lab.

6. Special vacuum distillation flasks to minimize contamination of the distillate caused by frothing of the boiling solution.

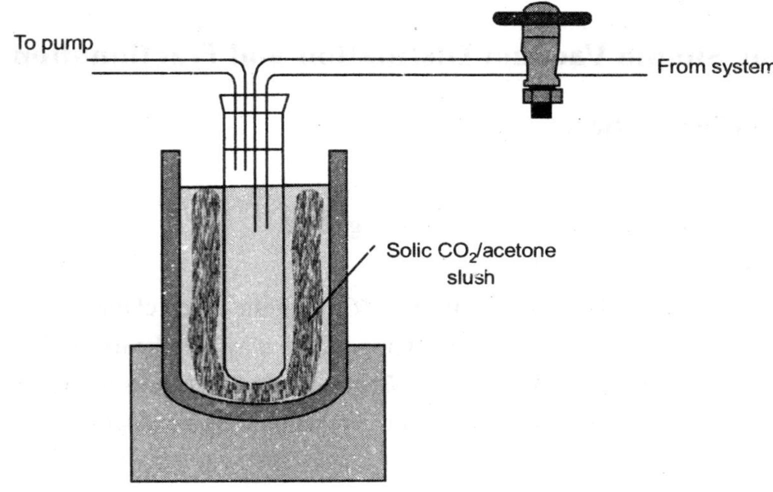

Fig. 2.20. Capillary gas-inlet tube

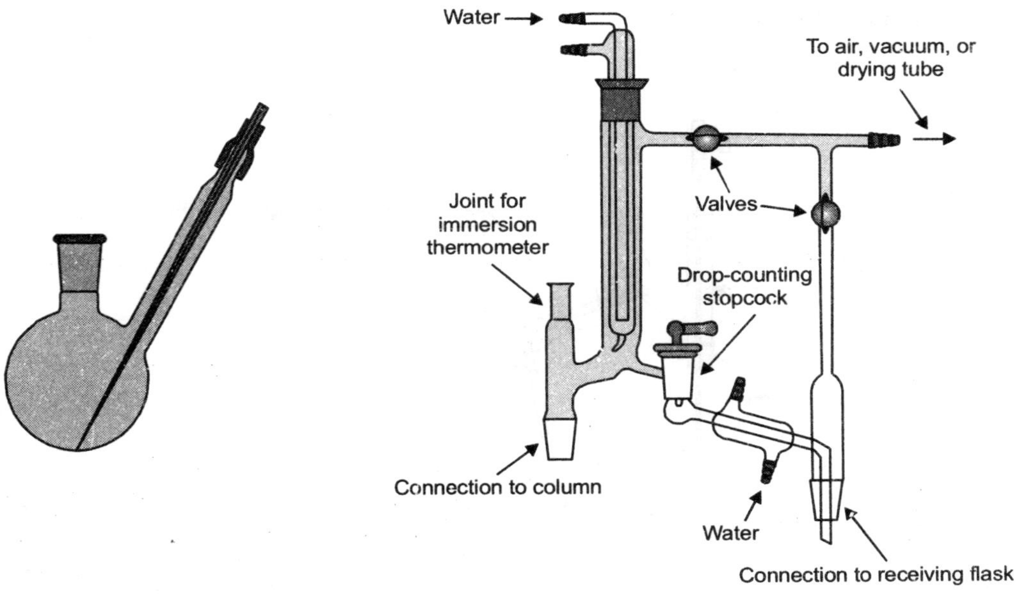

Fig. 2.21. The partial-takeoff distillation head and its component parts

7. Heating baths, electric mantles, and fusible alloy or sand baths.
8. Special distilling heads (Fig. 2.21) to permit removal of distillate fractions without interrupting the distillation.

Assemblies for Simple Vacuum Distillation and Fractionation

Simple Vacuum Distillation

Procedure

A Claisen flask for use in this process is shown in Fig. 2.22.

1. Fill the Claisen flask *(A)* one-third full.
2. Apply vacuum; adjust the capillary air inlet *(C)* with the pinch clamp *(D)*.
3. Heat bath to about *20°C* higher than the temperature at which the material will distill.
4. Cooling water flowing over the receiver *(B)* condenses the vapors to give a distillate.
5. A safety trap prevents any condensate from contaminating the suction pump or manometer.
6. When distillation is completed:
   (a) Remove the heating bath; allow the flask to cool.
   (b) Remove the capillary pinch clamp.
   (c) Cut off the cooling water.
   (d) Turn off the vacuum pump.

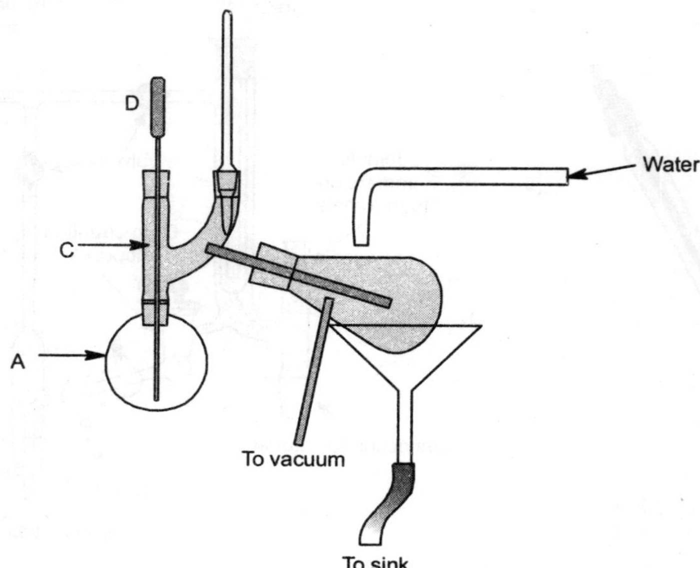

Fig. 2.22. The Claisen flask for vacuum distillation flask.

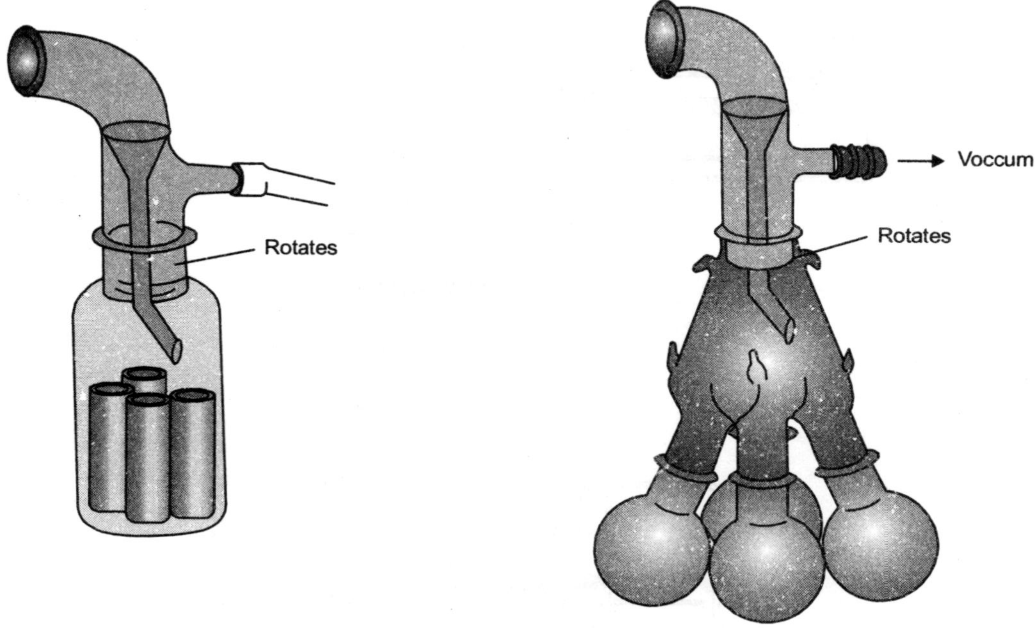

Fig. 2.23 and 2.24. The receivers for collecting of distillate fractions.

Steam Distillation

Principle

Steam distillation is a means of separating and purifying organic compounds by volatilization. The organic compound must be insoluble or slightly soluble in water. When steam is passed into a mixture of the compound and water, the compound will distill with the steam. In the distillate, this distilled compound separates from the condensed water because it is insoluble in water.

Most compounds, regardless of their normal boiling point, will distill by steam distillation at a temperature below that of pure boiling water.

Some high-boiling compounds decompose at their boiling point. Such substances can be successfully distilled at low temperature by steam distillation.

Steam distillation can be used to rid substances of contaminants because some water-insoluble substances are steam-volatile and others are not.

When it is desirable to separate non-volatile solids from high boiling organic solvents, steam distillation will remove all solvents (water-insoluble).

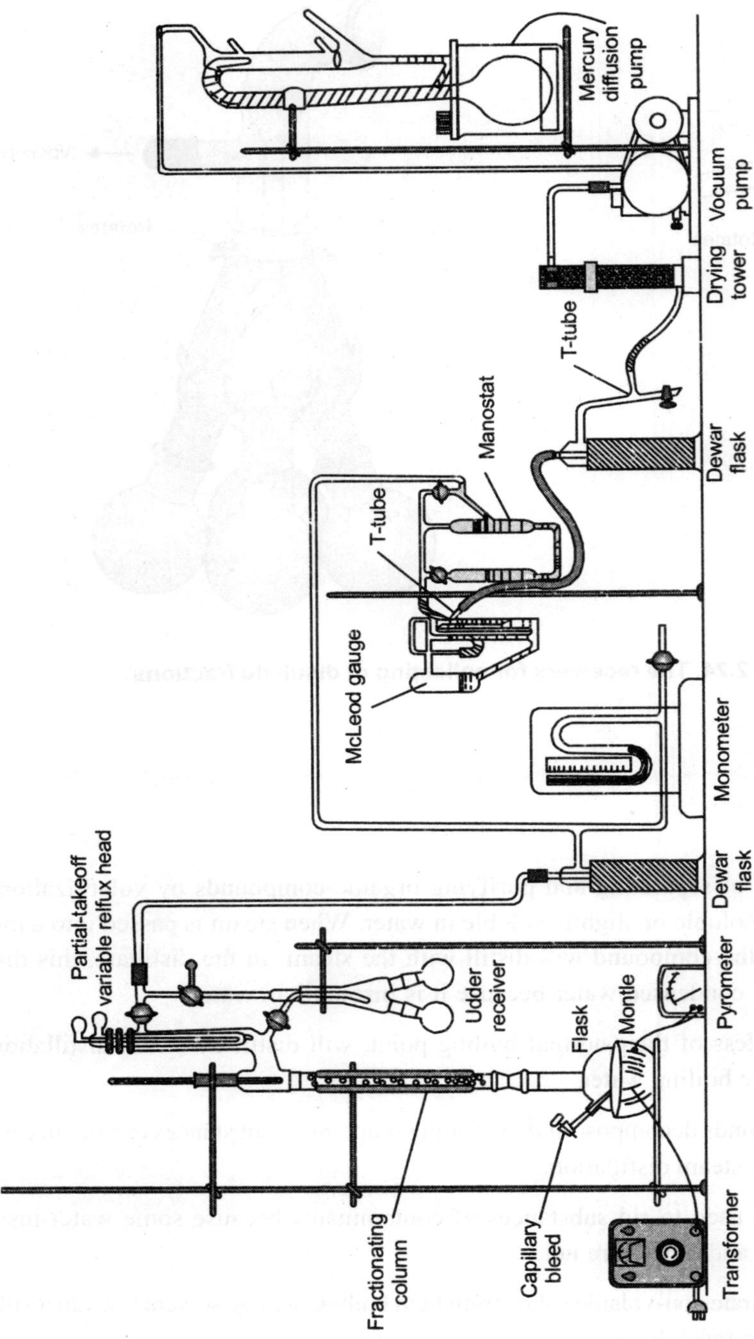

Fig. 2.25. The High Vacuum distillation set up.

Procedure

Place the compound or mixture in the distilling flask with a little water. Pass cooling water through the condenser (see Fig. 2.26). A Claisen flask may be subsituated for the round-bottomed flask. The Claisen still head helps to prevent an contamination of the distillate caused by spattering of the steam-distilled mixture. If there is no readily available source of piped steam, the steam can be generate. in an external steam generator (Fig. 2.27) and then passed into the mixture to the steam-distilled.

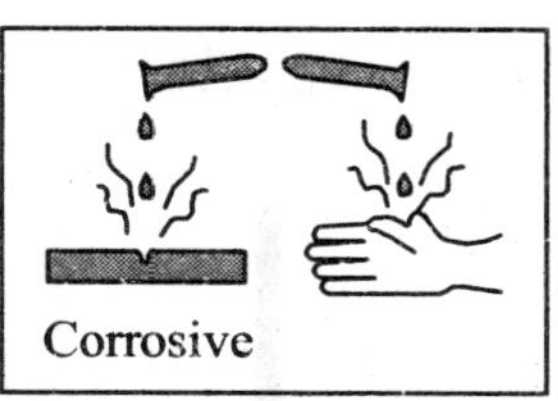

Corrosive

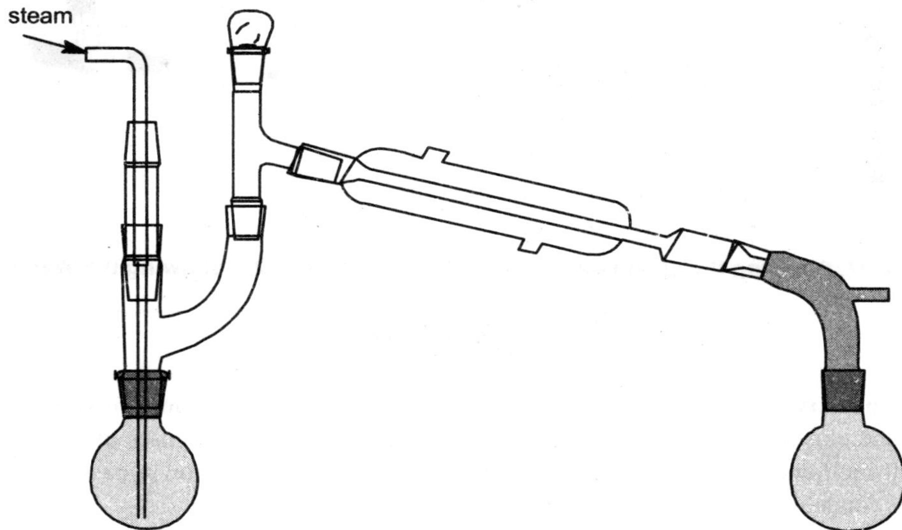

Fig. 2.26. The Set up for steam distillation.

NOTE

1. Always equip the steam generator with a safety tube to prevent explosions. If only a small amount of steam is needed, for instance, to rid a substance of a small amount of steam-volatile impurity, water can be combined with the material directly in the distilling flask; the flask is then directly heated with a Bunsen flame or any other suitable heat source. The long steam-inlet tube is replaced with a stopper.
2. Pass steam into the distilling flask with the steam outlet below the surface of the liquid. The distilling flask itself may be heated gently with a burner. If steam is available from a laboratory steam line, insert a water trap in the entering steam line to trap condensed water. Otherwise, the condensed water may fill up the distilling flask.
3. Continue passing steam into the flask until no appreciable amount of water-insoluble material appears in the condensate.

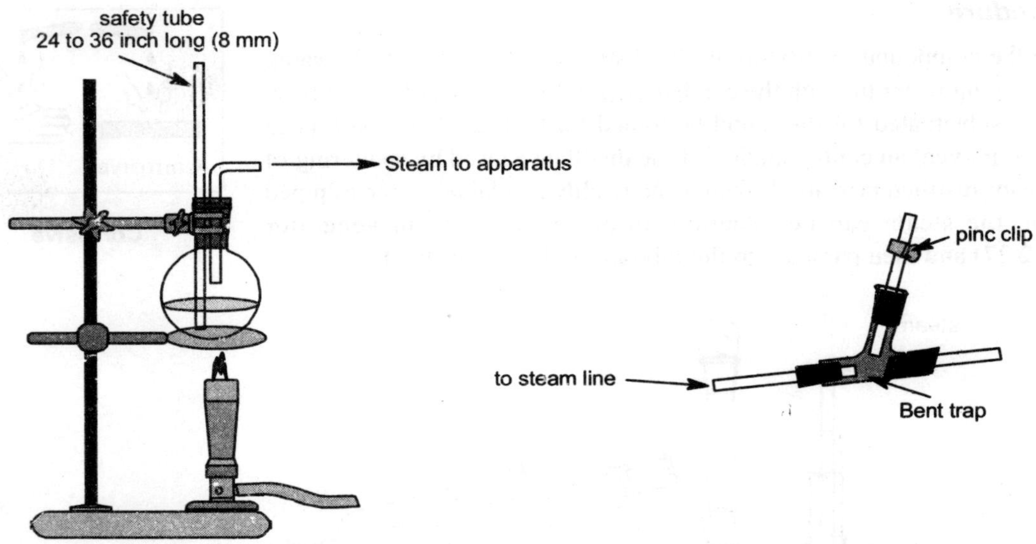

Fig. 2.27. Showing the steam generation. **Fig. 2.28. Showing the Water trap.**

DANGER

If the substance crystallizes in the condenser, it will close the tube. Steam pressure could build up when the tube closes and cause an explosion. Use care! Drain the condenser of cooling water. The crystals will melt and pass into the distillate. When the tube is clear, slowly pass the cooling water through the condenser.

Always disconnect the steam-inlet tube from the flask.

Steam Distillation with Superheated Steam

The use of superheated steam can increase the proportion of the low-vapour pressure component in the distillate, and, at the same time, reduce the amount steam condensate in the distilling flask. The distilling flask is surrounded by a heating bath, which is heated to the san temperature as the superheated steam; this minimizes any cooling of the steam, which could occur before it enters the flask. (See Fig. 2.29)

Refluxing

The reflux procedure allows you to heat a reaction mixture for an extended period of time without loss of solvent. The condenser, which is fixed in a vertical position directly above the heated flask, condenses all vapors to liquid. Because none of the vapors escape, the volume of liquid remains constant.

Reflux procedures are carried out in neutral, acid, or basic solution, depending upon the reaction.

Typical operations include hydrolysis-saponification of acid amides, esters, fat nitriles, substituted amides, and sulfonamides. Hydrolysis-saponification is used to split organic molecules (which were made by combination of two or more com-pounds) into the original compounds.

Experimental setups are shown in Figs. 2.31(*a*) to 2.31(*d*).

Procedure

1. The water inlet to the condenser is the lower one. The water outlet to the condenser is the upper one.
2. Fill the heating flask no more than half full; add a few boiling stones.
3. Turn on the cooling water.
4. Heat to reflux for the desired period of time.

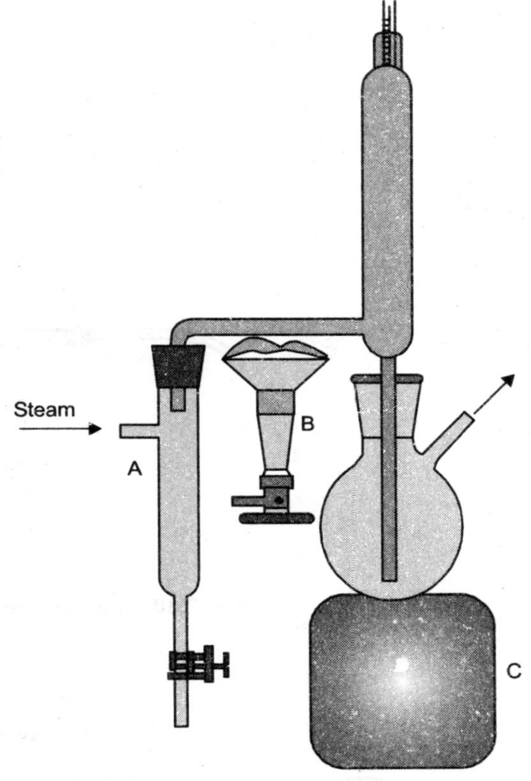

Fig. 2.29. Showing the Distillation set up for super heated steam.

Molecular Stills

Many organic substances cannot be distilled by any of the ordinary distilling methods because (1) they are extremely viscous, and any condensed vapours would plug up the distilling column, side arm, or condenser; (2) their vapours are extremely sus-ceptible to condensation.

To distill high-molecular-weight substances (molecular weights around 1300 for hydrocarbons and around 5000 for silicones and halocarbons) molecular stills are used. Molecular distillations differ from other distillations because:

1. All condensed vapour flows to the distillate receiver or collector.
2. Very low pressure (high vacuum) in the system favors vapourized molecules reaching the condensing surface without collision with other molecules to con-dense prematurely.
3. There is a very short distance between the surface of the evaporating liquid and the condenser surface. (See Fig. 2.32).

Laboratory models include a rotating still. Materials are fed at slightly above their melting point into the rotating still. They are distributed evenly and thinly over the heated evaporating surface. The sample distills, requiring a very short time, and its vapors condense to run into the collector. The degree of vacuum is controlled to collect the distillate effectively at the condenser, and the pressure can be as low as 1mm Hg.

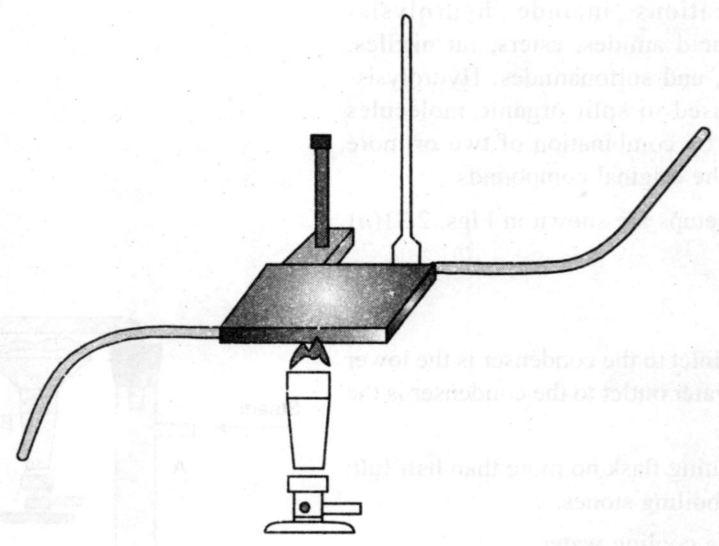

Fig. 2.30. The commercial metallic super heater.

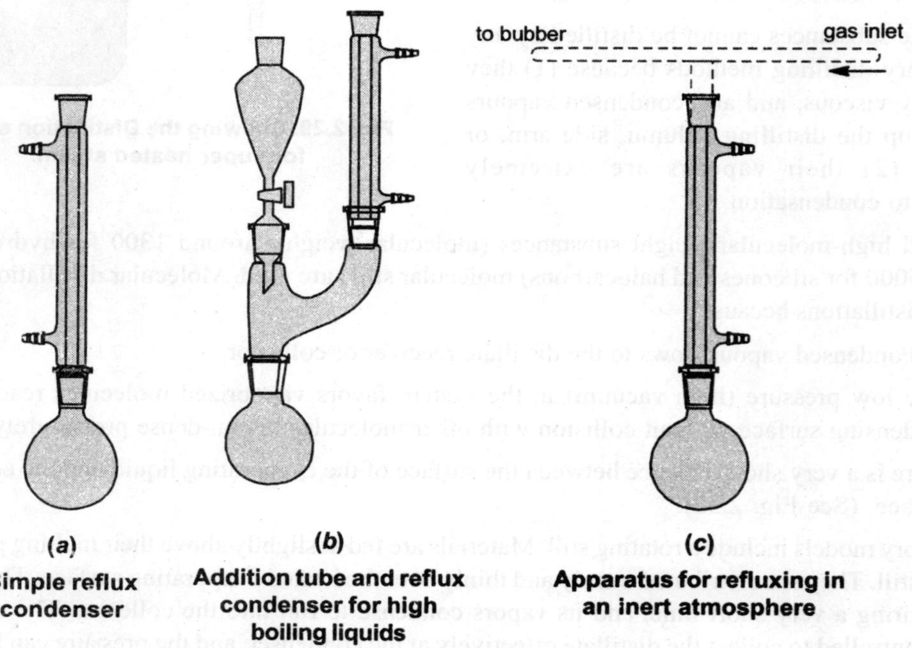

Fig. 2.31(*a*)

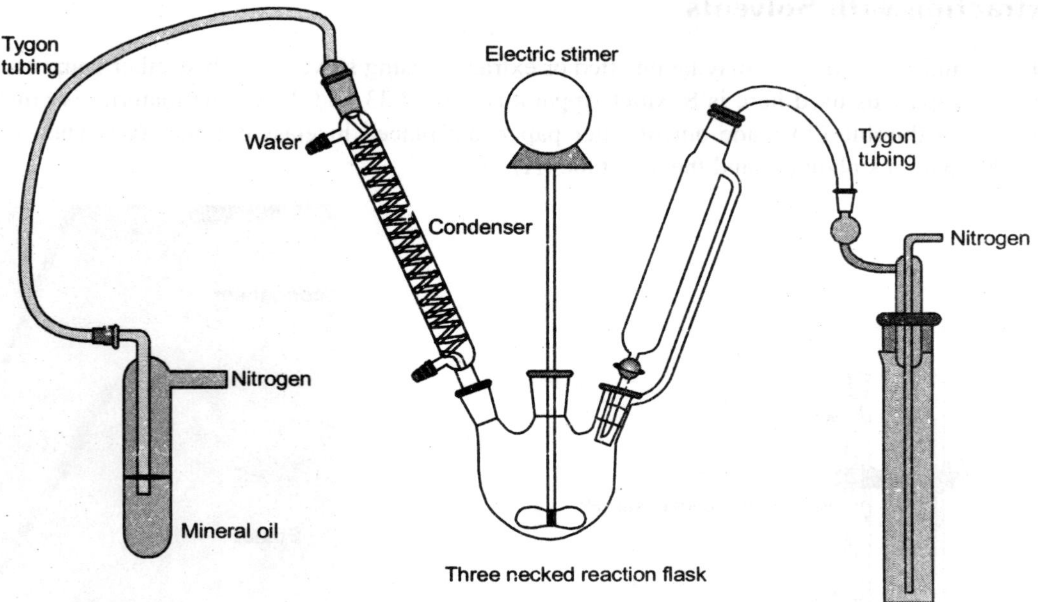

(*d*) Apparatus with mechanical fpr reflux reaction in an inert atmosphere.

Fig. 2.31(*b*)

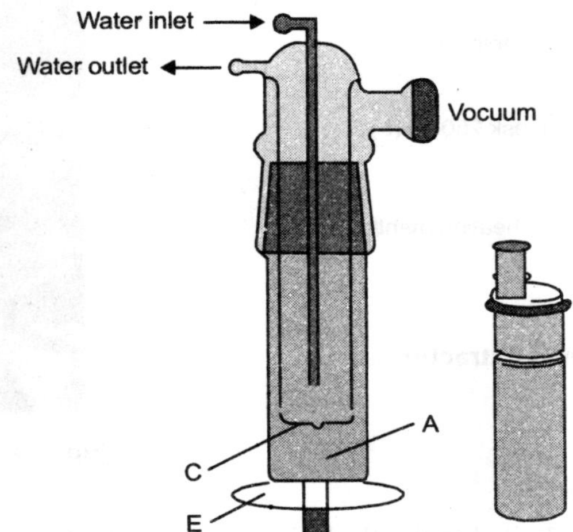

Fig. 2.32. Showing the simple molecular still made of glass A sample to be distilled, B cold finger condenser C, tip condenser to guide condensate into D; D, distillate collector, E source of heat

Extraction with Solvents

Fats oils and alkaloids etc., may be purified or extracted using solvents such as ether benzene, alcohol etc. The apparatus used here is Soxhlet Apparatus (Fig. 2.33 and 2.34) The material is crushed and placed in a thimble (T) made out of filter paper and placed in glass cylinder (C). The cylinder is provided with a siphon (S) and the side tube (A).

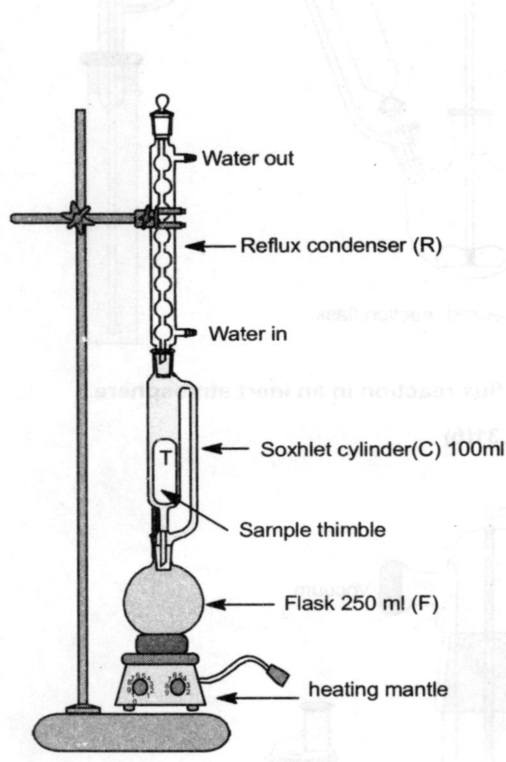

Fig. 2.33. The Soxhlet Extractor.

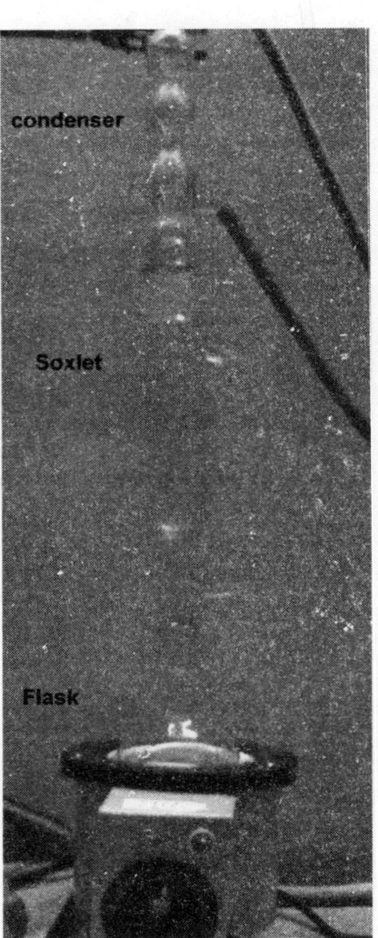

Fig. 2.34. The Soxhlet Extractor in author's lab

The solvent is carefully boiled in the flask F. The vapour rises through the side tube condenses in the upright reflux condenser (R) and drops down into the cylinder and therefore to the thimble and dissolves the sample, the dissolved sample solution accumulates in the cylinder (C) and then comes back to the flask (F) via the siphon (S) and then the solvent is redistilled and the sample is extracted and weighted.

Chemical Methods

Preparation for Derivation

An organic compound can also be isolated and purified by forming its solid derivative with another substance, which does not react with the impurities. The derivatives are then purified by recrystallization and decomposed to produce back the compound in the pure state. Methyl alcohol for example, may be purified by converting it into its solid ester methyl oxalate. This on hydrolysis with aqueous sodium hydroxide solution regenerates methyl alcohol, which may be distilled off and dried. Similarly an aldehyde can be converted into solid bisulphate compound and then decomposed by forming an insoluble salt and decomposing the latter with mineral acid.

Criteria for Purity

The purity of an organic substance is treated by the determination of its melting point in the case of a liquid. A pure substance has a sharp and definite melting point. This means that when the substance starts melting the whole of it melt at once at a fixed temperature in the case of the impure substance the melting temperature changes as the process continues.

In the case of those substances that are volatile in nature and also therefore does not give a sharp melting point, these cases the other parameters like density or refractive index are tested.

A rough idea of the purity of the organic solids is also obtained by the microscopic examinations.*

Since the determination of melting point is comparatively a simpler process, certain organic liquids are often tested through the formation of solid derivatives for example acetyl derivative melting at 113°C

Certain solids such as sugars, which do not have definite melting points, also form derivatives, which sharp melting points.

---

*This examination is done under electron microscope (the QUANTA) the name of the test is EDEX.

CHAPTER

3

Qualitative Organic Analysis

Introduction

In this chapter we will be discussing about the detection process of organic compounds. The elements commonly found in the organic compounds are mainly carbon hydrogen oxygen and nitrogen also sulphur, the halogen (chlorine bromine and iodine) and the metals and phosphorous sometimes.

Identification of Carbon and Hydrogen

All the organic compounds contain carbon hydrogen is also present in most of the organic compounds (except CS_2, CCl_4 etc.)

The detection of carbon is necessary only when it is not known whether the compound is organic is or not.

If the substance contains carbon, it gets charred or burns away with a smoky flame on heating alone or with sulphuric acid.

The best proof of the presence of carbon and hydrogen is obtained by completely oxidizing the organic compound by heating it with dry cupric oxide or lead chromate. The carbon and hydrogen of the organic compound are oxidized to CO_2 and H_2O respectively CO_2 is tested with lime water and H_2O with anhydrous copper sulphate changes to blue colour with water.

Reactions: -

$$\underset{\text{organic substance}}{[C + H_2]} + 3CuO \xrightarrow[\text{Heat}]{} CO_2 + H_2O + 3Cu$$

$$Ca(OH)_2 + CO_2 \rightarrow \underset{\substack{\text{calcium}\\ \text{carbonate}\\ \text{(milky)}}}{CaCO_3} + H_2O$$

$$\underset{\substack{\text{anhydrous}\\ \text{white}}}{CuSO_4 + 5H_2O} \rightarrow \underset{\substack{\text{hydrated}\\ \text{blue}}}{CuSO_4 + 5H_2O}$$

Experiment: A small quantity of the organic substance is mixed with excess of dry powdered cupric oxide. This mixture is then placed in an Borosil tube fitted with an delivery tube, whose other end is dipping in the lime water. (See Fig. 3.1).

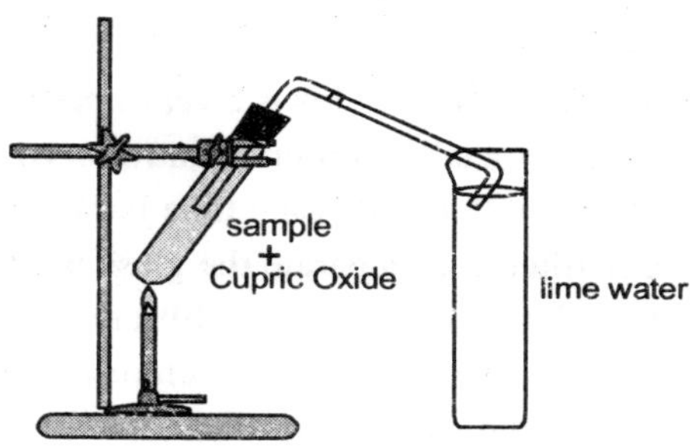

Fig. 3.1. The set up for detection of carbon in organic compounds.

The mixture is strongly heated if carbon is present it is oxidised to carbon dioxide and the lime water turns milky, If hydrogen is also present drops of water will appear on the upper walls of the tube, where they can be tested with anhydrous copper sulphate which turns like in the presence of water.

If the given organic substance is a gas or a volatile liquid the process is modified and the apparatus shown in the Fig. 3.1 is used for the detection of the carbon and hydrogen. In this vapours of the substance are passed over heated cupric oxide and the issuing gas are tested for moisture (H_2O) and carbon dioxide

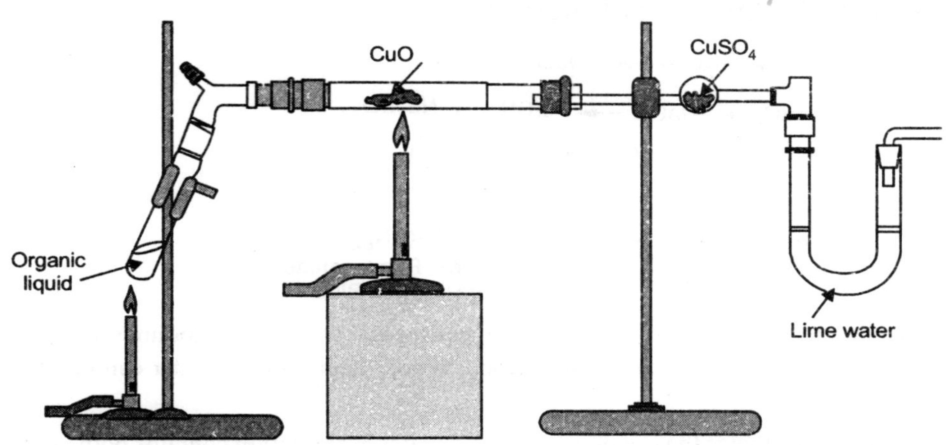

Fig. 3.2. the detection of carbon and hydrogen in the volatile organic liquid.

Detection of Nitrogen

Many organic compounds containing nitrogen when heated gives either a smell of burning hair or red fumes. Nitrogen is specifically detected by the following means.

Soda lime test: - The substance is treated with lime. The evaluation of ammonia gas proves the presence of nitrogen in organic compound. Ammonia gas may be tested. If the characteristic smell is found with Nessler's Reagent turns brown. This test for nitrogen is not very satisfactory and fails in the case of nitro-compounds azo-compounds etc. Therefore a negative result is obtained in the experiment, which may not mean the absence of the nitrogen.

Lassaigne's Test: - This is a very important and conclusive test for the nitrogen for the nitrogen. A small piece of metallic sodium is introduced with the sample in the ignition tube and the tube is heated until red hot and after this the tube is plunged into distill water kept in mortar and then grinded, the mixture is filtered to separate the glass beads and the filtrate which is known as ***Lassaigne's Filtrate*** and is used for analysis. To this filtrate a freshly prepared ferrous sulphate is added and the liquid is boiled and neutralized with hydrochloric acid and little ferric chlorine is added. A blue green colour or precipitate in the solution confirms the presence of nitrogen in the organic compound, the coloured compound id the Prussian blue.

Reactions: -

$$\underset{\text{organic substance}}{C + N + Na} \xrightarrow{\text{Fused}} \underset{\text{sodium cyanide}}{NaCN}$$

•

Reaction Equations of Nitrogen Estimation

$$2Na + 2H_2O \longrightarrow 2NaOH + 2H_2$$

$$FeSO_4 + 2NaCN \longrightarrow Fe(CN)_2 + Na_2SO_4$$

$$Fe(CN)_2 + 4NaCN \longrightarrow \underset{\text{Sodium Ferrocyanide}}{Na_4[Fe(CN)_6]}$$

$$Na_4[Fe(CN)_6] + 4FeCl_3 \longrightarrow \underset{\text{Ferric Ferrocyanide or Purssian Blue}}{Fe_4[Fe(CN_6]_3}$$

[it should remembered that Lassaigne's test fails in carbon causes and as diazo-compounds and methyl acetaldehyde. In such cases a number of potassium carbonate and magnesium powder can be used in place of sodium

Another point is to be noted that if the organic compound under examination contains both nitrogen and sulphur a blood red colour will be obtained in the above test, due to the formation of ferric sulphocyanide.]

$$C + N + S + Na \longrightarrow NaCNS$$

organic substance → sodium sulphocyanide

$$3NaCNS + FeCl_3 \longrightarrow Fe(CNS)_3 + 3NaCl$$

Ferric Sulphocyanide [blood red]

Lassaigne's test is used for testing sulphur phosphorous and halogens also.

Detection of Sulphur

The organic substance containing sulphur can be obtained in two processes.

Lassaigne's test: The substance is fused with sodium as in the case of nitrogen and Lassaigne's filtrate is tested for sodium sulphite.

(a) **Lead acetate and acetic acid**, when a black precipitated for the lead sulphide is produced

$$(CH_3COO)_2Pb + Na_2S \longrightarrow PbS + 2CH_3COONa$$

(b) Sodium nitroprusside solution: when deep violet coloration is produced

$$Na_2Fe(CN)_5(NO) + Na_2S \longrightarrow NA_3F_2(O = N - S - Na)(CN_5) \text{ (violet coloured compound).}$$

Oxidation

Sulphur in an organic compound can also be tested by fusing a little of the substance with excess of oxidizing mixture, i.e. a mixture of potassium nitrate an sodium carbonate. After allowing cooling the mass is extracted with water and filtered. The filtrate containing potassium and sodium sulphate is acidified with hydrochloric acid and a solution of barium chloride is added to it. A white ppt of barium sulphate is produced which is insoluble in inorganic acids, this proves that presence of sulphur in the substance.

Reactions:

$$\underset{\text{organic}}{(S)} + \underset{\text{oxidation mixture}}{KNO_3 + Na_2CO_3} \rightarrow K_2SO_4 + Na_2SO_4$$

$$K_2SO_4 + BaCl_2 \xrightarrow{HCl} \underset{\text{(white ppt)}}{BaSO_4} + 2KCl$$

$$NaSO_4 + BaCl_2 \xrightarrow{HCl} \underset{\text{(white ppt)}}{BaSO_4} + 2KCl$$

The presence of halogens is often proved by the following tests:

(1) ***Beilstein Test***: This is carried out by heating a copper wire in the flame until the colour of the flame is green, then the wire is dipped in the organic substance and again brought into the flame, the flame now gives a bluish colour, indicates halogen.

> This is not reliable because urea pyridine and its derivative does not contains any halogen also respond to this test due to the formation of copper cyanide.
> Also this test does not inform about the type of halogen.

(2) ***Lassaigne's Test:*** The organic substance is fused with halogen of the compound as in the case of nitrogen. When the halogen of the compound is converted into sodium halide NaCl NaBr or NaI. The Lassaigne's filtrate is then boiled with a little concentrated nitric acid to decompose the cyanide or silver nitrate solution is added to it when a precipitate is produced.

White precipitate of silver chloride soluble in ammonium hydroxide but insoluble in nitric acid indicates chlorine

$$NaCl + AgNO_3 \rightarrow AgCl + NaNO_3$$

Pale yellow precipitate of silver bromide, soluble only in excess of ammonium hydroxide but insoluble in nitric acid indicates bromine.

$$NaBr + AgNO_3 \rightarrow \underset{\text{pale yellow precipitate}}{AgBr} + NaNO_3$$

> Sodium cyanide and sulphide would be formed if the cyanide compound also contains nitrogen and sulphur. Their removed by boiling with HNO_3 in necessary as otherwise they would also give precipitate with $AgNO_3$.

Deep yellow precipitate of silver iodide, insoluble in ammonium hydroxide and nitric acid indicates iodine.

$$NaI + AgNO_3 \rightarrow \underset{\text{deep yellow precipitate}}{AgI} + NaNO_3$$

Distinction Between Bromine and Iodine

The Lassaigne's filtrate after boiling with conc. nitric acid is then treated with carbon di sulphide or carbon tetrachloride and a few ml of strong chlorine water. The mixture is thoroughly shaken. If CS_2 or CCl_4 layer turns brown if bromine is present and if turns violet iodine is present.

$$2NaBr + Cl_2 \rightarrow 2NaCl + Br_2 \text{ (soluble in CS or } CCl_4\text{) gives brown colour.}$$

$$2NaI + Cl_2 \rightarrow 2NaCl + I_2 \text{ (soluble in CS or } CCl_4\text{) gives violet colour.}$$

The halogen in an organic compound can also be detected by fusing it with quick lime, the residue is extracted with water and filtered. Calcium halide ($CaCl_2$ or CaB_2 and CaI_2) formed remains in the filtrate and is identified with $AgNO_3$ solution in the used way.

Detection of Phosphorous

The organic substance is fused with sodium peroxide. The fused mass is extracted with water. The filtrate containing sodium phosphate is boiled with an excess of nitric acid and ammonium molybdate. A yellow colour or a precipitate of ammonium phosphomolybdate indicates the presence of phosphorus.

Reactions:

$$\text{(1) } 2Na_2PO_4 + 3(NH_4)_3MoO_4 \rightarrow 2(NH_4)_3PO_4 + 3Na_2MoO_4$$

$$\text{(b) } Na_2MoO_4 + 2HNO_3 \rightarrow H_2MoO_4 + 2NaNO_3$$

$$\text{(c) } H_2MoO_4 \rightarrow H_2O + MoO_3$$

$$\text{(d) } (NH_4)_3PO_4 + 12MoO_3 + 6H_2O \rightarrow \underset{\text{ammonium phosphomolybdate}}{(NH_4)_3PO_4 + 12MoO_3 + 6H_2O}$$

Detection of Oxygen

There is no definite method of detecting the presence of oxygen compounds. If the substance on heating alone in a test tube gives water vapours oxygen may be present,. The presence of oxygen is however proved by the results of quantitative analysis. The presence of (– O –) aldehyde.

(– CHO) ketone (>C = O) amide (– $CONH_2$) etc. group in an organic compound also proves the presence of oxygen in it.

Detection of Metals

The substance containing metals are strongly ignited to destroy all organic matter. The residual metallic oxide or carbonate is then identified by the usual methods of inorganic quantitative analysis.

QUANTITATIVE ANALYSIS

Estimation of the Element in an Organic Compound

The estimation of the common elements present in an organic compound is known as (quantitative organic analysis) it is carried out as follows: -

Estimation of Carbon an Dhydrogen

Liebig's Method of Combustion

All organic compounds (except the stable carbonates and cyanide of the alkali metals and the alkaline earth metals) are decomposed when they are brought into contact with red-hot cupric oxide. The carbon of the compound is converted into carbon dioxide and hydrogen into water. By taking a known weight of a substance and collecting and weighing the products of combustion i.e. CO_2 and H_2O percentage of carbon and hydrogen may be readily determined. (See Fig. 3.3)

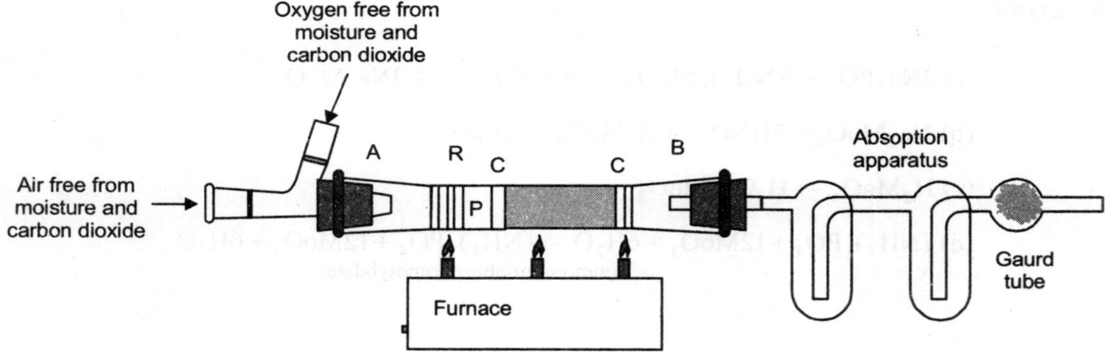

Fig. 3.3. The set up Liebig's Method.

The combustion is carried out in a glass tube (A, B) of Fig. 3.3 about 90 cm long and 12-15 mm diameter. About two-thirds of the tube (C, C is filled with cupric oxide kept in position by two pieces of oxidised copper rolls E, E). Before the experiment is started the tube is strongly heated in a combustion furnace. Dry air free from CO_2 is admitted into the tube. The air may be freed of moisture and CO_2 by passing it through conc. sulphuric acid in a water and soda lime in 'U' tube (not shown in the fig). The moisture traces of organic matter and CO_2 are driven off the tube empty section of the tube is cooled and the free end (B). The empty section of the tube is cooled and the free end then connected through a rubber cork. To an accurately weight 'U' tube containing fused calcium chloride the other end of the 'U' tube is connected to another 'U' tube containing soda lime this can be replaced by the introduction of potash bulb (Fig. 3.4) containing about 40% potassium hydroxide.

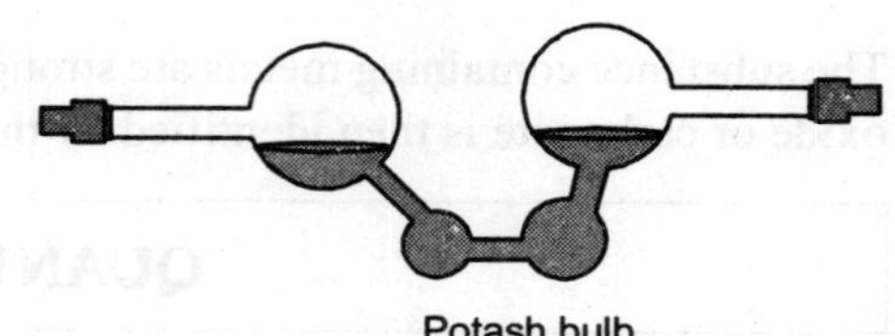

Fig. 3.4. The Potash bulb

The free end of the soda lime tune is connected to the guard tube filled with a mixture of fused calcium chloride and soda lime. This serves to prevent the entry of moisture and form CO_2 from the atmosphere.

All connections must be perfectly airtight. About 0.2 g of the organic substance is dried and accurately weighted in the small porcelain boat P. This boat is then introduced into the combustion tube. All connections must be perfectly airtight. A freshly ignited role of copper (R) is placed behind organic liquid is to be analyzed is weighted out in a thin glass bulb which is then placed in the boat.

A slow current of purified air is passed through the combustion tube, which gradually heated, from the both ends proceedings towards the boat. The heating is continued until the whole of the organic substance in the boat has burnt away. As soon as the whole tube has been raised dull red heat the current of air is turned off and a stream of pure oxygen is passed in order to burn away any remaining organic matter and to oxidize the copper, which has been formed by the reduction of some cupric oxide. Finally, air is again passed to expel oxygen from the apparatus. The burners are turned off one by one and the apparatus is allowed to cool.

The experiment takes about 2-3 hours to complete. The $CaCl_2$ tube and the soda lime tube is disconnected, closed and weighted. The increase in weight of $CaCl_2$ tube gives the weight of H_2O and the increase in weight of soda-lime tune gives the weight of CO_2 produced.

Knowing the weight of the organic compound and the weights of the CO_2 and H_2O formed during the combustion the percentage of carbon and hydrogen is easily calculated.

The molecular weight of Carbon dioxide is 44 and it consist of 12 g of carbon

$$\therefore \text{ The percentage of Carbon} = \frac{12}{44} \times \frac{\text{weight of } CO_2}{\text{weight of organic compound}} \times 100$$

The molecular weight of H_2O is 18 and it contains of 2 g of hydrogen

$$\therefore \text{ The percentage of Hydrogen} = \frac{2}{18} \times \frac{\text{wt of } H_2O}{\text{wt of compound}} \times 100$$

Precaution:

1. The air or oxygen flow through the combustion tube should be regulated at 2-3 bubbles per second.
2. If the substance contains nitrogen, a roll of bright copper gauze is placed near the right end of the combustion tube. This roll is kept red hot during combustion. This reduces the oxides of nitrogen, which may be produced during the experiment and which would otherwise be dissolved in calcium chloride tube.
3. If the substance contains hydrogen silver gauze is used instead of copper gauze. Silver gauge used instead of copper gauze, silver gauze will prevent them from passing it into the adsorption apparatus.
4. If the substance contains halogen, sulphur or phosphorous fused lead chromate is used in place of cupric oxide.

$$2PbCrO_4 \rightarrow 2Pb + Cr_2O_3 + 5(O)$$

Example: 0.11330 g of substance gave 0.6118 g of CO_2 and 0.1315 g of H_2O. Calculate the percentage of carbon and hydrogen in the substance.

Solution: 44 g of CO_2 contains 12 g of carbon

$\therefore$ 0.6118 g of CO_2 conatins $\dfrac{0.6118 \times 12}{44} = .1668$ g of C

Hence % of C = $\dfrac{.1668 \times 100}{0.1830} = 91.19$ **Ans.**

Again 18 g of H_2O conatins 2 g of H

$\therefore$ 0.1315 g of H_2O contains $\dfrac{0.1315 \times 2}{18} = 0.0146$ g of H

Hence, the percentage of H = $\dfrac{0.0146 \times 100}{0.1830} = 7.978$

Example: 0.20 g of an organic compound containing carbon hydrogen and oxygen give on combustion 0.040 of H_2O and 0.195 g of CO_2 calculate the percentage of the compound.

Solution:

Percentage of Carbon $= \dfrac{12}{44} \times \dfrac{.195}{.20} \times 100 = 26.59$

Percentage of Hydrogen $= \dfrac{2}{18} \times \dfrac{\cdot 04}{.20} \times 100 = 2.22$

And the percentage of oxygen by diffence 100 − $(26.59 + 2.22) = 71 \cdot 19$.

Estimation of Nitrogen

Nitrogen estimation is done by the following methods:

1. ***Dumas's Method***: In this method a known weight of an organic compound is heated in a combustion tube with cupric oxide and a bright metallic copper spiral in the atmosphere of CO_2. The organic compound decomposes and produces CO_2 H_2O and nitrogen gas. The products of decomposition are collected in a nitro meter tube containing a strong solution of potassium hydroxide; the carbon dioxide and water are absorbed while the nitrogen gas gets collected in the tube. When the whole of the substance has completely decomposed, the volume of the nitrogen is measured and its temperature and pressure are noted, Its volume is then reduced to N.T.P and the weight of this volume from the fact that the weight of 22.4 litres of nitrogen at N.T.P is 28 g*. From the weight the percentage of nitrogen can be easily calculated.

   The apparatus is shown in Fig. 3.5.

   (*a*) It consists of a Borosil glass flask: charged with either magnesium carbonate on heating this CO_2 is produced.

---

*It is to be noted that 1 cc of nitrogen weighs 0.00126 at N.T.P.

$$\text{Sodium Bicarbonate} \xrightarrow{\text{heat}} \text{Sodium Carbonate} + \text{Water} + \text{Carbon Dioxide}$$
$$2NaHCO_3 \xrightarrow{\text{heat}} Na_2CO_3 + H_2O + CO_2$$

$$\text{Magnesium Carbonate} \xrightarrow{\text{heat}} \text{Magnesium Oxide} + \text{Carbon Dioxide}$$
$$MgCO_3 \xrightarrow{\text{heat}} MgO + CO_2$$

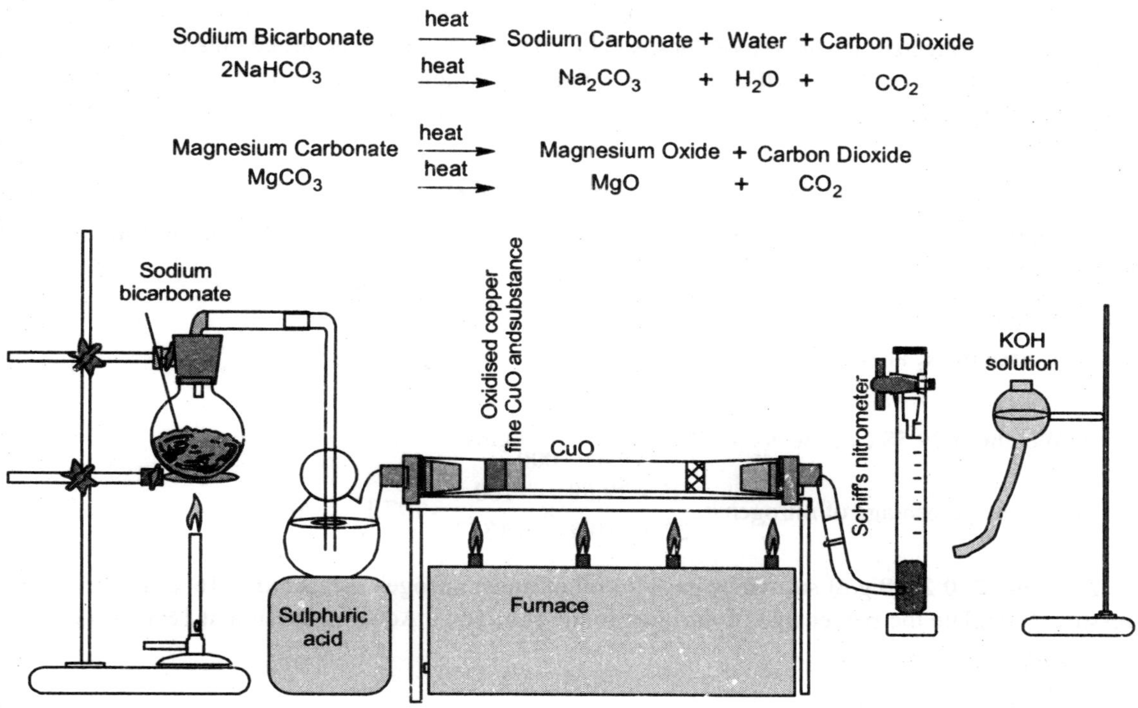

Fig. 3.5. The Set up for nitrogen estimation by Dumas's Method

(*b*) **A bubbler tube containing conc. sulphuric acid:** Carbon dioxide is dried by passing it through this bubbler. The bubbler also serves to show the rate at which the gas is passing.

(*c*) **A combustion tube:** open at both ends and filled with oxidised copper coil a boat filled with copper oxide and a spiral or bright copper.

(*d*) **Schiff's Nitro meter:** A graduated glass tube with a mercury trap at the base and concentrated solution of potassium hydroxide.

A known weight of the organic substance (say .2 g) is mixed with powdered cupric oxide and the mixture placed in the boat. The boat is then shipped into the combustion tube. A small piece of copper spiral is placed at the right end of the combustion tube. It served to reduce any oxide of nitrogen formed during the combustion. The apparatus is then set in order

Carbon dioxide, produced by heating sodium bicarbonate or magnesium carbonate, placed in a Borosil flask is passed trough the combustion tube in order to remove all the air inside the tube and then the combustion tube consisting the CuO is heated first, when a test portion of the gas is collected in the nitrometer is completely absorbed by KOH, all the air is considered to be displaced. The current of CO_2 is now stopped and the nitrometer is filled. The boat containing the organic compound and fine CuO is now heated. The organic substance decomposes and the gaseous products of combustion (CO_2, H_2O

and N_2) leaving the combustion tube enter the nitrometer when CO_2 and H_2O are observed and the nitrogen gas is collected, when no more gas is evolved the CO_2 stream is turned on and the nitrogen in the combustion tube is passed into the nitrometer is noted by bringing the KOH solution to the temperature and pressure are also noted and the volume of percentage of nitrogen is calculated as is explained in the following examples:

Dumas's method for estimation nitrogen is of universal application.

***Example*: 0.1877 gm of an organic compound gave combustion 29.58 cc of dry nitrogen at NTP calculate the percentage of nitrogen in the compound.**

***Solution*:**

22.4 litres of N_2 at NTP weighs 28 gms

$$\therefore \text{ Weight of 29.58 c of the } N_2 \text{atNTP} = \frac{28 \times 29.58}{22.4 \times 1000} \text{gms}$$

$$\text{Hence the percentage of nitrogen} = \frac{28 \times 29.58}{22.4 \times 1000} \times \frac{100}{0.1877} = 197.1 \quad \text{Ans}$$

***Example 2*: 0.2248 g of substance gave 76 ml of moist nitrogen measured at 16°C and 753.5 mm pressure. Calculate the percentage of nitrogen in the substance. Aqueous tension at 16°C is 13.5 mm.**

***Solution*:**

V_1 = 76 ml

V_2 = vol at NTP

P_1 = 753.5 –13.5 = 740 mm

T_1 = 16 + 273 = 289°A

P_2 = 760 mm

T_2 = 273°A

Convert the volume to NTP by applying the formula

$$\frac{P_1V}{T_1} = \frac{P_2V_2}{T_2}$$

$$\frac{740 \times 76}{289} = \frac{760V_2}{273}$$

$$\therefore \quad V_2 = \frac{740 \times 76 \times 273}{249 \times 760} = 69.90 \text{ ml}$$

$$\text{wt of 69.90 cc of } N_2 = \frac{28 \times 69.9}{22400} = 0.873 \text{ g}$$

$$\text{Hence, the percentage of } N_2 = \frac{0.0873}{0.2248} \times 100 = 38.87 \quad \textbf{Ans.}$$

Kjeldahl's Method

A known weight (usually .2 to .3 g) of the substance is mixed with about 20 cc of concentrated sulphuric acid and about 10 g of solid potassium sulphate. This dark coloured mixture is then heated in long necked round bottom flask called Kjeldahl's flask (Fig. 3.6).

Potassium sulphate is added just to raise the boiling point of the sulphuric acid. A little copper sulphate potassium permanganate mercury, potassium dichromate manganese dioxide or selenium can also be added during the heating, to catalyze the decomposition. The heating is continued till contents of the flask are colourless (See Fig. 3.6).

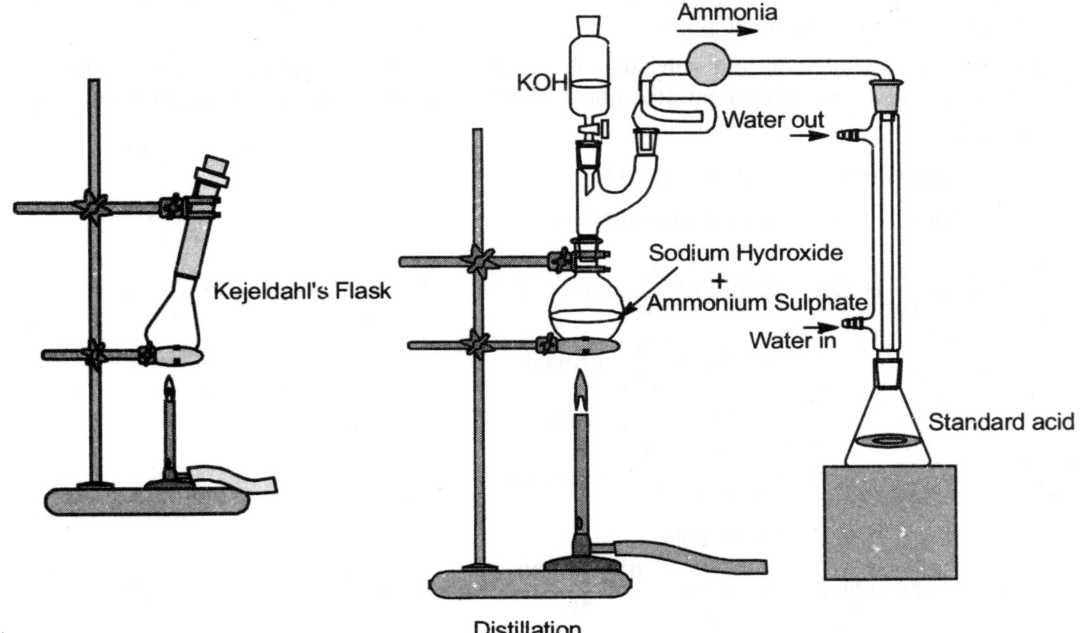

Fig. 3.6.The Kjeldahl's Method

Carbon and hydrogen of the organic compound are converted to CO_2 and H_2O and escape during heating. Nitrogen of the compound is converted to ammonia, which with H_2SO_4 forms ammonium sulphate, and hence the solution in the flask becomes clean and colourless.

$$\underset{\text{Organic Substance}}{\{C+H+N\}} + H_2SO_4 \xrightarrow{\text{Heat strongly}} CO_2 + H_2O + NH_3$$

$$2NH_3 + H_2SO_4 \rightarrow \underset{\text{Ammonium Sulpahte}}{(NH_2)_2SO_4}$$

The reaction mixture is now distilled with excess of sodium hydroxide is specially designed apparatus (Fig. 3.6). The liberated ammonia is absorbed in known excess and standard H_2SO_4 and HCl

$$(NH4)_2SO_4 + 2NaOH \xrightarrow{\text{Distil}} Na_2SO_4 + NH_3 \uparrow + 2H_2O$$

$$NH_3 + HCl \rightarrow NH_4Cl$$

After complete absorption the excess acid is titrated back with standard NaOH or KOH. From this the amount of absorbed ammonia and hence that of nitrogen in the organic substance can be calculated.

This process is not applicable for the nitro compounds, pyridine or compounds containing azo group (–N=N–)

Kjeldahl's method gives good results with amines ammonium compounds amino acids, proteins and related compounds. It is, therefore, employed particularly in agricultural laboratories for the analysis of foodstuffs, manures and other products.

Example: 0.28 g of an organic substance is Kjeldahlized. The ammonia evolved requires 25 cc of N/10 sulphuric acid for complete neutralization. Calculated the percentage of nitrogen in the compound.

Solution:

Wt of organic compound taken = 0.280 g

Since equinormal solution are equivolumetric

$$\therefore \text{ 25 cc of } \frac{N}{10} H_2SO_4 \equiv \text{25 cc of } \frac{N}{10} \text{ ammonia}$$

$$\equiv \text{25 cc of } \frac{N}{10} \text{ nitrogen}$$

$$\equiv \text{2.5 cc of N Nitrogen}$$

$$= \frac{2.5 \times 14}{1000} \text{ g of Nitrigen}$$

$$= 0.35 \text{ gms}$$

$$\text{Hence the percentage of Nitrigen} = \frac{.035 \times 100}{.28} = 12.5 \quad \textbf{Ans.}$$

Example 2: 0.442 g of an organic substance was treated according to Kjeldahl's method. The ammonia evolved was absorbed in 50 ml of NH_4SO_2 the residual acid required 65.6 ml of $\frac{N}{2}$ alkali solution. Find the percentage of nitrogen in the compound.

Solution:

Wt of the substance = 0.4422 g

Total acid taken = 50 ml of N H_2SO_4

Alkali used for back titration = 65.5 ml $\frac{N}{2}$ Alkali

= 32.75 ml N H_2SO_4

That is acid unused = 32.75 ml of NH_2SO_4

∴ Acid used = (50 – 32.75) ml H_2SO_4

The acid used for neutralizing ammonia

∴ 17.25 ml of N H_2SO_4

= 17.25 ml of N H_2SO_4

= 17.25 ml of N nitrogen

$= \frac{17.25 \times 14}{1000}$ g nitrogen

= .2415

Hence, the percentage of nitrigen

$= \frac{.2415}{.492} \times 100$

= 54.61 **Ans.**

Estimation of Halogen

Carius Method

A known weight of the organic compound is taken in a small ignition tube. This tube is then slipped carefully into a double walled hard glass tube about 40 cm long called Carius tube or bomb tube containing 4-5 cc of fuming nitric acid a few crystals of silver nitrate (see Fig. 3.7). The open end of the tube is drawn out and sealed. The tube is then wrapped in thin asbestos paper and heated to 200°C-250 °C in an iron "Bomb tube".

After this operation the tube is cooled and the drawn out and broken carefully. The precipitate of silver halide, which is thus formed, is filtered washed and dried and finally weighted. From the weight of the precipitate the percentage of the halogen is calculated.

Percentage of Chlorine

$$= \frac{35.5 \times \text{Wt of the precipiate of AgCl} \times 100}{(108+35.5) \times \text{weight of the Organic Compound}}$$

Percentage of Bromine

$$= \frac{80 \times \text{Wt of the precipiate of AgBr} \times 100}{(108+80) \times \text{weight of the Organic Compound}}$$

Percentage of Iodine

$$= \frac{127 \times \text{Wt of the precipitate of AgI} \times 100}{(108+127) \times \text{weight of the Organic Compound}}$$

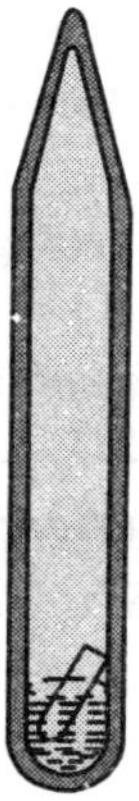

Fig. 3.7. Bomb Tube

***Example*:** 0.151g of a substance when heated with HNO_3 and $AgNO_3$ in Carius method gave 0.139 g of AgBr. Calculate the percentage of bromine in organic substance.

***Solution*:** 188 g of silver bromide contain 80 g of Br

$\therefore$ 134 g of AgBr will contain $\frac{80}{188} \times .134$ g of Br

Weight of the substance taken = 0.158 g

Hence the percentage of Bromine = $\frac{80}{188} \times \frac{.134}{.151} \times 100 = 37.51$

Estimation of Sulphur

Carius Method

Sulphur in an organic compound can also be estimated by Carius Method.

A known weight of the sample is heated with conc. Nitric acid in Carius tube as in the estimation of halogens. No $AgNO_3$ is used in this case. Nitric acid oxidizes the sulphur of the organic compound to sulphuric acid. The contents of the Carius after cooling are washed out and treated with sulphuric acid and excess of barium nitrate solution. A white ppt of barium sulphate is this produced. The precipitate is filtered washed and dried and weighted. From the weight of $BaSO_4$ the percentage of sulphur can be calculated

$$H_2SO_4 + Ba(NO_3)_2 \longrightarrow BaSO_4 + 2HNO_3$$

Oxidation

A known weight of an organic sample containing sulphur is strongly heat with an oxidizing mixture of sodium and sodium peroxide. Sulphur is thus oxidised to sodium sulphate. The hot mass is boiled with water and filtered. The filtrate containing Na_2SO_4 is acidified with hydrochloric acid and $Ba(NO_3)_2$ or $BaCl_2$ added to it. The white ppt of $BaSO_4$ is so formed is filtered washed dried and weighted the percentage of sulphur is then calculated as percentage of sulphur.

$$\text{Percentage of Sulphur} = \frac{32 \times \text{ of } BaSO_4 \text{ ppt} \times 100}{233 \times \text{ wt of organic compound}}.$$

***Example*:** 0.2518 g of organic substance containing sulphur, after suitable treatment gave 0.2638 g of $BaSO_4$. Calculate percentage of sulphur in the substance [Ba = 137 and 0 = 16].

***Solution*:**

$BaSO_4 = 137 + 32 + 64 = 233$

i.e., 233 g of $BaSO_4$ contains 32 g of sulphur

$\therefore$ 0.2633 g of $BaSO_4$ contain $\frac{32 \times .2638}{233}$ of sulphur.

Weight of the organic substance = 0.2518 g

Hence, percentage of sulphur $= \frac{32}{233} \times \frac{.2638}{.2518} \times 100$

$= 14.39$ **Ans.**

Estimation of Phosphorous

The estimation of phosphorous is also carried out in the same way as of sulphur. The phosphorous is oxidised to phosphoric acid or soluble phosphate, which is then treated with magnesia mixture. The precipitate of magnesium ammonium phosphate ($MgNH_4PPO_4$) so obtained is filtered washed and dried and ignited, when magnesium pyrophosphate is obtained

$$2MgNH_4PO_4 \xrightarrow{\text{ignite}} \underset{\text{Magnesium Pyrophate}}{Mg_2P_2O_7} + 2NH_3 + H_2O$$

From the weight of magnesium pyrophosphate, the percentage of phosphorous is calculated.

$$\text{Percentage of phosphorous} = \frac{62 \times \text{ wt of } Mg_2P_2O_7 \times 100}{222 \times \text{ wt of organic substance}}$$

Example: 0.6 g of an organic substance containing phosphorous was treated in Carius tube and phosphoric acid formed was precipitated with magnesia mixture as $MgNH_4PO_4$, which on subsequent ignition produced 0.999 g of $Mg_2P_2O_7$. Calculate the percentage of phosphorous in the organic compound.

[Mg=24, P =31, and O=16]

Solution:

$Mg_2P_2O_7 = 48 + 62 + 112 = 222$

That is 222 g of $Mg_2P_2O_7$ contains 62 g of phosphorous

$\therefore$ .999 g of $Mg_2P_2O_7$ conatins $\frac{62 \times .999}{222}$ g of P

wt of organic substance is = 0.6 g

Hence, the percentage of phosphorous $= \frac{62}{222} \times \frac{.999}{.6} = 46.5\%$

Estimation of Oxygen

A method of estimation of oxygen directly has been developed. A known weight of substance is heated in a current of hydrogen and the vapour evolved is passed over heated metallic nickel. The oxygen in the compound is wholly converted into H_2O which is absorbed in $CaCl_2$ and weighted.

CHAPTER

4

Structure of Organic Compounds

Introduction

After the determination of the molecular formula the next step for the diagnosis of the organic compound is to assign a structural formula to it, which is finally confirmed by its synthesis.

The formula, which gives the arrangement of, the atoms of different elements present in a molecule of the compound, is known as the *graphic* or *constitutional formula.*

It is a common phenomenon in organic chemistry that the same molecular formula may rearrangement more than one different properties. This is because the different atoms in a molecule can arrange themselves n more than one ways to form different structures. Consider for instance the molecular formula C_2H_6O. Keeping in mind that carbon is tetra valent oxygen divalent and hydrogen monovalent, these are only two possible ways in which two carbon six hydrogen atoms and one oxygen atom can be arranged as shown below

```
    H    H            H                  H
    |    |            |                  |
 H—C——C—OH      H—C————O————C—H
    |    |            |                  |
    H    H            H                  H
   ethanol
      I                  dimethyl ether
                               II
```

In structure one the two carbon atoms are directly linked together whereas in structure II the carbon atoms are linked trough oxygen atoms. Structure I is given to a compound called ethyl alcohol and the structure II is for dimethyl ether. Alcohol and ether are two different types of compounds. They are obtained by different methods and show different properties and reactions.

Such organic compound which have different structural formulae but same percentage composition and therefore the same empirical formula are termed as isomers or isomerides and the property is called isomerism.

As the number of the carbon atoms in the molecule increases the number of possible isomers will also increase. The formula C_5H_{12} has three isomers and C_6H_{14} has as many five isomers. Among the different structural formula, the one, which agrees best with the reaction of the compound, is selected as the correct fornmula of the compound.

Hence it is very important to determine the structural formula for a compound in order to make an intelligent study of its methods of preparation and properties.

The study of the structural formula is based upon the following facts

1. ***Tetra-valency of Carbon Atom***: The carbon atom is usually tetravalent. This is an arrangement with its potion in the periodic table. Carbon is placed in the IV group of the periodic table and has therefore electrons in the centre most orbit the valency shell of its atomic structure. Thus it has remarkable tendency to complete its outermost valency shell by sharing electrons rather than giving or accepting electrons. The sharing of electrons which always takes place in pairs and that is responsible for the contribution of carbon with other atoms through co-valences or co-valent bondings.

   The tetra-valency of the carbon atom is also proved by the fact that it can combine with four atoms of monovalent elements, groups or radicals. The valences are usually included by small lines and the valency electrons may be represented by either dots or crosses.

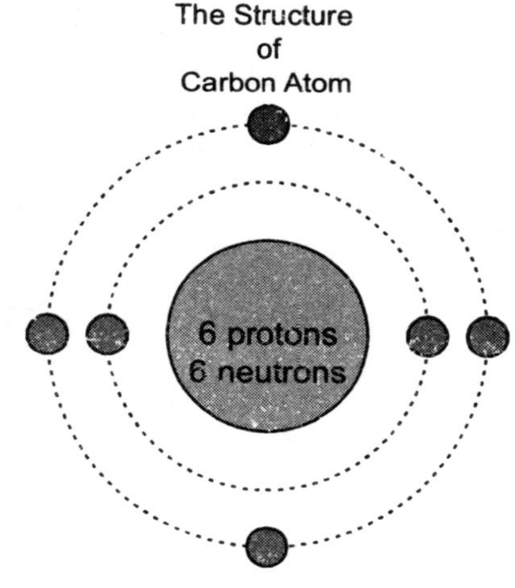

The Structure of Carbon Atom

The tetravalent carbon atom is represented in the following structural templates below:

—C— (with bonds above and below)   —C≡   =C=

The electronic representation of the carbon atom is

ˣCˣ (with crosses above and below)   ˣC⁑ˣ   ˣCˣ (crosses at four corners)

or

·Ċ· (with dots above and below)   ·C: :C:

The crosses or the dots represent the valency of the carbon atom.

Similarly the structural and the electronic formula for methane (CH_4) and carbon tetrachloride (CCl_4) etc can be written as shown:

In case of electronic formulæ for methane and carbon tetrachloride, the crosses are the valence electrons of the carbon atom and the dots represent the valence electrons of hydrogen and chlorine atoms.

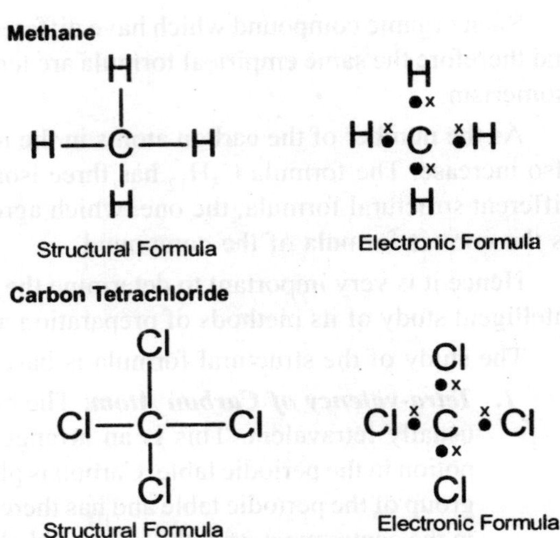

2. ***Equivalence of the Four Carbon Valences:*** The four valences of carbon are equivalent to one another. This can be shown by replacing any one of the four hydrogen atoms of methane by same mono-valent atom or group of atoms, when the sane mono-substitution product is obtained. This means that mono-substituted derivative means that a mono-substituted derivative of methane exists in only one form i.e. isomerism is not met with in this case.

Henry replaced each hydrogen by nitro group and obtained same compound nitro-methane in all the four cases

iodomethane $\xrightarrow{AgNO_3}$ 1-nitromethane

iodomethane $\xrightarrow{KCN}$ acetonitrile $\xrightarrow{H_2O}$ acetic acid $\xrightarrow{Cl_2}$ chloroacetic acid $\xrightarrow{AgNO_3}$ nitroacetic acid $\xrightarrow{CaO,\ Heat}$ 2 -nitromethane

1-nitromethane is identical with 2-nitromethane. Hence valences numbered 1 and 2 are identical same arrangement applies to 3 and 4.

3. ***Linkage of Carbon Atom***: The carbon atom possess a remarkable property containing with other carbon atoms to form chain or rings.

These compounds in which carbon atoms are linked together by single valency or bond are called saturated carbon compounds. They then form derivatives by substitution and are not very reactive compound. Methane ethane propane etc. are all saturated compounds.

methane ethane propane

Carbon atoms can also unite with other carbon atoms by double or triple bonds or linkages.

ethylene acetylene

The presence of the double or the triple bond is the sign of unsaturation. These compounds are therefore called **unsaturated organic compounds** and show characteristic reactions. They form addition compounds and are more reactive than saturated compounds

Electronic formula

Double or triple bond in an organic compound is a sign of weakness rather than strength

NOMENCLATURE SYSTEM OF ORGANIC COMPOUNDS THE INTERNATIONAL UNION OF PURE AND APPLIED CHEMISTRY METHOD AND THE GENEVA METHOD

Homology

Organic compounds are divided into two different classes

I. The open chain compounds

II. Closed chain compounds or the cyclic compounds

Open chain compounds, these compounds are those in which the carbon atoms are in straight or branched chain e.g.

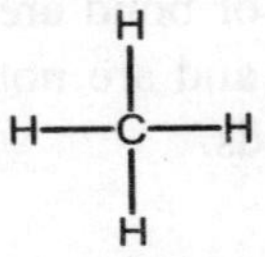

methane

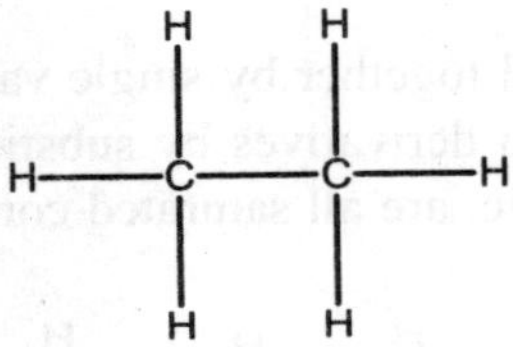

ethane

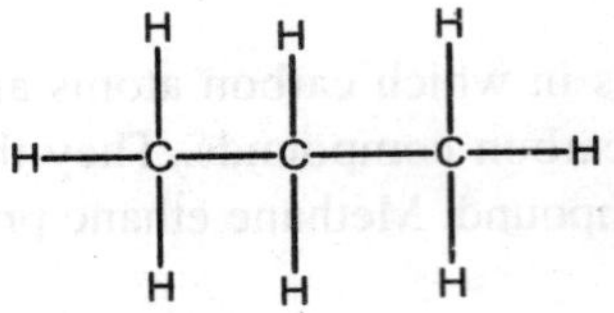

propane

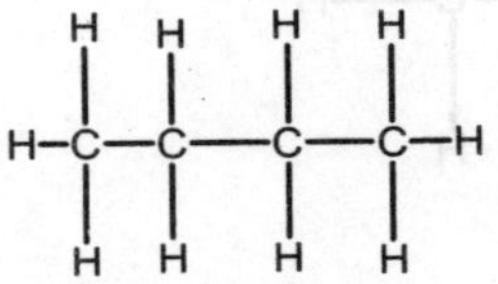

butane

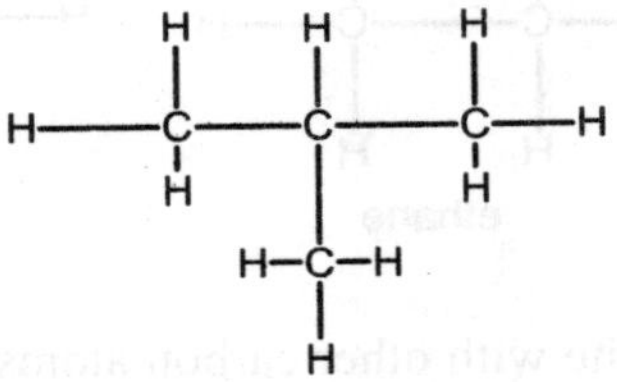

isobutane

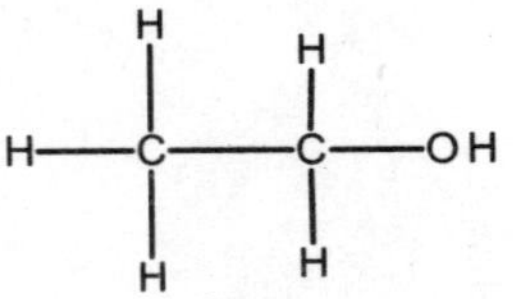

ethanol

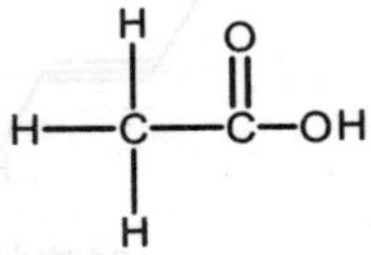

acetic acid

A large number of important open-chain carbon compounds are found in animal and vegetable fats and hence are known as fatty or aliphatic compounds. These compounds can be derived from methane and therefore are called methane derivatives. Fatty or aliphatic compounds form a different section of organic chemistry.

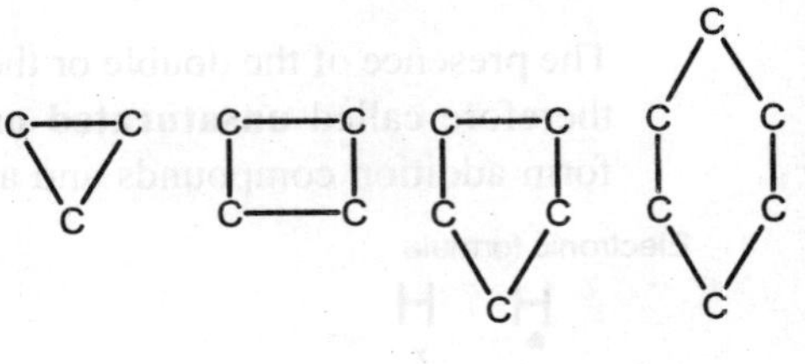

If in addition to carbon some other polyvalent elements such as O, S, N is also present in the ring the compounds are termed as hetrocyclic.

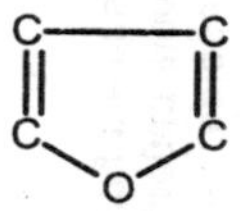

furan 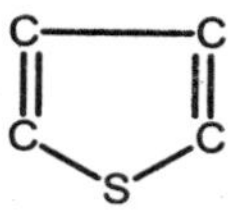 thiophene 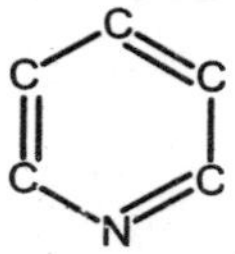 pyridine

Among the carbocyclic compounds those containing six carbon atoms possess a pleasant aromatic odour (aroma) and are called ***Aromatic Compounds***, the parent substance of this family is benzene and hence they are called benzene derivatives

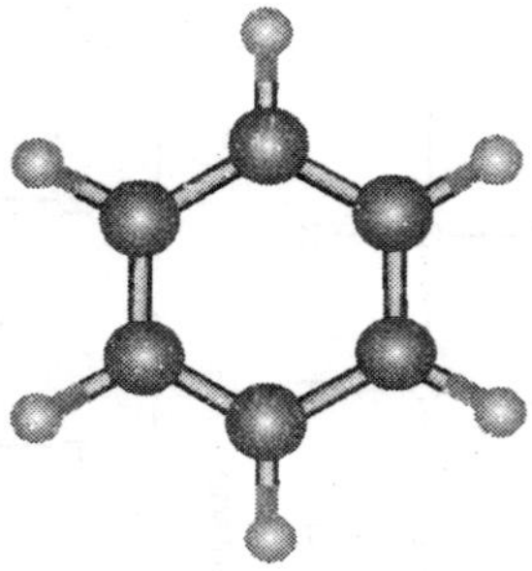

3-D model of Benzene

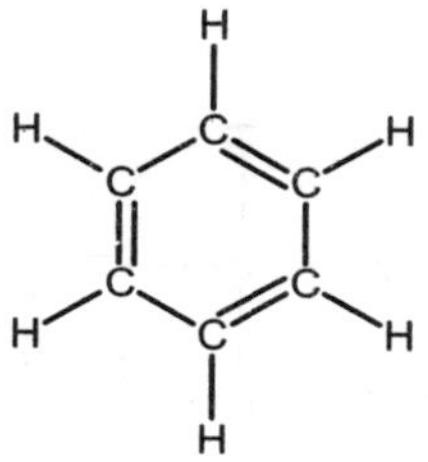

Benzene

Aromatic compounds are dealt in a separate section of this book. In order to make study of organic compounds simpler and comparative, both aliphatic and aromatic compounds have been divided into various families. The members of each family can be represented by a general formula. The various important families of the aliphatic class, with their general formulae, common characteristic groups and the first few members are given in Table 4.1.

Homologous Series

In the study of the organic compounds it is mandatory to group together all the compounds with similar structures and properties. To take an example, when the paraffins are arranged according to numbers of carbon atoms continued in the molecule, they form a series in which adjacent numbers differ by $-CH_2$

Methane CH_4
} CH_2
Ethane C_2H_6
} CH_2
Propane C_3H_8
} CH_2
Butane C_4H_{10}
} CH_2
Pentane C_5H_{12}
} CH_2
Hexane C_6H_{14}

Table 4.1. Families of aliphatic class with their general formulae, characteristic group

Family	General Formula	Characteristic Group	Examples	Remarks
Hydrocarbon	—	—	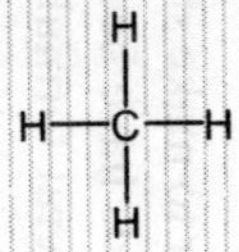 Methane 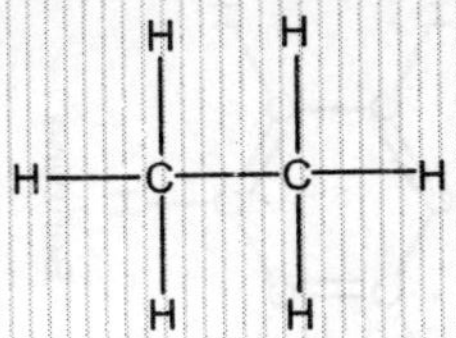 Ethane 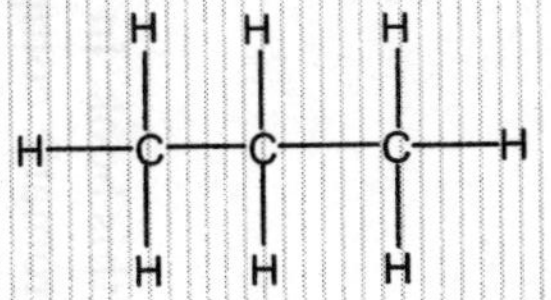	There are two classes 1. Saturated and 2. Unsaturated
Saturated Hydrocarbon	C_nH_{2n+2} (n=1, 2.....n)	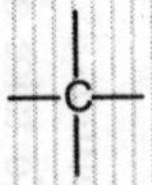 $-\overset{\vert}{\underset{\vert}{C}}-$	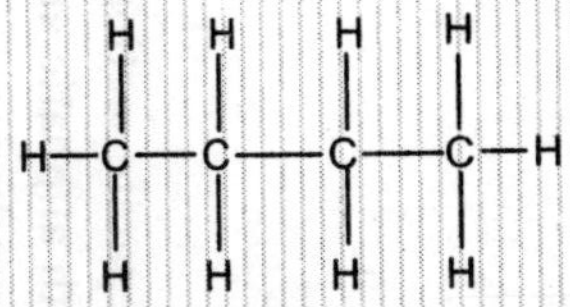 Butane 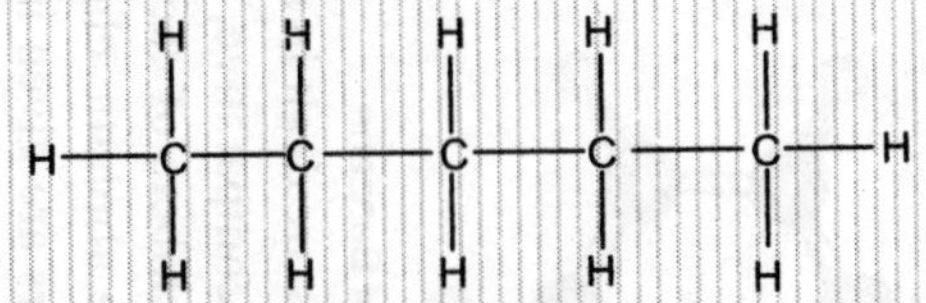 Pentane	(a) Succeeding member are formed by the replacement of a hydrogen atom of the preceding member CH_3 group
Unsaturated Hydrocarbon	—	—	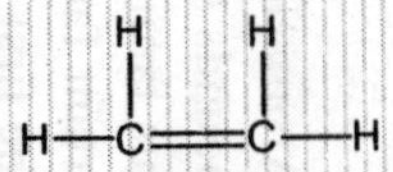 Ethylene 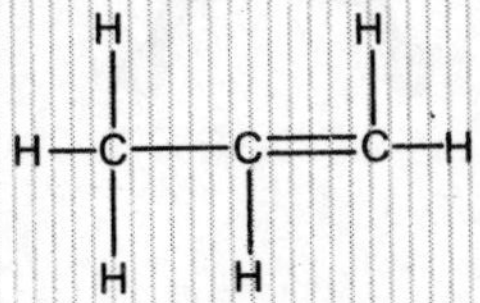 Propylene	Their names are in (ane) Chemically are not very active

(1) Olefins	C_nH_{2n} (n=1, 2....n)	$-\overset{\vert}{C}=\overset{\vert}{C}-$	$H-CH_2-CH_2-CH=CH_2$ Buthylene	(i) They are two types olefins and acetylenes (ii) They contain two hydrogen atoms less than the corresponding paraffin (iii) At least two carbon atoms one associated with double bon(=)
(2) Acetylene	C_nH_{2n+2} n=2, 3 .....n	—C=C—	$H-C\equiv C-H$ Acetylene; $H-CH_2-C\equiv C-H$ Prop-1-yne	(c)Their names either 'eyene' or 'ene' (d) First member (c) isn not known (e) Chemically they are reactive.
Halogen compounds	————	————	$H-CH_2-Cl$ Methyl chloride; $H-CH_2-CH_2-Cl$ Ethyl chloride	(a) They contain two h atoms less than the corresponding olefins (b) At least 2 carbon atoms are untied by a triple bond (C≡) (c) Their names end in either ylene or ine or yne (d) Chemically they are active
(a) Mono halogen compounds	$C_n+H_{2n+}X$ R-x (x=Cl. Br, I) $R=CH_3$. C_2H_5	$-\overset{\vert}{\underset{\vert}{C}}-X$		They are various types mono di tri tetra etc. (a) They are derived from paraffins by replacing one hydrogen atom by a halogen atom. (b) They are very reactive compound.

(Contd...)

Table 4.1—(Contd...)

Family	General Formula	Characteristic Group	Examples	Remarks
(a) Alkylene Halides	$C_xH_{2x-}X_2$	$-\overset{\mid}{C}(X)-\overset{\mid}{C}(X)-$	$H-CH(Cl)-CH(Cl)-H$ Ethyl dichloride	The two halogen atoms are different C atoms
(b) Alkaledine Halides	$C_nH_{2n}X_2$	$-C-C(X)-X$	$H-CH_2-CH(Cl)-Cl$ 1,1-dichloroethane	The two halogen atoms are attached to one and the same C atom
(c) Five halo compounds a s		$C(X)(X)-X$	$H-C(Cl)(Cl)-Cl$ Chloroform; $H-C(Br)(Br)-Br$ Bromoform; $H-C(I)(I)-I$ Iodoform	These tri-halogen derivatives are obtained from methane by replacing 3 H toms by 3 halogen atoms
(d) Tetra halogen Compounds	CX_n n=4	$X-C(X)(X)-X$	$Cl-C(Cl)(Cl)-Cl$ Carbon Tetrachloride	This is obtained by replacing all the 4-atoms by chlorine atoms
3 alkohols		————	$H-CH_2-CH_2-CH_2-OH$ Propan-1-ol	(a) They contain OH groups and re of various types monohydric dihydric tri-hydric (b) Their names end with "ol"

Monohydric Alcohol	$C_nH_{2n+1}OH$	$-\overset{\vert}{\underset{\vert}{C}}-OH$	$H-C(OH)(H)-C(H)(H)-C(H)(H)-H$ Isopropyl Alcohol $H-C(H)(H)-OH$ Methanol $H-C(H)(H)-C(H)(H)-OH$ Ethanol	They are derived form the corresponding paraffins by replacing one hydrogen atom by –OH group. They are further divided into three classes: (*i*) Primary (*ii*) secondary alcohols, and (*iii*) tertiary alcohols R is either the hydrogen or nay alkyl group such as – $CH_3 - C_2H_5$ etc.
(1) Primary Alchohols	$C_nH_{2n+1}OH$ $R-CH_2OH$	$-\overset{H}{\underset{H}{C}}-OH$	H, CH_2OH; CH_2CH_2OH CH_2CH_3OH	
(2) Second alcohols	$C_nH_{2n+1}OH$ R_2CH-OH	$-\overset{H}{\underset{\vert}{C}}-OH$	$H-C(OH)(H)-C(H)(H)-C(H)(H)-H$ Isopropyl Alcohol Dimethyl Carbinol $H-C(H)(H)-C(H)(H)-C(H)(H)-C(H)(H)-H$ Butane	They are obtained from Methyl alcohol or carbinol by the substitution of the two alkyl groups for two atoms of hydrogen.

(Contd...)

Table 4.1—(Contd...)

Family	General Formula	Characteristic Group	Examples	Remarks
(3) Tertiary Alcohol	$R_3 \rightarrow OH$ (three R attached to C–OH)	$-\overset{\vert}{\underset{\vert}{C}}-OH$ Hydroxyl Group	$H_3C-\overset{H_3C}{\underset{HO}{C}}-CH_3$ $H-\overset{H}{\underset{H}{C}}-\overset{H-\overset{H}{C}-H}{\underset{OH}{C}}-\overset{H}{\underset{H}{C}}-H$ Trimethyl carbinol or tertiary isobutyl alcohol	They are obtained from the carbinol by replacing the three hydrogen atoms by the three alkyl group
(B) Dihydric alcohols	$C_2H_{2n}(OH)_2$	$-\overset{\vert}{\underset{OH}{C}}-\overset{\vert}{\underset{OH}{C}}-$	$H-\overset{H}{\underset{OH}{C}}-\overset{H}{\underset{OH}{C}}-H$ Ethylene glycol	The two hydroxyl groups (– OH) are attached to two different carbon atoms.
		$-\overset{\vert}{\underset{OH}{C}}-\overset{\vert}{\underset{OH}{C}}-\overset{\vert}{\underset{OH}{C}}-$	$H-\overset{H}{\underset{OH}{C}}-\overset{H}{\underset{OH}{C}}-\overset{H}{\underset{OH}{C}}-H$ Glycerol	The three –OH groups are attached to three different carbon atoms.

(IV) Ester	C_XH_{2X+1}–O–C_XH_{2X+1}	—O—	H_3C—O—O—H Dimethyl Ether H_3C–CH_2–O–CH_2–CH_3 Diethyl ether H_3C—O—CH_2–CH_3 Ethyl methyl ether Mixed ether	If the two alkyl groups are the same the ether is a simple ether i.e. Dimethyl ether and diethyl ether. If the two alkyl groups are different as in the methyl ethyl ether is known as mixed ether.
(V) Aldehyde	R—C(=O)—H	—C(=O)—H	H—C(=O)—H Formaldehyde H_3C–C(=O)–H Acetaldehyde H_3C—CH_2—C(=O)—H Propionaldehyde	(a) R is the general formula may be an alkyl group or hydrogen (b) Their names end with 'al'

(Contd...)

Table 4.1—(Contd...)

Family	General Formula	Characteristic Group	Examples	Remarks
(VI) Ketones	$R_2C{=}O$	$>C{=}O$ ketonic group	$(H_3C)_2C{=}O$ Dimethyl ketone; $(H_5C_2)(H_3C)C{=}O$ Methyl ethyl ketone	(a) They like ethers are also of two types simple and mixed (b) The names of ketones ens with 'one'.
(VII) Acids (*a*) Mono carboxylic	$R{-}COOH$	$-COOH$ (Carboxylic Group)	$H{-}COOH$ Formic acid; $H_3C{-}COOH$ Acetic acid; $H{-}CH_2{-}CH_2{-}COOH$ Propionic acid	R-may be an alkyl group or hydrogen
(*b*) Dicarbo oxylic acid	———	———	$HOOC{-}COOH$ Oxalic acid; $HOOC{-}CH_2{-}COOH$ Malonic acid; $HOOC{-}CH_2{-}CH_2{-}COOH$ Succinic acid	The two carboxylic groups may be attached to the same or different carbon atoms

(c) tri carboxylic			$HOOC-CH(OH)-CH(OH)-COOH$ Tartaric acid $HOOC-CH_2-C(OH)(COOH)-CH_2-COOH$ Citric Acid	The three carboxylic group are attached to three different carbon atoms
(VIII) Ester	$R-C(=O)-OR'$	$C-C(=O)-OR'$	$H-C(=O)-OCH_3$ Methyl formate $CH_3-C(=O)-OCH_3$ Methyl acetate $CH_3-C(=O)-OC_2H_5$ Ethyl acetate $CH_3-CH_2-C(=O)-OCH_3$ (H_3CO) Methyl Propionate	R is the hydrogen or an Alkyl group
9 Acid chloride	$R-C(=O)-Cl$	$-C(=O)-Cl$	$H-C(=O)-Cl$ Formyl chloride $H_3C-C(=O)-Cl$ Acetyl chloride $H_3C-CH_2-C(=O)-Cl$ Propanoyl chloride	They are obtained from the corresponding fatty acids by replacing –OH of the carbonyl group by a chlorine atom

(Contd...)

Table 4.1—(Contd...)

Family	General Formula	Characteristic Group	Examples	Remarks
10 Acid Amide	$R{-}C(=O){-}NH_2$	$-C(=O){-}NH_2$	$H{-}C(=O){-}NH_2$ Formamide; $H{-}CH_2{-}C(=O){-}NH_2$ Acetamide; $H{-}CH_2{-}CH_2{-}C(=O){-}NH_2$ Propionamide	—OH of the group is replaced by $-NH_2$ group or —C of the acid chloride by NH_2 group
11 Acid anhydride	$R{-}C(=O){-}O{-}C(=O){-}R$	$-C(=O){-}O{-}C(=O){-}$	$H_3C{-}C(=O){-}O{-}C(=O){-}CH_3$ Acetic anhydride; $H_5C_2{-}C(=O){-}O{-}C(=O){-}C_2H_5$ Propanoic anhydride	They are obtained by removing a molecule of water from two molecules of an acid.
12 Amines (a) Primary amines	$R{-}NH_2$	$-C{-}NH_2$ (amino group)	$H{-}CH_2{-}NH_2$ Methylamine; $H{-}CH_2{-}CH_2{-}NH_2$ Ethylamine; $H{-}CH_2{-}CH_2{-}CH_2{-}NH_2$ Propylamine	They are of three types (i) primary (ii) secondary and (iv) tertiary amines
(b) secondary amines	R_2NH	$-N(H)-$ (imino group)	$H_3C{-}NH{-}CH_3$ *N, N*-dimethylamine; $H_5C_2{-}NH{-}C_2H_5$ *N, N*-dimethylamine	Formed by the replacement of one H-atom of the NH_3 by the alkyl group or by the H-atom of paraffin %NH_2 group

(c) Tertiary amines	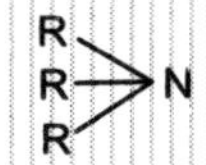 $R_3 \rightarrow N$ (R, R, R → N)	 $\geq N$	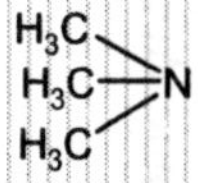 $(H_3C)_3N$ *N,N,N*-trimethylamine; 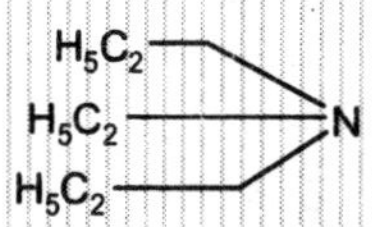 $(H_5C_2)_3N$ *N,N,N*-triethylamine	All the 3H hydrogen atoms of NH_3 are replaced by three alkyl groups same or different for simple or mixed amines
13 Alkyl cyanides	$R—C \equiv H$	$—C \equiv N$	$H_3C—C \equiv N$ Acetonitrile methyl cyanide; $H_5C_2–C \equiv N$ Ethyl cyanide propiononitrile	(i) they may be regarded as derived from the carboxylic acid by the replacement of the –COOH group by the –CN group (ii) The valency of N is three
14 Alkyl isocyanides or Acid isonitrile	$R—M—N \equiv C$	$—N \equiv C$	$H_3C—\overset{+}{N} \equiv C^-$ Isocyanomethyane methyl isocyanide; 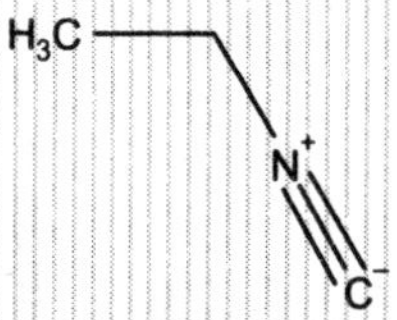 $H_3C—CH_2—\overset{+}{N} \equiv C^-$ Ethyl isocyanide pripino isocyanide	They are isometric with alkyl cyanides. The valency of N is five.
15 Alkyl nitrites	$R—O—N \equiv O$	$—O—N \equiv O$	$H_5C_2—O—N=O$ ethyl nitrite	These compounds are derived from the hydrocarbons by replacing one H atom by nitrite (O–H=O) or 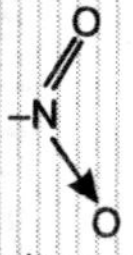 $–N(=O) \rightarrow O$ nitro group
16 Nitro compounds	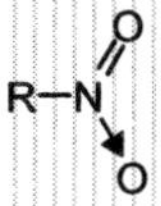 $R–N(=O) \rightarrow O$	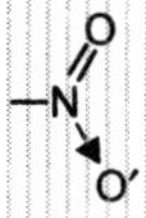 $–N(=O) \rightarrow O'$	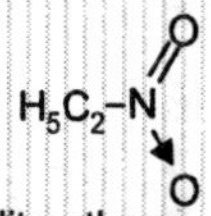 $H_5C_2–N(=O) \rightarrow O$ Nitroethene	In nitrates N is trivalent and in nitro compounds N is pentavalent. They are isomeric compounds.

Such a series the members of which are similar in constitution and therefore also in chemical properties is termed as Homologous series and the several members are called homologous to each other. This property of virtue of which organic compounds from a homologous series is termed as homology. Paraffins, alcohols acids amines etc all form homologous series.

Characteristics of Homologous Series

1. A homologous series can be represented by a general formula that is general formula for paraffins is C_xH_{2x+2} where x is the number of homologue
2. The two adjacent members differ by a difference namely $-CH_2$
3. The members of a homologous series can be prepared by some general reaction called the general method of preparations.
4. The chemical properties of the members of a series are almost similar while their physical properties like boiling points, melting points solubility as specific gravities show a regular gradation from the member to member.

Nomenclature of Organic Compounds

The Geneva system of naming the organic compounds was decided at the meeting of International Chemical Congress held at Geneva in 1892. It is based upon the molecular structure of organic compounds. This system later on in 1930 was revised by International Union of Pure and Applied Chemistry

Paraffins

The first members have special names

```
Methane          H
                 |
               H—C—H
                 |
                 H

Ethane           H   H
                 |   |
               H—C———C—H
                 |   |
                 H   H

Propane          H   H   H
                 |   |   |
               H—C———C———C—H
                 |   |   |
                 H   H   H
```

Butane

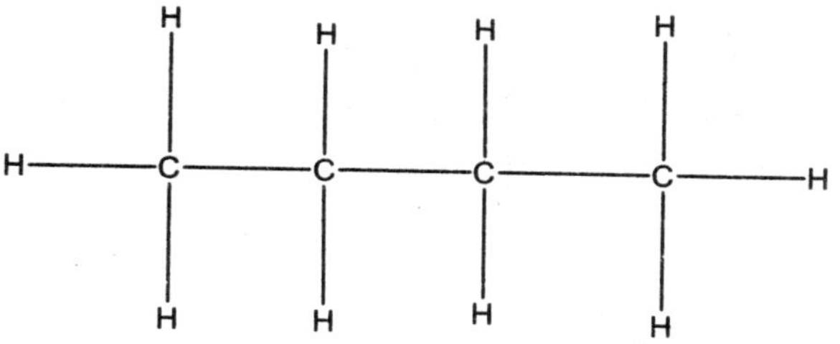

```
     H    H    H    H
     |    |    |    |
 H - C -- C -- C -- C - H
     |    |    |    |
     H    H    H    H
```

The higher hydrocarbons are named by putting paraffin indicating the number of carbon atom. Thus,

Pentane
(5 - C atoms)

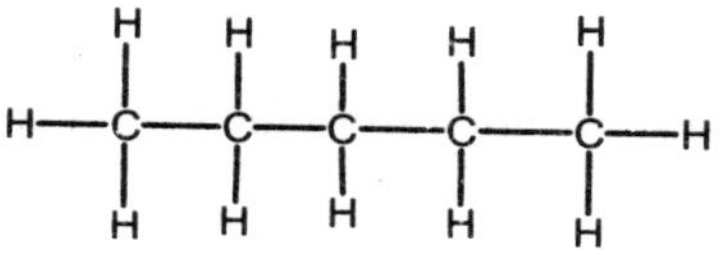

```
    H   H   H   H   H
    |   |   |   |   |
H - C - C - C - C - C - H
    |   |   |   |   |
    H   H   H   H   H
```

Hexane
(6 - C atoms)

```
    H   H   H   H   H
    |   |   |   |   |
H - C - C - C - C - C - CH3
    |   |   |   |   |
    H   H   H   H   H
```

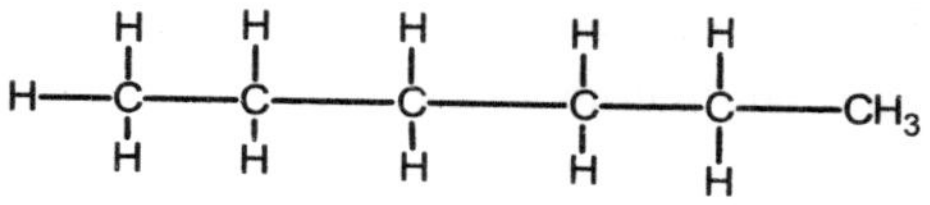

and so on.

The hydrocarbon with carbon chains are known as normal hydrocarbon. While these with the branched chains are called iso-hydrocarbons.

Butane

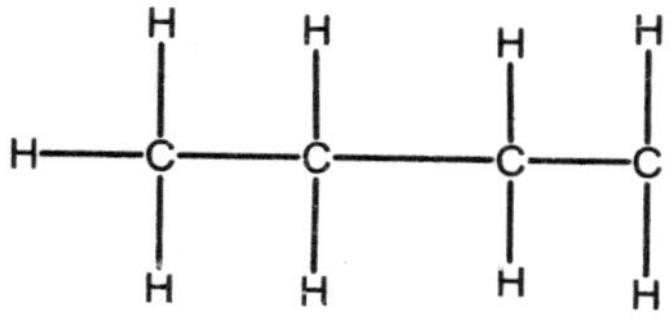

```
    H   H   H   H
    |   |   |   |
H - C - C - C - C
    |   |   |   |
    H   H   H   H
```

Isobutane

```
    H   H   H
    |   |   |
H - C - C - C - H
    |   |   |
    H   |   H
    H - C - H
        |
        H
```

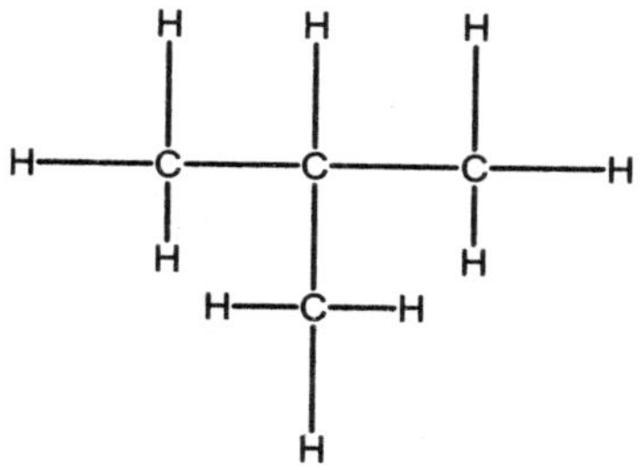

Pentane

```
    H   H   H   H   H
    |   |   |   |   |
H - C - C - C - C - C - H
    |   |   |   |   |
    H   H   H   H   H
```

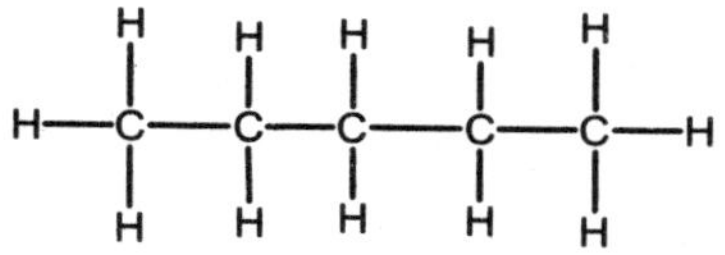

Isopentane

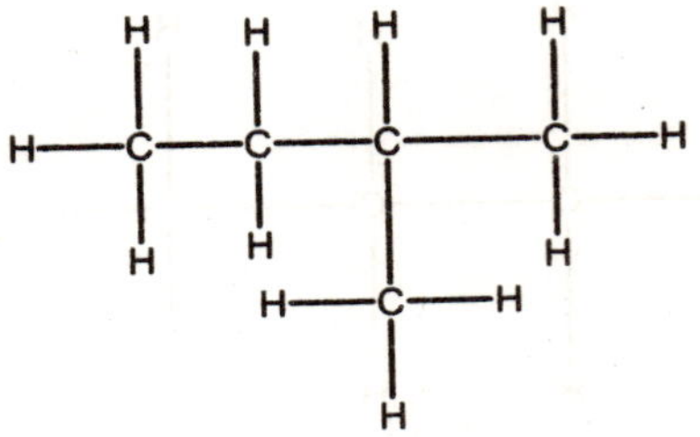

The names of all the paraffins.

Olefines or Alkenes

The names of the olefines are obtained from the names of the corresponding paraffins by changing the termination(ane) to either ‘ylene’ or ‘ene’.

The first member corresponding to methane is not known. However, the formula for the member should be CH_2 = Methylene or methane.

Now some more important members are:

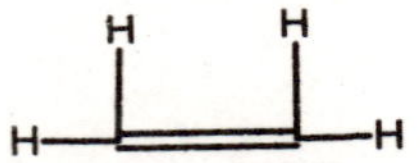

Ethylene or ethene

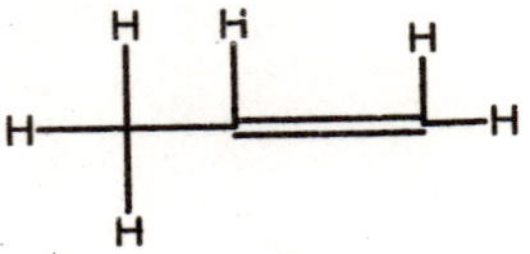

Propylene or propane or methyl ethane

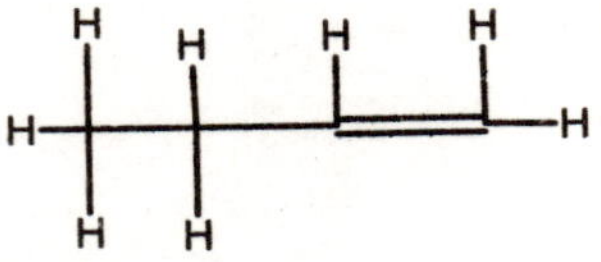

Normal butylenes or butane or ethyl ethylene

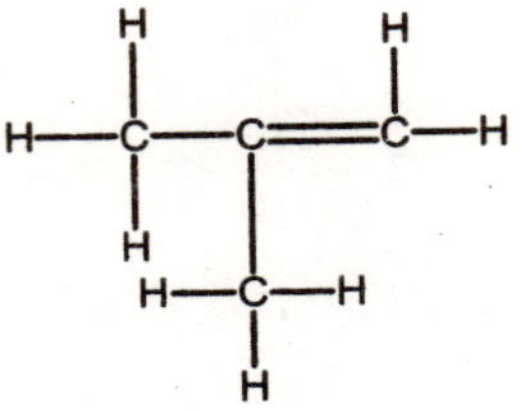

Iso-Butylene or unsymmetrical dimethyl ethylene

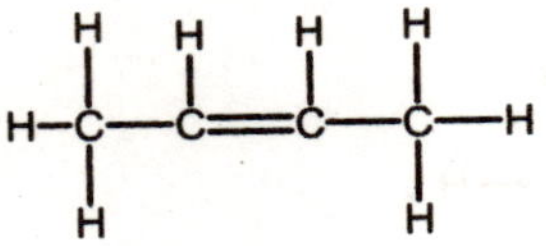

Symmetrical dimethyl ethylene

and so on.

Acetylene

The names of the members of this family are derived from the names of the corresponding paraffins by the changing the terminations 'ane' to either 'ine' or 'yne'

$H-C \equiv C-H$ acetylene ethyne or et

Halogen Compounds

(I) Allyl Halides: (*a*) they are named after the alkyl groups that they contain:

(*b*) they are also normal as halogen substituted paraffins.

CH_3Cl, Methyl chloride, or chloromethane

C_2H_5I, Ethyl Iodide, or Iodoethane

C_3H_7Br, Propyl Bromide or Bromo-propane

(II) Alkylene Halide: They are named as

(*a*) Substitution products of paraffins

(*b*) Additions products of olefines

CH_2Cl_2 di Chloro = methane or Methylene chloride

(1) H_2C—Cl
(2) H_2C—Cl   1, 2-dichloroethane or ethylene chloride

$[CH_3, CHCl_2]$ 1,1 dichlorothane or ethylidene

Alcohol

Alcohols are named as under:

(*a*) The monohydric alcohols are named after the alkyl group that they contain.

(*b*) The names of the alcohols are derived from the names of the corresponding paraffins by changing the terminal 'e' into 'ol'.

(*c*) Alcohols may also be regarded as derived from methyl alcohol or carbinol CH_3OH.

H–C(H)(H)–OH   Methyl Alcohol or methanol, or Ethyl Carbinol

H–C(H)(H)–OH   Ethyl Alcohol or ethanol or Methyl Carbinol

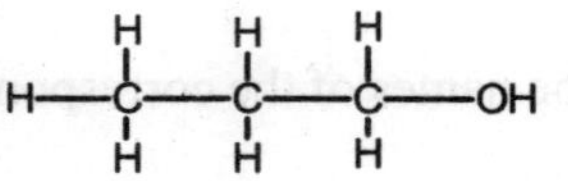

N propyl Alcohol or propanol.

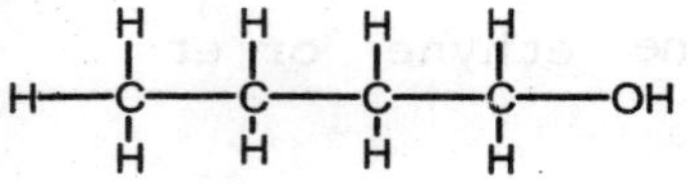

N-butyl alcohol or butenol or Propyl Carbinol (primary alcohol)

Some more examples are

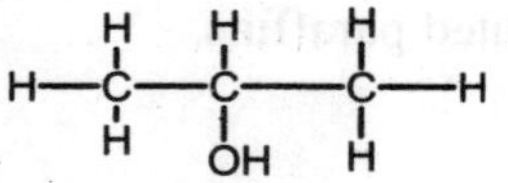

Isopropyl Alcohol or Dimethyl carbinol (secondary alcohol)

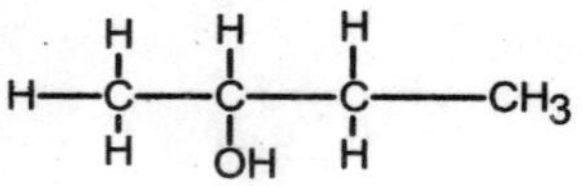

Isobutyl alcohol or Methyl Ethyl Carbinol

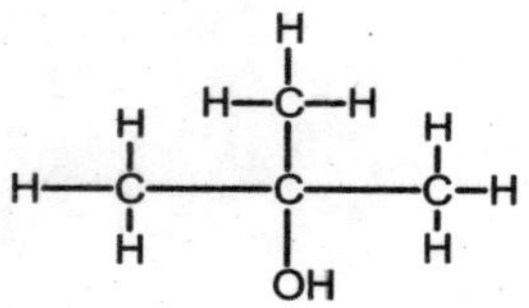

Isobutyl alcohol or Tri-methyl carbinol (tertiary alcohol)

Ethers

They are named after alkyl group they contain

CH_3–O–CH_3	Dimethyl Ether
C_2H_5–O–C_2H_5	Diethyl Ether
CH_3–O–C_2H_5	Methyl Ethyl Ether

Aldehydes

They are three systems for naming these aldehydes:

(a) Aldehydes are named after the acids they giver on oxidation.

(b) Aldehydes are also name after the alcohols from which they may be obtained by oxidation.

(c) The names of aldehydes may also be derived from those of the parent hydrocarbons by replacing the terminal 'e' by '**al**'.

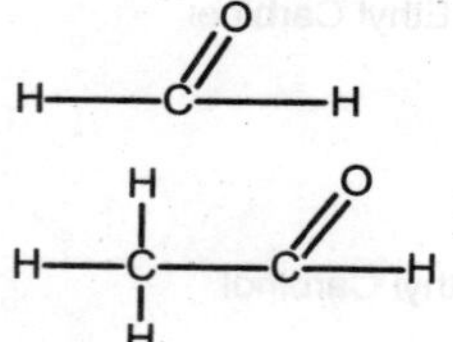

Formaldehyde, Methaldehyde or Methanal

Acetaldehyde Ethaldehyde or Ethanal

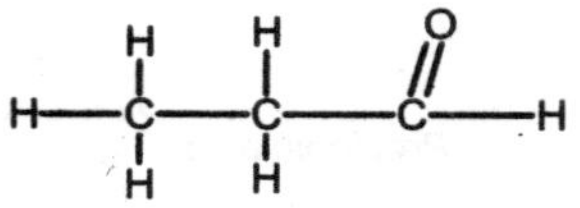

Propionaldehyde, Propaldehyde, or Propanal

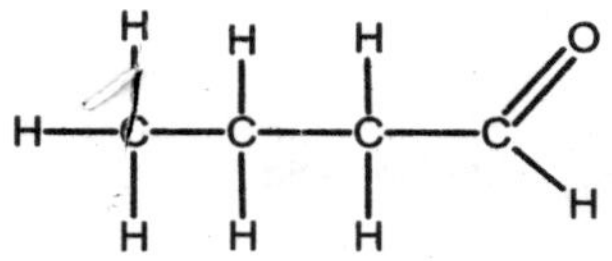

Butyraldehyde, Butyraldehy de or Butanal

Ketone

Ketones are named as under:

(a) Ketones are named after the acids from which they may be obtained.

(b) Ketones are also named after the allyl groups that they contain.

(c) The names of ketones may also derive from those of the parent hydrocarbon by replacing '**e**' by '**one**'

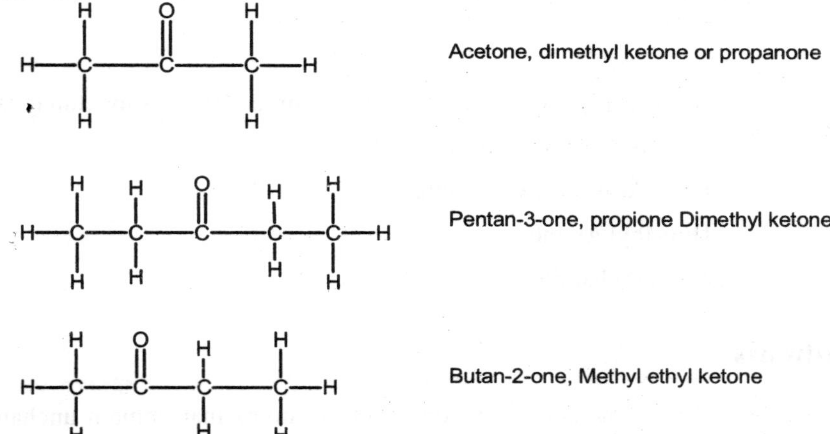

Acetone, dimethyl ketone or propanone

Pentan-3-one, propione Dimethyl ketone

Butan-2-one, Methyl ethyl ketone

Acids

They have special names which includes the source from which they can be isolated. The names of the acids end in '**ic**'

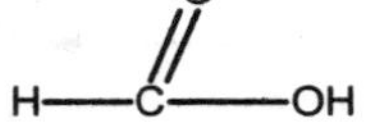

Formic acid

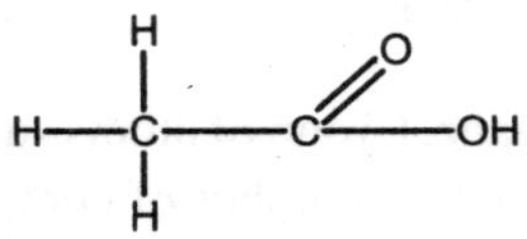

Acetic acid

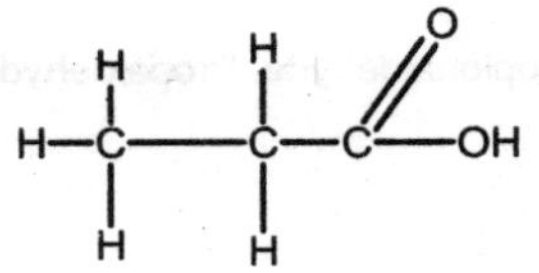

```
   H   H   O
   |   |  //
H—C——C——C——OH
   |   |
   H   H
```

Propionic acid

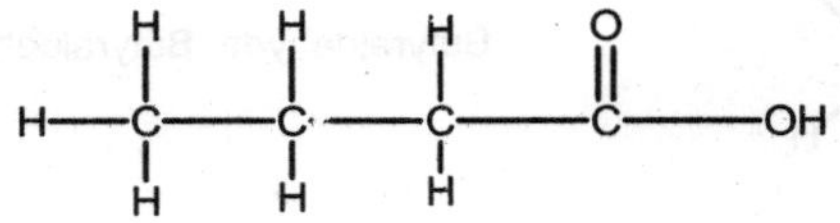

```
   H    H    H    O
   |    |    |    ||
H——C————C————C————C————OH
   |    |    |
   H    H    H
```

Butyric acid

Esters

Esters are named after the acids from which they are obtained.

Acids	Esters	Names
HCOOH(Formic acid)	$HCOO{-}CH_3$	Methyl Formate
CH_3COOH (Acetic acid)	CH_3COOCH_3	Ethyl Acetate
C_2H_5COOH (Propionic acid)	$(C_2H_5COOCH_3)$	Methyl Propionate

Amines

They are named either after the alkyl group that they contain or as Amino-substituted paraffins.

$CH_3{-}NH_3$	Methyl amine or Ammino methane
$C_2H_5{-}NH_2$	Ethyl Amine or Amino ethane
$(CH_3)_2NH$	Dimethyl Amine
$(CH_3)_3NH$	Trimethyl amine

Organic Radicals

Inorganic chemistry we often come across certain group of atoms than remain unchanged during a whole series of reactions Methane $-CH_4$ for instance may be converted into methyl alcohol $-CH_3OH$ and this compound may be converted into methyl chloride $-CH_3Cl$ latter may again be transferred into methyl alcohol $-CH_3OH$.

In all these reactions the CH_3 that is methane remains unchanged and behaves like a single atom

$$\underset{\text{Methane}}{CH_3H} \xrightarrow{\text{Chlorine}} \underset{\text{Methyl Chloride}}{CH_3Cl} \xrightarrow{\text{Water}} \underset{\text{Methyl Alcohol}}{CH_3OH} \xrightarrow{\text{Phosphorous Pentachloride}} \underset{\text{Methyl Chloride}}{CH_3Cl}$$

Ethane $\longrightarrow C_2H_6$ also behaves exactly like methane. Group of atoms such as CH_3 and C_2H_5 which act like the single atoms and which enter unchanged into a number of compounds are called radicals.

They may be mono-valent, di-valent etc. Radicals CH_3–C_2H_5– are mono-valent and are called alkyl groups. These radicals are obtained from the saturated hydrocarbon Methane CH_4 ethane C_2H_6 etc. By removing one atom of hydrocarbon from the molecule of each. Their distinctive names are also derived from the corresponding hydrocarbons by changing the termination '**ane** to '**yl**'

Examples:

Hydrocarbon	Alkyl groups
C_xH_{2n+2}	C_nH_{2n+1}
R—H	R—
CH_4	CH_3—
Methane	Methyl
C_2H_6	C_2H_5
Ethane	Ethyl
C_3H_8	C_3H_7
Propane	Propyl
C_4H_{10}	C_4H_{10}
Butane	Butyl

The divalent radicals are known as Alkylenes the names end with ene

CH_2 Methylene or methane
C_2H_4 Ethylene or ethane

Similarly the trivalent radicals are written with the termination '**ine**'

Such as

CH ≡ Methine
C_2H_3 ≡ Ethine
C_3H_5 ≡ Propine

Other important radicals are:

–OH Hydroxyl

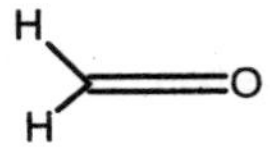

H, H, =O Carbonyl

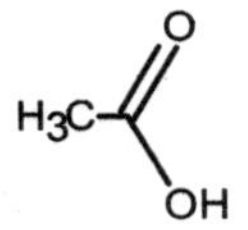

H_3C–C(=O)OH Carboxyl

H—≡N Cyanogen or cyanide

H_3C–C(=O)–OH Acetyl

H_3C–C(=O)–OH Aldehyde

H_2N—H Amino

HN═NH Imino

O═N–O^- Nitro

It is very important to have a clear idea of the above reactions of the above-mentioned groups as to learn how they determine the properties of the molecules, all the reactions of the compounds depend on those of its constituent groups.

IUPAC Rules for Naming Alkanes from their Structural Formulas

1. **If the molecule is a continuous-chain alkane with no branches, use a prefix to indicate the number of carbon atoms in the chain and add the suffix *-One* to indicate that it is an alkane or saturated hydrocarbon. The prefixes for the first 10 hydrocarbons are*:**

C_1 meth	C_3 prop	C_5 pent	C_9 non
C_2 ethy	C_4 but	C_6 hex	C_{10} Dec

2. **If the molecule has one or more branches attached to the continuous chain, do the following:**

   (*a*) **Find the longest continuous chain of carbon atoms and write the name for this chain using the appropriate prefix (see Rule 1) followed by the suffix *-ane*. Note from the following example that the longest continuous chain need not appear in a straight line.**

   Example :

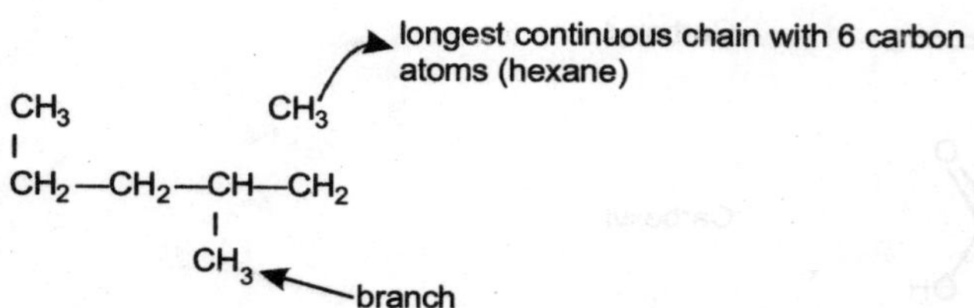

1 A device for remembering these prefixes is Mary Eats Peaches, But Paul Has Had only Nine. Donuts. *meth- eth- prop- but- pem- hex- hept- oct- non- dee

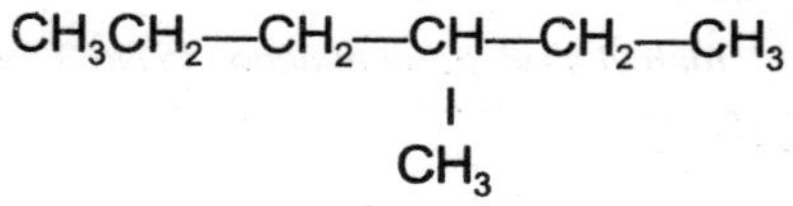

$$CH_3CH_2—CH_2—\underset{\displaystyle CH_3}{\underset{|}{CH}}—CH_2—CH_3$$

hexane

(*b*) Number the carbon atoms in the longest continuous chain con-secutively, starting with the carbon atom at the end of the chain that is nearer to the branch or branches attached to the continuous chain.

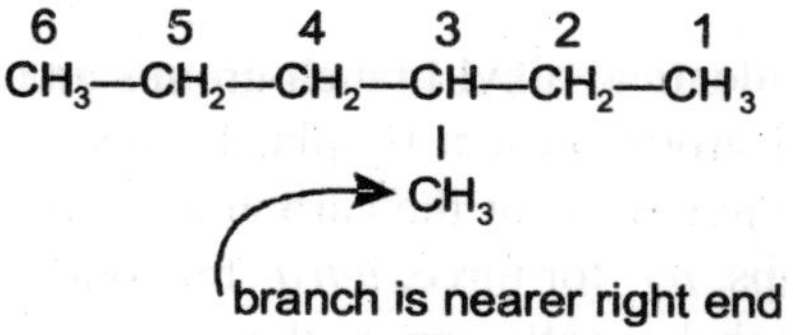

$$\overset{6}{CH_3}—\overset{5}{CH_2}—\overset{4}{CH_2}—\underset{\displaystyle CH_3}{\underset{|}{\overset{3}{CH}}}—\overset{2}{CH_2}—\overset{1}{CH_3}$$

branch is nearer right end

not

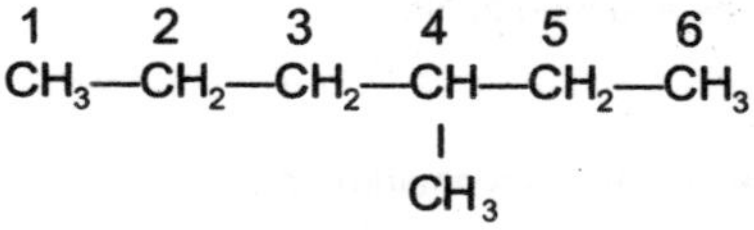

$$\overset{1}{CH_3}—\overset{2}{CH_2}—\overset{3}{CH_2}—\underset{\displaystyle CH_3}{\underset{|}{\overset{4}{CH}}}—\overset{5}{CH_2}—\overset{6}{CH_3}$$

(*c*) Name each branch by giving the number of the carbon atom in the longest continuous chain to which it is attached, followed by a hyphen, the name of each branch. Table and the name of the longest continuous chain from Rule 2a. Branches are called *alkyl groups* and are attached to the continuous chain by removing a hydrogen atom from an alkane. Branch alkyl groups are named by changing the suffix for the parent alkane from *-ane* to *-yl,* as shown for several of the more common alkyl groups in Table.

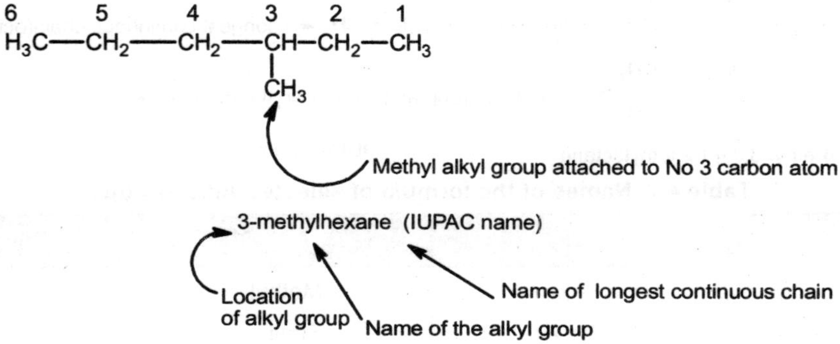

$$\overset{6}{H_3C}—\overset{5}{CH_2}—\overset{4}{CH_2}—\underset{\displaystyle CH_3}{\underset{|}{\overset{3}{CH}}}—\overset{2}{CH_2}—\overset{1}{CH_3}$$

Methyl alkyl group attached to No 3 carbon atom

3-methylhexane (IUPAC name)

Location of alkyl group

Name of the alkyl group

Name of longest continuous chain

(*d*) When two or more alkyl groups are attached as branches to the longest continuous chain at different locations, give the number of the carbon atom to which each group is attached, list them in alphabetical order, and follow each name with a hyphen until the last alkyl group is named (for example, ethyl before methyl).

methyl alkyl group attached to No 3 carbon atom

$$\begin{array}{l} \quad\quad\quad\quad\quad CH_2 \\ H_3C{-}CH_2{-}CH{-}CH{-}CH_2{-}CH_2{-}CH_3 \leftarrow \text{longest continuous chain} \\ \quad\quad\quad\quad\quad\quad\quad\quad CH_2 \\ \quad\quad\quad\quad\quad\quad\quad\quad CH_3 \end{array}$$

Ethyl alkyl group attached to No 4 carbon atom

(IUPAC name)

4-ethyl-3 methylheptane

not 3-methyl-4ethylpropane

(*e*) If there are two or more identical alkyl branch groups attached to the carbon chain, indicate the number of each alkyl group present by placing one of the following prefixes before its name before showing the numbers for the carbon atom or atoms to which they are attached: *di-* for two identical groups, *tri-* for three, *tetra-* for four, and *penta-* for five. In listing these multiple-branch groups alphabetically, ignore these prefixes and use the first letter in the name of the alkyl group (Rule 2d).

to methyl groups attached to No 2 carbon atom

$$\begin{array}{c} CH_3 \\ H_3C{-}C{-}CH_3 \\ CH_3 \end{array} \leftarrow \text{longest continuous chain(propane)}$$

2,2 dimethyl propane IUPAC name

indicates 2 methyl groups are attached to the No 2 carbon atom

3 methyl group branches with 1 on No 4 carbon atom and 2 on No 3 carbon atom

$$\begin{array}{l} \quad\quad\quad\quad\quad\; CH_3 \quad CH_3 \\ H_3C{-}CH_2{-}C{-}C{-}CH_2{-}CH_2{-}CH_2{-}CH_3 \leftarrow \text{longest continuous chain(octane)} \\ \quad\quad\quad\quad\quad\; CH_3 \quad CH_2 \\ \quad\quad\quad\quad\quad\quad\quad\quad CH_3 \end{array}$$

1 ethyl group attached to No 4 carbon atom

4-ethyl-3,3,4-trimethyloctane IUPAC name

Table 4.2. Names of the formula of selected Alkyl groups

Structural Formula of Alkyl group	Name of the Alkyl group
$H_3C{-}$	Methyl
$H_3C{-}CH_2{-}$	Ethyl
$H_3C{-}CH{-}CH_2{-}$	Propyl
$H_3C{-}CH{-}CH_3$	Isopropyl
$H_3C{-}CH_2{-}CH_2{-}CH_2{-}$	Butyl

Writing Structural Formulas for Alkanes from their Names. Now reverse the process and write the structural formula for an alkane from its IUPAC name. The following rules can be used for this procedure.

Rules for Writing Structural Formulas for Alkanes from their IUPAC Names

1. Use the last part of the name to identify the number of carbon atoms in the longest continuous chain. Write a carbon skeleton with this number of carbon atoms and number the carbon atoms in this chain consecutively-

   Example: Write the structural formula for 2,2,4-trimethylpentane.

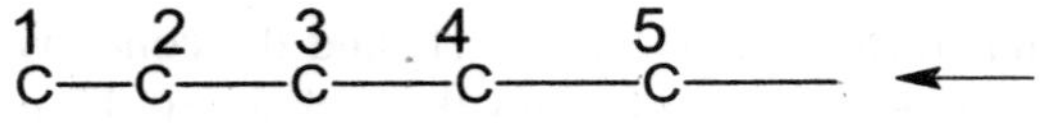

   ```
   1   2    3    4    5
   C—C——C——C——C——   ←   longest continuous chain with 5
                           carbon atoms
   ```

2. Use the first part of the name to determine the type and number of alkyl branch groups that are to be attached to the continuous chain.

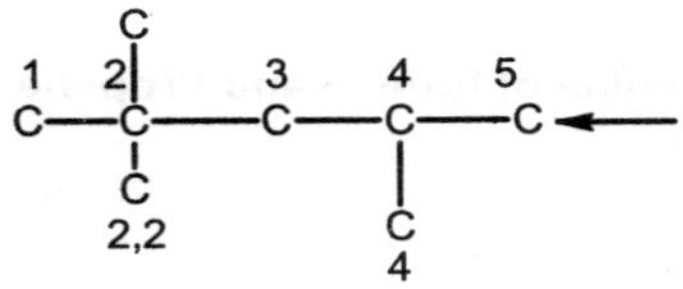

   ```
          C
   1    2|     3     4     5
   C——C——C——C——C ←—— attach 2 methyl groups to the No2
         |           |         carbon atom (2,2 and 1 methyl group to the No.4 carbon
         C           C         atom
        2,2          4
   ```

3. Add enough hydrogen atoms to have each carbon four covalent bonds.

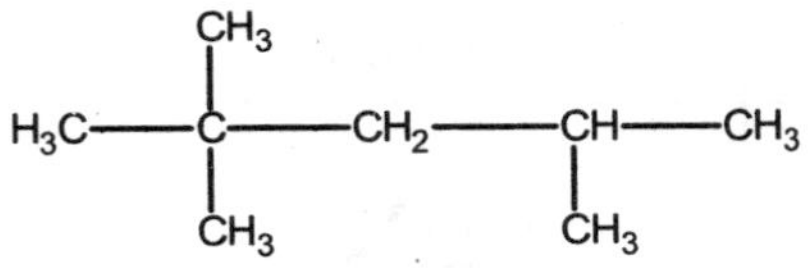

   ```
             CH3
              |
   H3C——C——CH2——CH——CH3
              |          |
             CH3        CH3
   ```

   2,2,4-trimethylpentane

Alkene, Alkyne, and Aromatic Hydrocarbons

Structures of Alkenes

There are three types of unsaturated hydrocarbons that contain fewer hydrogen atoms than the alkane with the same number of carbon atoms:

(1) *alkenes* that contain at least one double covalent bond betwcen carbon atoms (%C=C%),

(2) *alkynes* that contain at least one triple covalent bond between carbon atoms (-C-C-), and

(3) *aromatic hydrocarbons* that are ring compounds that include benzene and all other compounds with structures and chemical properties similar to benzene.

The double and triple covalent bonds are *functional groups* that determine most of the chemical behavior of an unsaturated hydrocarbon. As a result, unsaturated hydrocarbons are much more chemically reactive than saturated hydrocarbons or alkanes.

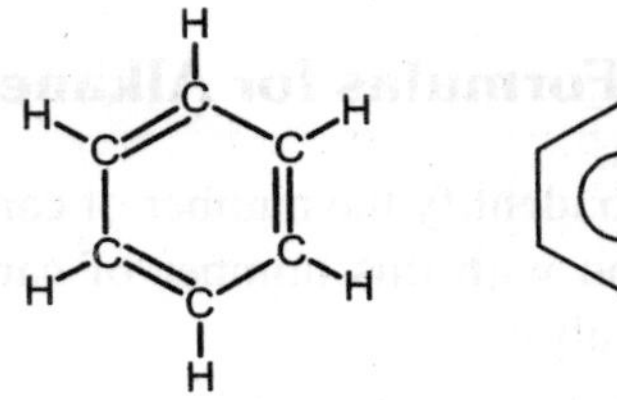

The *alkenes* —C=C— are members of a homologous series with the general molecular formula (C_nH_{2n})' the simplest alkene has the molecular formula $C_2H_{2\times2}$ = C_2H_4 and the condensed structural formula (CH_2=CH_2). Notice that it contains two less hydrogen atoms than ethane (C_2H_6 or CH_3-CH_3), the alkane with two carbon atoms. To indicate the presence of a double bond, the ending *-ene* is substituted for the ending *-ane* in an alkane. Thus, the IUP AC name for the simplest alkene is *ethene,* although it is often called by its common name *ethylene.* The second member of this series has the molecular formula $C_3H_{2\times3}$ = C_3H_6 and the condensed structural formula CH_3- CH=CH_2. Its IUPAC

Table 4.3 : IUPAC Name and Structural Formulae of Ethene and Propene

IUPAC name	Ethene	Propene
Common name	Ethylene	Propylene
Molecular Formula (C_nH_{2n})	C_2H_4	C_3H_6
Condensed formula	CH_2=CH_2	CH_3—CH=CH_2
Expanded structural Formula	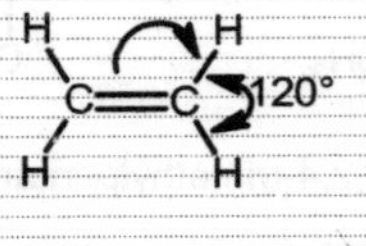 120°	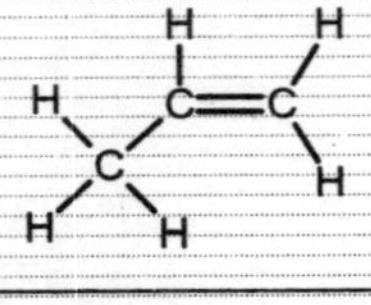
Ball and stick model	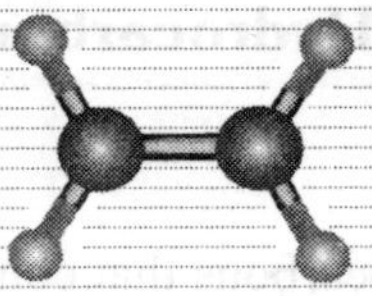	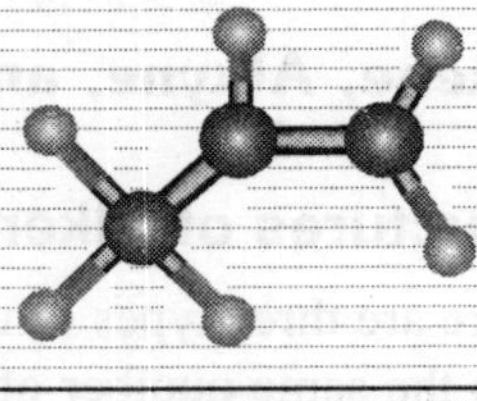
Space fill model	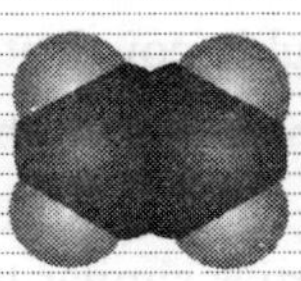	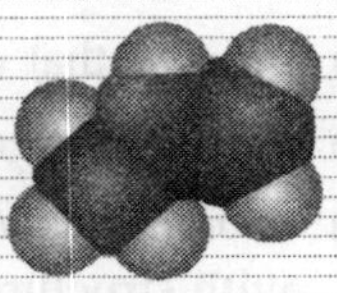

name is *propene* and its common name is *propylene.* More detailed structural formulas for ethene and propene are shown in Table 4.3. Alkenes COD-taining four or more carbon atoms have structural isomers that increase in number with each additional carbon atom.

Ethene (ethylene) is one of the most important commercially produced chemicals in the United States. It is used to make the plastic polyethylene, ethylene glycol that is the major component of most brands of anti-freeze used in automobile radiators, and a number of other important chemicals. Propene (propylene) is also an important industrial chemical. It is used to make polypropylene plastics, isopropyl alcohol (rubbing alcohol), and a number of other useful substances.

Structures of Alkynes

The *alkynes* (-C- C-) are members of another homologous series with the general molecular formula C_nH_{2n-2}'. The simplest alkyne has the molecular formula

$$C_2H_{(2\times 2)-2} = C_2H_2$$

and the condensed structural formula CHa" CH.

Notice that it contains four less hydrogen atoms than ethane (C_2H_6) and two less hydrogen atoms than ethene (C_2H_4). To indicate the presence of a triple covalent bond, the ending *-yne* is substituted for the ending *-ane* in an alkane. This means that the IUP AC name for the simplest alkyne is *ethyne,* although it is usually called by its common name of *acetylene.* It is a flammable and explosive gas that burns with a bright flame. When ethyne (acetylene) is burned with oxygen in an oxyacetylene torch, it gives off so much energy that it can be used to cut and weld metals.

IUPAC name	Ethyne	2-Butyne
Common name	Acetylene	Dimethylacetylene
Molecular Formula(C_nH_{2n})	C_2H_2	C_4H_6
Condensed formula	$CH\equiv CH$	$H_3C-C\equiv C-CH_3$
Expanded structural Formula	$H-C\equiv C-H$	H—C(H)(H)—C≡C—C(H)(H)—H
Ball and stick model		
Space fill model		

Alkynes in general are highly reactive molecules Acetylene is so reactive that it can be used as a starting material for the industrial production of most simple organic compounds. These intermediates can in turn be used to make acrylic and vinyl plastics synthetic fibers, and a variety of other chemical products. The second member of the alkyne series has the molecular formula $C_3H_{(2 \times 3) - 2} = C_3H_4$ and the condensed structural. formula $CH_3C{=}CH$. Its IUPAC name is *propyne.* More detailed structural formulas for two alkynes are given in the table below:

Naming Alkenes and Alkynes

The following IUPAC rules are used to name alkenes and alkynes.

IUPAC Rules for Naming Alkenes and Alkynes

1. Select the longest continuous chain of carbon atoms that includes the double or triple covalent bond. Name this chain as you would an alkane but change the ending *-ane* to *-ene* for an alkene and to - *yne* for an alkyne.
2. Number the carbon chain starting at the end closest to the first double or triple bond. Place the number corresponding to the first carbon atom in each double or triple bond in front of the alkene or alkyne name from Rule 1, and separate the number and name with a hyphen.

Examples:

$H_3CCH_2{\cdot}CH{=}CH_2$ — but-1-ene

$\overset{1}{H_3C}\overset{2}{CH}{=}\overset{3}{C}\overset{4}{HCH_3}$ — but-2-ene

$H_3CC{\equiv}CCH_3$ — but-2-yne

3. Name branch chains the same way they are named in alkanes by indicating the number of the carbon atom to which they are attached followed by a hyphen and the name of the alkyl group.

$H_3C{-}CH{=}CHCH(CH_3)CH_3$ — 4-methylpent-2-ene

$\overset{8}{H_3C}{-}\overset{7}{C}(CH_3)_2{-}\overset{6}{CH_2}{-}\overset{5}{CH}(CH_3){-}\overset{4}{C}{\equiv}\overset{3}{C}{-}\overset{2}{CH_2}{-}\overset{1}{CH_3}$ — 5,7,7-trimethyloct-3-yne

The Process of Wrtting the Structural Formula from IUPAC Name

To write the structural formula from the IUPAC name for an alkene or alkyne, this process is reversed. For example, suppose you are asked to write the condensed structural formula for

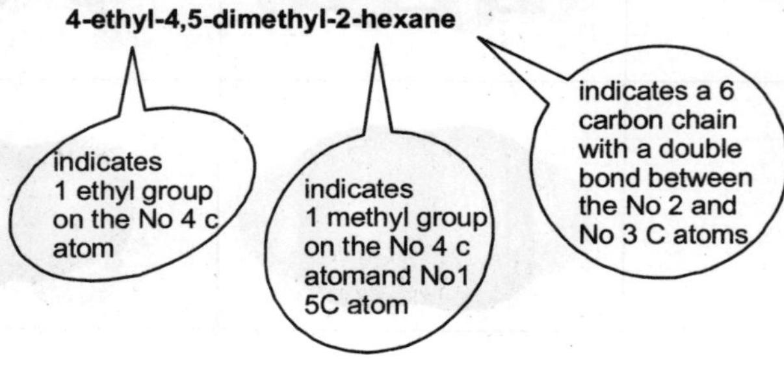

Write a carbon skeleton with six carbon atoms, place a double bond between the No. 2 and No. 3 carbon atoms, and place an ethyl group on the No. 4 carbon atom and one methyl group on the No. 4 carbon atom and one on the No. 5 carbon atom. Then add hydrogen atoms so that each carbon has four bonds.

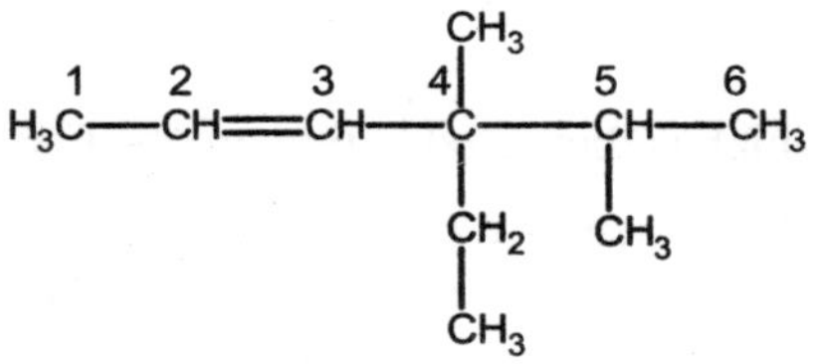

Aromatic Hydrocarbons

Aromatic hydrocarbons are hydrocarbons containing one or more benzene rings. Originally the word *aromatic* (derived from *aroma)* was used to describe the rather strong and sometimes pleasant odour of these compounds. Some aromatic compounds, however, have vile odours and others have very little odour. The simplest and most important aromatic hydro-carbon is *benzene* with the molecular formula C_6H_6 and the following structural formula:

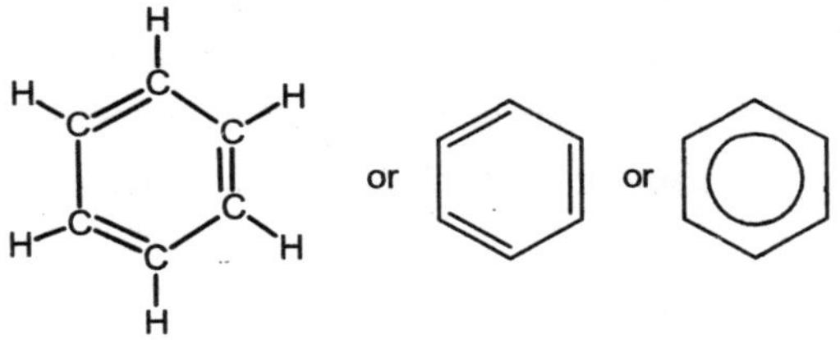

Benzene and many other aromatic hydrocarbons are obtained from coal tar, a sticky black mixture produced when bituminous (soft) coal is heated in the absence of air. Benzene is widely used in the chemical industry as a solvent in common products such as glues, paints, and thinners and as a starting material for making a variety of organic compounds. When inhaled, however, its fumes can cause nausea and damage to the central nervous system, blood cells, and liver. Inhaling a fairly large amount of benzene can cause death. It is also known to cause cancer in humans and some animals. For these reasons, in recent years the use of benzene has been restricted and subject to much stricter safety regu-lations.

Naming Benzene Compounds

Alkyl groups and functional groups. Such as the halogens (-F, -CI, -Br, and -I) and others can be attached to one or more carbon atoms on a benzene ring by substituting them for a hydrogen atom. For example, substituting a methyl group for one of the hydrogen atoms on the benzene ring gives the compound *methylbenzene,* better known by its common name *toluene,* with the structure.

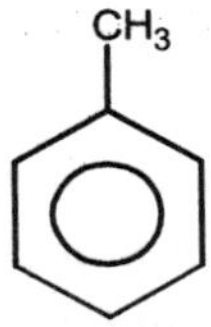

Methylbenzene or toluene

Notice that its IUP AC name is obtained by giving the name of the methyl group followed by the word *benzene.* Toluene is an important industrial solvent that has replaced benzene for many purposes because it is less toxic than ben-zene. It is also used as a starting material to make other aromatic compounds. Substitution of an ethyl group for a hydrogen atom on the benzene ring gives *ethyl benzene* with the structure.

CH_3
CH_2

Ethylbenzene

It is used to prepare styrene, which is then used to prepare the plastic polystyrene.

When two groups are attached to a benzene ring, two different naming methods are used to indicate their relative positions on the benzene ring. The simplest method is to number the six carbon atoms in the ring and use these numbers to indicate the location of the groups on the benzene ring in the same way they are located on a continuous chain of carbon atoms in an alkane, alkene, or alkyne. An older and more commonly used method is to use the prefixes *ortho-* (abbreviated *o-), meta-* (abbreviated *m-),* and *para-* (abbreviated *p-)* to show the relative position of the second group on the benzene ring. In an *or tho- (0-)* compound the second group is located on the carbon atom next to the one containing the first group. In a *meta- (m-)* compound it is located two carbon atoms away from the first group, and in a *para- (p-)* compound it is located three carbon atoms away.

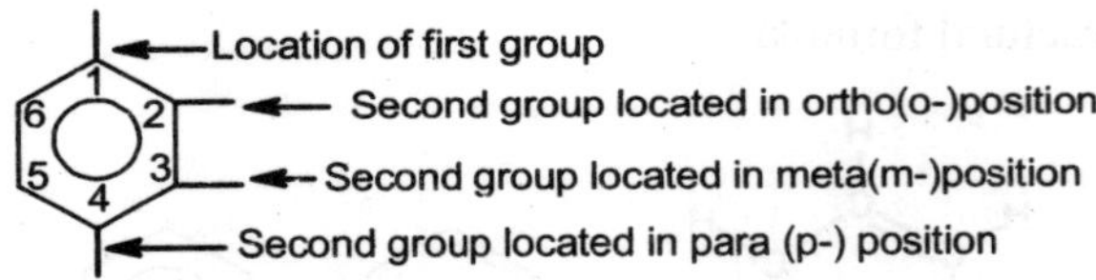

Examples of both systems for naming benzene compounds containing two substituted group are:

CH_3, CH_3	CH_3, CH_3	CH_3, CH_3
1,2-dimethylbenzene or *o*-dimethylbenzene	1,3-dimehylbenzene or *m*-dimethylbenzene	1,4-dimethylbenzene or *p*-dimethylmenzene

Chemical Reactions of Hydrocarbons

Major Types of Hydrocarbon Reactions

In this section we will learn about four major types of reactions that some or most hydrocarbons can undergo:

(1) *combustion redox reactions* (alkane, alkene, alkyne, and aromatic hydrocarbons),
(2) *substitution reactions* (alkane and aromatic hydrocarbons),
(3) *addition reactions* (alkenes and alkynes), and
(4) *addition polymerization reactions* (alkenes).

In general, the alkanes are fairly unreactive. However, they can undergo combustion and substitution reactions under the appropriate reaction condi-tions. The alkenes and alkynes are much more reactive because of the presence of the

$$-\overset{|}{C}=\overset{|}{C}- \text{ and } -C\equiv C-$$

functional groups. Under the appropriate reaction conditions, they can undergo all four of the major types of organic reactions listed above but rarely undergo substitution reactions. Aromatic hydrocarbons in general are less reactive than alkenes and alkynes but more reactive than alkanes. They tend to undergo combustion and substitution reactions. Look at examples of each of the four types of organic reactions.

Combustion Redox Reactions

In the presence of excess oxygen, all types of hydrocarbons can undergo complete combustion redox reactions to form carbon dioxide and water. Because such combustion reactions release large amounts of heat, many hydrocarbons (especially alkanes such as methane, propane, butane, and octane) are used as fuels. Some examples of combustion reactions follow:

Alkane: $CH_4(g) + 2O_2 \longrightarrow CO_2(g) + 2H_2O(g) + 801\,kJ/mol\,CH_4$
membrane

Alkene: $CH_2 = CH_2(g) + 3O_2(g) \longrightarrow 2CO_2(g) + 2H_2O(g) + 1323\,kJ/mol\ C_2H_4$
ethene

Alkyne: $2\,CH \equiv CH(g) + 5O_2(g) \longrightarrow 4\,CO_2(g) + 2H_2O(g) + 740\,kJ/mol\,C_2H_2$
ethyne

Aromatic: $2C_6H_6(l)$ or [benzene ring] $15O_2(g) \longrightarrow 12CO_2(g) + 6H_2O(g) + 3136kJ/mol\,C_6H_6$
benzene

Substitution Reactions

In a substitution reaction one of the hydrogen atoms in a hydrocarbon is replaced by another type of atom such as a halide (- F, - Cl) - Br, or - I) or a group of atoms such as a nitro group (- NO_2). With alkanes this type of reaction requires either high temperatures or exposure *to* ultraviolet or visible light. Once initiated, however, such reactions occur very rapidly.

An example of a substitution reaction for an alkane is

Alkane: $H-\underset{H}{\overset{H}{\underset{|}{\overset{|}{C}}}}-H(g) + Cl_2(g) \xrightarrow{\text{UV light}} H-\underset{H}{\overset{H}{\underset{|}{\overset{|}{C}}}}-Cl\ (g) + HCl(g)$

Methane — Chloromethane

If enough Cl_2 is present, Cl atoms can also be substituted for the other hydrogen atoms on methane to produce

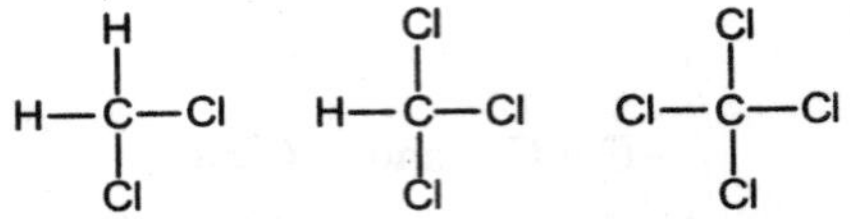

$$CH_2Cl_2 \qquad CHCl_3 \qquad CCl_4$$

Alkenes and alkynes can also undergo substitution reactions, but they are very difficult to carry out. This is because the double and triple bonds between carbon atoms in such molecules are so reactive that addition rather than sub-stitution reactions usually take place. By contrast, most aromatic hydrocarbons readily undergo substitution reactions.

Aromatic:

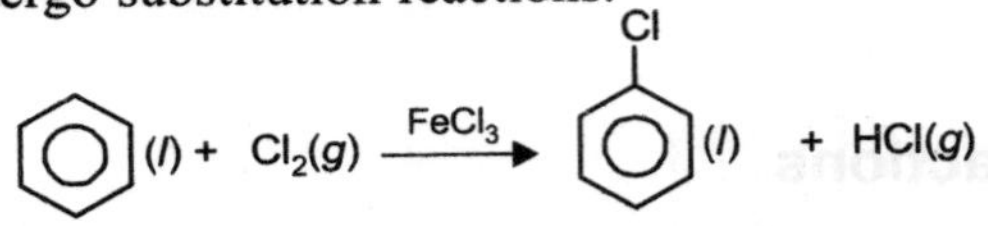

$$C_6H_6(l) + Cl_2(g) \xrightarrow{FeCl_3} C_6H_5Cl(l) + HCl(g)$$

Benzene Chlorobenzene

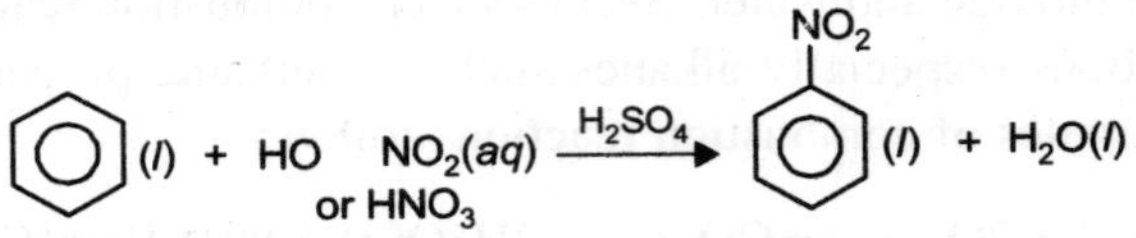

$$C_6H_6(l) + HO\ NO_2(aq) \text{ or } HNO_3 \xrightarrow{H_2SO_4} C_6H_5NO_2(l) + H_2O(l)$$

Benzene Nitric acid Nitrobenzene

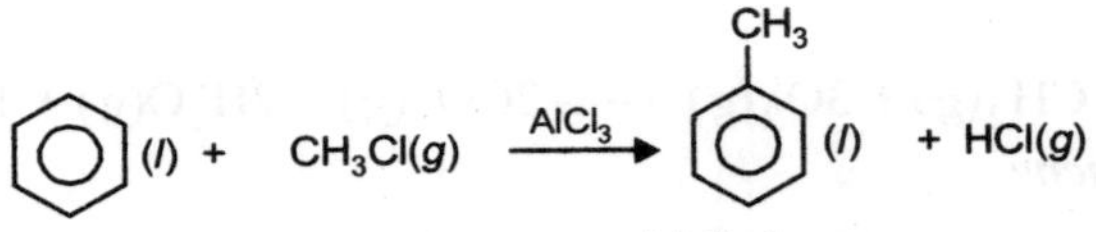

$$C_6H_6(l) + CH_3Cl(g) \xrightarrow{AlCl_3} C_6H_5CH_3(l) + HCl(g)$$

Benzene Chloromethane Methylbenzene

Addition Reactions

Most reactions of alkenes and alkynes are addition reactions in which two or more atoms of elements such as H, Cl, or Br are added to each of the carbon atoms in a double or triple bond. Examples follow:

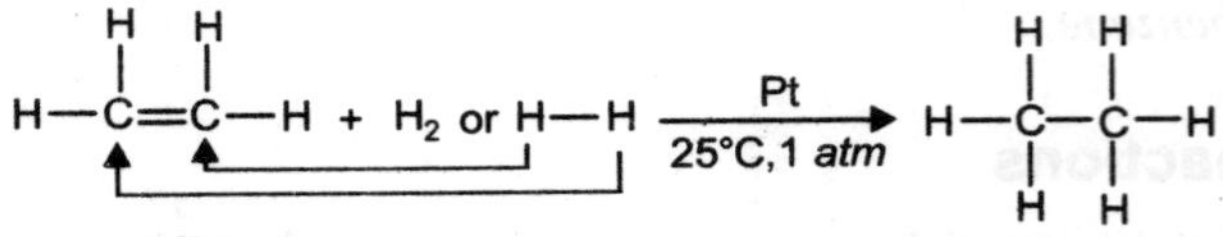

$$H_2C{=}CH_2 + H_2 \text{ or } H{-}H \xrightarrow[25°C,1\ atm]{Pt} H_3C{-}CH_3$$

Ethene Ethane

Alkenes:

$$H_2C{=}CH_2 + Br_2 \text{ or } Br{-}Br \rightarrow BrH_2C{-}CH_2Br$$

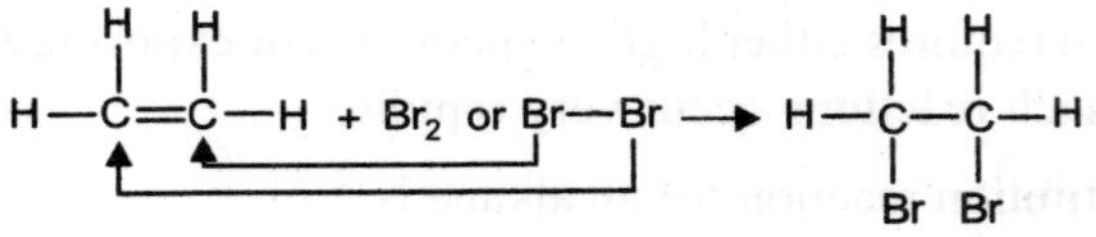

Ethene 1,2-dibromoethane

Alkynes:

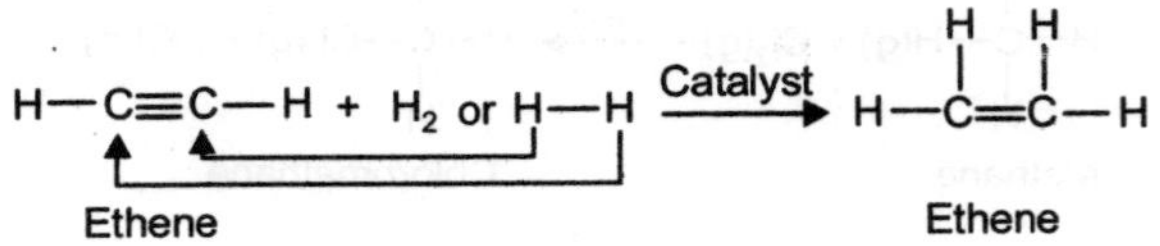

$$H{-}C{\equiv}C{-}H + H_2 \text{ or } H{-}H \xrightarrow{Catalyst} H_2C{=}CH_2$$

Ethene Ethene

With more H_2' the ethene produced by this reaction could also undergo an addition reaction to produce ethane (CH_3CH_3).

$$H-C\equiv C-H + Br_2 \text{ or } Br-Br \longrightarrow H-\underset{Br}{C}=\underset{Br}{C}-H$$

Ethyne *1,2-dibromoerhene*

A second addition reaction would produce 1,1,2,2-tetrabromoethane.

$$H-\underset{Br}{\overset{Br}{C}}-\underset{Br}{\overset{Br}{C}}-H$$

Addition Polymerization Reactions

Your life and much of your life-style depends on a class of compounds known as polymers. Polymers are giant molecules formed by linking together many small molecular units called monomers. During the past 80 years chemists have created a new multibillion dollar industry based on synthetic organic polymers such as vinyl, nylon, polyesters, PVC (polyvinyl chloride), polystyrene, polyethylene, Plexiglas, Lucite, Teflon, and synthetic rubber.

There are different types of polymers depending on the type of chemical reaction involved. Alkenes and alkene derivatives can form a class of polymers called *addition polymers.* In an addition polymerization reaction, as each two unsaturated monomers combine, they undergo an addition reaction in which their carbon-carbon double bonds are converted to carbon-carbon single bonds.

The simplest addition polymer is *polyethylene,* today's most widely used plastic. Polyethylenes are cheap and they can be molded into almost any shape, extruded into fibres or filaments, and precipitated or blown into films. Polyethylenes are made by the simple end-to-end addition of ethene, or ethylene, to form chains of 350 to 1400 monomer units:

$$\text{Petroleum} \longrightarrow H-\underset{H}{\overset{H}{C}}-\underset{H}{\overset{H}{C}}-H \xrightarrow[\text{(cracking)}]{-2H} \left(\begin{matrix}H & & H\\ & C=C & \\ H & & H\end{matrix}\right) \xrightarrow[\text{High temperature}]{\text{Catalyst and pressure}}$$

Ethane — Ethane or ethylene monomer

$$\cdots-\underset{H}{\overset{H}{C}}-\underset{H}{\overset{H}{C}}-\underset{H}{\overset{H}{C}}-\underset{H}{\overset{H}{C}}-\underset{H}{\overset{H}{C}}-\underset{H}{\overset{H}{C}}-\underset{H}{\overset{H}{C}}-\cdots \quad \text{or} \quad \left(-\underset{H}{\overset{H}{C}}-\underset{H}{\overset{H}{C}}-\right)_n$$

Polyethylene — Repeating unit formula

Actually, the carbon atoms join at a slight angle to form a kinked structure.
So that carbon skeleton of polyethylene looks like

When one or more of the hydrogen atoms on the ethylene monomer are replaced with an alkyl side chain (R) or some functional group (X) such as a chlorine atom (–Cl), a vinyl monomer is created

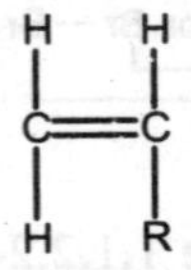

vinyl monomer

When vinyl monomers polymerize, vinyl polymers are produced for example, substituting a methyl group for a hydrogen on ethene gives the monomer propene or propylene. Polymerization yields the vinyl polymer polypropylene, which is used to make plastic milk bottlers, films and indoor-outdoor carpettings.

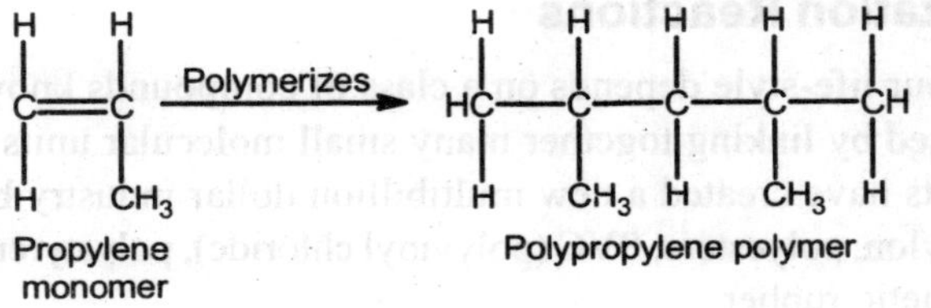

Polymerizes

Propylene monomer

Polypropylene polymer

or

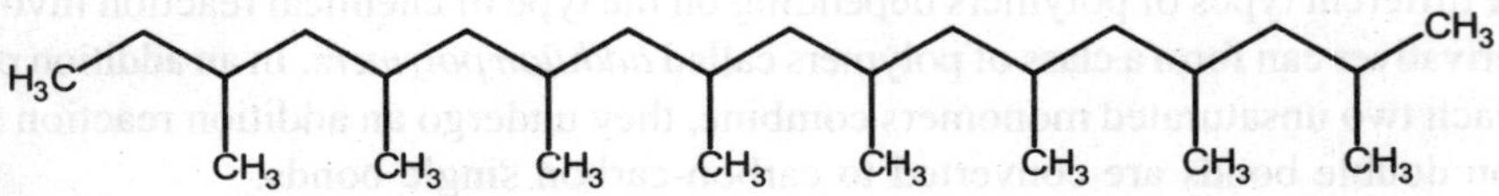

when a chlorine atom is the x group, the monomer is chloroethene (vinyl chloride). It can be polymerized to yield a plastic called polyvinyl chloride (PVC).

polymerizes

Vinyl chloride monomer

Polyvinyl chloride (PVC) polymer

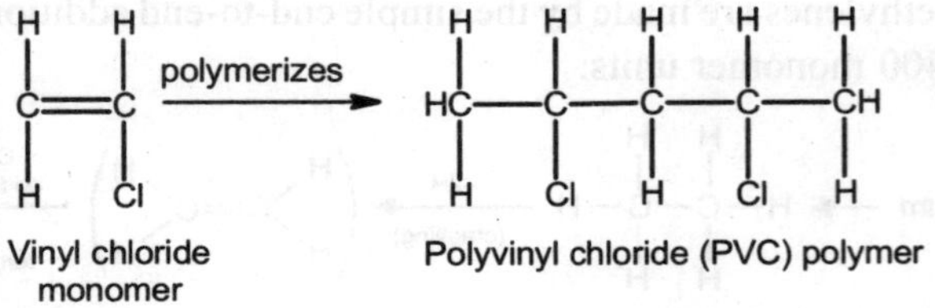

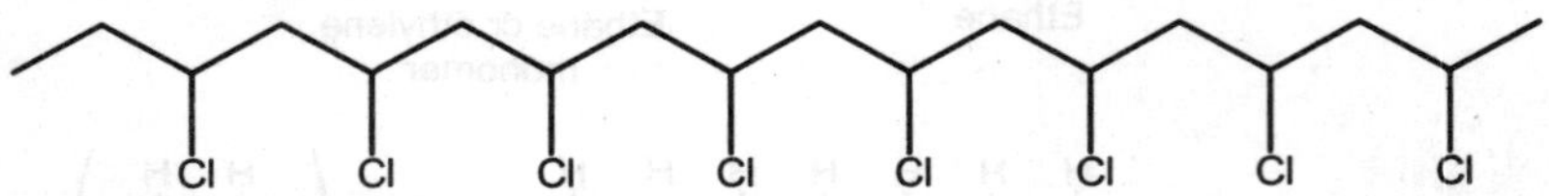

Other vinyl polymers are listed in the Table 4.4.

Hydrocarbon Derivatives

Functional Groups

We have seen that alkyl groups can replace one or more hydrogen atoms on a hydrocarbon chain or ring. Such hydrogen atoms can also be replaced by other atoms, such as chlorine (-Cl), or by groups

Table 4.4 : Some Vinyl Polymers

Monomer	Monomer Formula	Polymer Formula and Name	Trade Names	Uses
Ethene (ethylene)	$CH_2=CH_2$	$-CH_2-CH_2-CH_2-CH_2-CH_2-CH_2-$ Polyethelene	Polythene	Bags, films, toys, and molded objects; electrical insulation
Propene (propylene)	$CH_2\ CH(CH_3)$	$-CH_2-CH(CH_3)-CH_2-CH(CH_3)-CH_2-CH(CH_3)-$	Prolene, Vectra, Herculon	Bottles, films, outdoor and indoor carpets, bucket seats
Chloroethene (vinyl chloride)	$CH_2=CHCl$	$-CH_2-CHCl-CH_2-CHCl-CH_2-CHCl-$	PVC, Koroseal	Floor tiles, phonograph records, raincoats, garden hose, packaging, car tops
1, 1-Dichloroethene (vinylydene chloride)	$CH_2=CCl_2$	$-CH_2-CCl_2-CH_2-CCl_2-CH_2-CCl_2-$	Saran	Food wrap

(Contd...)

Table 4.4—(Contd...)

Monomer	Monomer Formula	Polymer Formula and Name	Trade Names	Uses
Tetrafluoroethene	$F_2C{=}CF_2$	$-CF_2-CF_2-CF_2-CF_2-CF_2-CF_2-$ Polytetrafluoroethene	Teflon	Pan coatings, electrical insulation, gasket bearings
Acrylonitrile	$H_2C{=}CH(C{\equiv}N)$	$-CH_2-CH(C{\equiv}N)-CH_2-CH(C{\equiv}N)-CH_2-CH(C{\equiv}N)-$ Polyacrilonitrile	Acrilan, Cresian, Orlon	Rugs, fabrics
Methyl methacrylate	$H_2C{=}C(CH_3)-C({=}O)-O-CH_3$	$-CH_2-C(CH_3)(C({=}O)OCH_3)-CH_2-C(CH_3)(C({=}O)OCH_3)-CH_2-C(CH_3)(C({=}O)OCH_3)-$ Polymethyl methacrylate	Lucite, Plexiglas, Perspex	Windows, surface coatings, latex paints, colourful signs, molded items such as combs and brushes
Styrene	$H_2C{=}CH(C_6H_5)$	$-CH_2-CH(C_6H_5)-CH_2-CH(C_6H_5)-CH_2-CH(C_6H_5)$ Polystyrene	Styron, Lustron, Styrofoam	Food and drink coolers, insulation, plastic furniture, radio cabinets, piping, battery cases, building material

of atoms, such as the hydroxyl group (-O-H). Atoms or groups of atoms that can be bonded to a carbon atom on the end of a carbon chain or between two carbon atoms within a carbon chain in an organic compound and that can determine much of the molecule's chemical properties are called *functional groups.*

Once attached, the functional group serves as the active reaction site for the molecule. The rest of the molecule, excluding the functional group, typically consists of one or more hydrocarbon alkyl groups. Because these alkyl groups are fairly unreactive, they are often represented by the letter R when the functional group is attached to the end of an organic molecule. For example, the molecule can be represented as R-Cl, where R stands for all the molecule except the-Cl functional group (Table 4.5).

A brief summary of some of the major functional groups was given earlier in gives a more detailed summary of these major functional groups. Notice that in cases such as ethers where the functional group is between two carbon atoms, the alkyl groups on each side of the functional group are represented by R and R′ if they are different. If the chemistry of a particular functional group is understood, another compound containing the same functional group will normally have similar chemical properties. In the remainder of this section you will learn more about the structures, names, and properties of molecules containing the major functional groups shown in

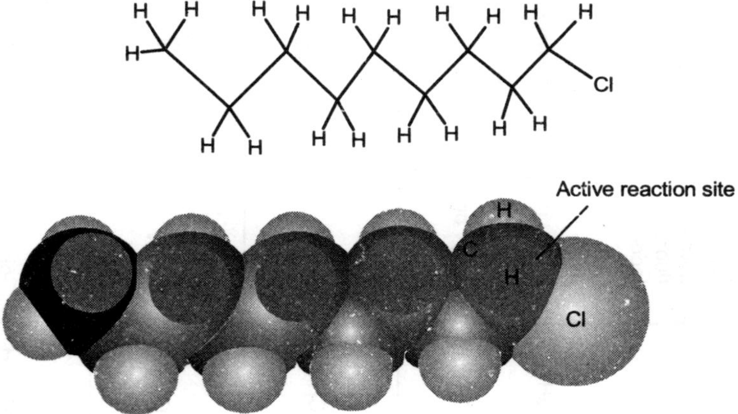

Organic Halides

Replacing a hydrogen atom in a hydrocarbon with a chlorine atom or other halogen atom from Group VIIA of the periodic table yields a hydrocarbon derivative called an organic halide. The simplest organic halides can be formed by substituting one or more F, Cl, Br, or 1 atoms for hydrogen on methane, with Cl being the most common halide functional group. One of the simplest organic halides is CIt3Cl. Its common name is *methyl chloride* and its IUP AC name is *chloromethane.*

To write the IUPAC name for such compounds, place an *o* after the first syllable of the halogen name *(Fluoro-, Chloro-, Cromo-,* and *Iodo-)* and then add the name of the alkane from which the compound is derived. The carbon atom or atoms to which each halogen atom is attached are numbered in the same way used to indicate alkyl side chains on hydrocarbons. Other examples of organic halides are given in Table 4.6.

Table 4.5 : Some Major Functional Groups Found in Organic Compounds

Type of Compound	Structural Formula	Condensed Formula	Prefix or Suffix for Group Name	Example		
				Condensed Structural Formula	IUPAC Name	Common Name
Halide	R—Cl (or — Br, —F, —I)	RCl	Chloro- (or bromo-, fluoro-, iodo-)	$CH_3—CH_2—Cl$	Chloroethane	Ethyl chloride
Alcohol	R—O—H	ROH	-ol)	$CH_3—CH_2—O—H$	Ethanol	Ethyl alchol
Ether	R—O—R′	ROR′	-oxy- or ether	$CH_3—O—CH_3$	Methoxymethane	Dimethyl ether
Aldehyde	R—C(=O)—H	RCHO	-al	$CH_3—C(=O)—H$	Ethanal	Acetaldehyde
Ketone	R—C(=O)—R′	RCOR′	-one	$CH_3—C(=O)—CH_3$	Propanone	Acetone
Carboxylic acid	R—C(=O)—O—H	RCOOH	-oic acid	$CH_3—C(=O)—O—H$	Ethanoic acid	Acetic acid
Ester	R—C(=O)—O—R′	RCOOR′	-oate	$CH_3—C(=O)—O—CH_3$	Methyl ethanoate	Methyl acetate

Alcohols

If a hydrogen atom attached to a carbon skeleton is replaced with an -O-H or *hydroxyl functional group,* the resulting compound is called an alcohol. The -o-o or hydroxyl group in alcohols should not be confused with the hydroxide ion (OH–) in NaOH and other inorganic bases. The hydroxyl group is covalently bonded to a carbon atom and does not give a basic solution when the compound is placed in water. In contrast, the OH - ion in NaOH is ionically bonded to a Na + ion and forms a basic solution in water.

Table 4.6 : Examples of Organic Halides

Structural Formula	Names		Typical Uses
	IUPAC	Common	
Cl H—C—Cl Cl	Trichloromethane	Chloroform	Gcod solvent, general anesthetic
Cl Cl—C—Cl Cl	Tetrachloromethane	Carbon tetrachloride	Good solvent for non-polar molecules
F F—C—Cl Cl	Dichlorodifluoromethane	Freon-12	Solvent and refrigerant fluid
Br F H—C—C—F Cl F	2-Bromo-2-chloro-1, 1, 1-Trifluoroethane	Halothane	General inhalation anesthetic
Cl (benzene ring) Cl	1, 4-Dichlorobenzene	p-Dichlorobenzene	Ingredient in some types of mothballs
Cl Cl H Cl H H Cl H H Cl Cl H	Hexachlorocyclohexane	Lindane	Insecticide

The simplest alcohol is *methanol* with the common name *methyl alcohol.* Its other common name, wood alcohol, comes from the fact that it was first made by heating wood in the absence of air.

The common name for an alcohol is obtained by giving the name of the alkyl (R) group attached to the -OH group followed by the word *alcohol* (methyl alcohol). The IUP AC name for a simple alcohol is obtained by changing the last letter in the name of the parent hydrocarbon to *-ol.* Thus, the alcohol produced from methane (CH_4) is methanol (CH_3OH). When necessary, the position of the -OH group on a carbon chain is indicated in the usual manner by giving the number of the carbon atom to which it is attached.

Thus, $CH_3CH_2\underset{\displaystyle OH}{\underset{|}{C}}HCH_2CH_3$ is called 3-pentanol because the –OH group is attached to the No.3 carbon atom in a five-carbon chain.

Methanol is a deadly poison. Excessive inhalation of its vapour or ingestion of only a few millilitres can dissolve the fatty sheath around the optic nerve and cause permanent blindness. A little more can kill when the enzymes in the liver convert methanol to highly toxic formaldehyde.

The next alcohol, with an –OH group replacing one of the hydrogen atoms in ethane (C_2H_6), has the IUPAC name *ethanol* and the structure shown in below. It is commonly called *ethyl alcohol* or "alcohol."

H H
H—C—C—O—H or CH_3CH_2OH
H H

Showing the formulas of the ethanol

Ethanol is also found in alcoholic beverages. In this case, it is produced by the fermentation of starches or sugars in fruits or vegetables in reactions that are catalyzed by enzymes found in yeast.

$$\underset{\textit{glucose (sugar)}}{C_6H_{12}O_6(S)} \xrightarrow{\textit{engymes in years}} \underset{\textit{ethanol}}{2CH_3CH_2-OH(l)} + 2CO_2(g) + \text{energy}$$

Alcoholic beverages produced by fermentation include *undistilled* beverages such as beer, champagne, and wine that typically contain from 3.5% to 12% by volume ethanol unless fortified by addition of extra ethanol. Alcoholic beverages such as whiskey, rum, gin, brandy, and vodka contain about 40% to 50010 by volume ethanol and are obtained in these higher concentrations by *distilling* fermented solutions with lower ethanol concentrations.

Ethanol used in alcoholic beverages is heavily taxed by the government and is an important source of revenue. It is not a crime to produce ethanol for consumption, but producing "bootleg" alcohol without paying the tax on it is a crime: Tax-free ethanol is sold for laboratory and industrial uses. To avoid its illegal use as a beverage, it is usually *denatured* by adding a small amount of a poisonous substance such as methanol or gasoline. This makes it unfit to drink without affecting its chemical properties.

Despite the fact that ethanol is the least toxic of the continuous-chain alcohols, its widespread use and abuse causes more deaths each year than any other chemical. Contrary to popular belief, ethanol acts not as a stimulant but as a depressant of the central nervous system. Extensive long-term use of ethanol (chronic alcoholism) causes (1) destruction of liver cells, swelling of the liver, and permanent damage to the liver (cirrhosis); (2) malnutrition by reducing the appetite and the ability of the body to digest foods properly and produce certain essential vitamins; and (3) low blood sugar (hypoglycemia).

In a three-carbon chain there are two possible sites for attaching an –O–H or hydroxyl group. Putting it on either end gives I-propanol (or n-propyl alcohol):

H–C(H)(H)–C(H)(H)–C(H)(H)–O–H or $H_3C—CH_2—CH_2—OH$

1-propanol or n-propyl alcohol

Attaching to the middle carbon atom gives 2 propenol (isopropyl alcohol)

H–C(H)(H)–C(H)(O–H)–C(H)(H)–H or $\overset{1}{H_3C}—\overset{2}{CH}(OH)—\overset{3}{CH_3}$

2-propanol or isopropyl alcohol

which is used as rubbing alcohol and as an antiseptic and disinfectant. Both 1-propanol and 2-propanol have the molecular formula C_3H_5O and are examples of structural isomers.

When a hydroxyl group is attached directly to a benzene ring, the resulting compound is called *phenol.*

Phenol

Phenol is used industrially to prepare several kinds of plastics and dyes. When mixed with water, it forms a solution known as *carbolic acid* that can be used as a germicide to clean surgical instruments and hospital linens. It is no longer applied to the skin to kill germs because it can burn and severely damage the skin.

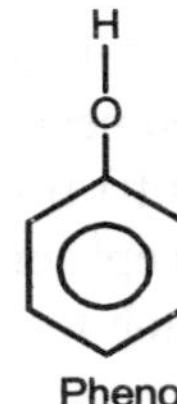

Phenol

Ethers

An ether is an organic compound that contains the

>C—O—C<

or R-O-R' functional group, where Rand R′ are hydrocarbon groups that can be derived from saturated, unsaturated, or aromatic hydrocarbons. The simplest ether contains two methyl groups (CH_3-) on each side of an oxygen atom. Most ethers are usually called by their non-IUPAC names. These names are obtained by naming the hydrocarbon group attached to the oxygen atom preceding it by the prefix *di-* if the groups are identical and then adding the word *ether.* Thus, the compound in the figure given below is usually called *dimethyl ether.* Its IUP AC name is *methoxymethane.*

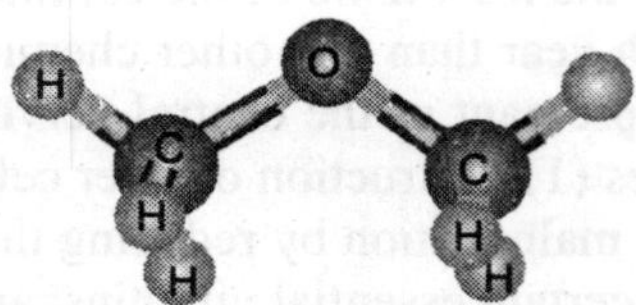

The IUPAC name for an ether is obtained by (1) locating the longest hydro-carbon group attached to the oxygen atom and giving it the name of the corre-sponding alkane, (2) changing the *-yl* ending for the other hydrocarbon group to *-oxy* (for example, CH_3-O is called *methoxy),* and (3) combining the name from Step 2 with that from Step 1. Thus, CH_3-O-CH_3 is called *methoxy-methane* and CH_3-O-$CH_2CH_2CH_3$ is called *methoxypropane.*

$$H-\overset{H}{\underset{H}{C}}-O-\overset{H}{\underset{H}{C}}-H \qquad H_3C-O-CH_3$$

Figure showing the Structure of the simplest ether molecule. It is usually called *dimethyl ether* but its IUPAC name is *methoxymethane.*

Two other common ethers are methoxyethane or methyl ethyl ether and ethoxyethane or diethyl ether.

CH_3–O–CH_2–CH_3
Methoxyethane or Methyl ethyl ether

CH_3–CH_2–O–CH_3–CH_3
Ethoxyethane or diethyl ether

Ethoxyethane (diethyl ether) is the most well-known ether and is often called simply *ether.* It is sometimes used as an anesthetic. However, this use has declined somewhat because it is highly flammable and causes aftereffects such as nausea and vomiting. Methoxypropane (methypropyl ether) is currently more widely used as an anesthetic because it is relatively free of after-effects.

All ethers are highly flammable and form explosive mixtures with air. Thus, they should never be used near a flame or a possible source of a spark. For example, striking a match as far away as 15 m (50 ft) from an open bottle of ethoxyethane (diethyl ether) has caused an explosive laboratory fire. Except for their flammability, ethers are generally unreactive. This makes them excellent solvents in which to carry out organic reactions.

Aldehydes and Ketones

Aldehydes and ketones are compounds containing a carbonyl group

$$R-\overset{O}{\overset{\|}{C}}-H$$

In an aldehyde, the carbonyl group is attached to the end of a hydrocarbon group to give the general formula

$$R-\overset{O}{\overset{\|}{C}}-H \text{ (Figure).}$$

In a ketone, it is found between two hydrocarbon groups to give the general formula

$$R-\overset{\overset{\displaystyle O}{\|}}{C}-R'$$

The IUPAC name for an aldehyde is given by locating the longest continuous chain containing the aldehyde group and changing the last letter in its alkane name to *-al.* The same procedure is followed for ketones except that the last letter in the alkane corresponding to the longest continuous chain attached to the carbonyl group is changed to *-one.*

Thus, the simplest aldehyde has the formula

$$H-\overset{\overset{\displaystyle O}{\|}}{C}-H$$

and is called *methanal.* You may know this compound from biology lab by its common name, *formaldehyde.* Formalin, a 40% by volume solution of formaldehyde (methanal) in water is used to preserve biological specimens. It is also used as an embalming fluid and to manufacture certain plastics and synthetic fibers. If your clothes are crease resistant, you can usually thank methanal. Unfortunately, some people are allergic to methanal (freed from clothes by perspiration) and get rashes if they wear permanent press clothes. The structure of another aldehyde, propanal, is shown in Figure below.

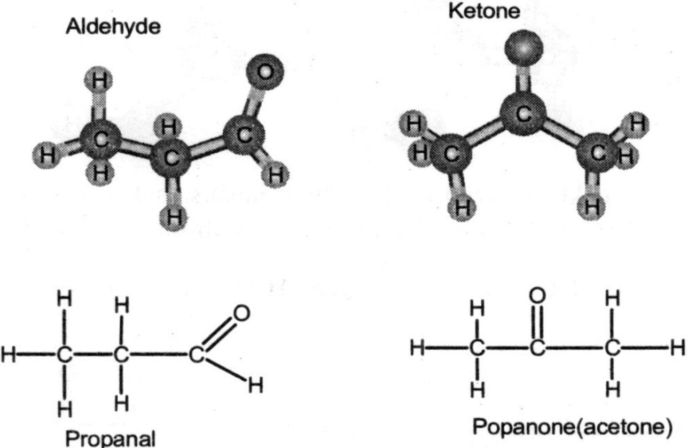

The simplest ketone is *propanone*. It is usually called by its non-IUPAC name, *acetone.* If you smell the sweetish odour of acetone on an unconscious person's breath, call a doctor at once. Such a smell usually means that the person is in a diabetic coma. If you don't know what acetone smells like, ask your instructor to let you smell some. Acetone is used around laboratories to clean and dry glassware. After a piece of glassware is washed with water, any remaining water droplets can be rinsed out with acetone, which then evaporates very quickly. Acetone, however, is flammable and must be used with care. It is also widely used as a solvent in the manufacture of plastics, lacquers, paint removers, explosives, and some drugs.

Carboxylic Acids

A carboxylic acid is an organic compound that contains a functional group called carboxyl group,

$$-\overset{\displaystyle O}{\overset{\|}{C}}-O-H$$

attached to the end of a hydrocarbon chain and represented by

$$R-\overset{\displaystyle O}{\overset{\|}{C}}-O-H$$

or RCOOH. The IUPAC name for a carboxylic acid is given by dropping the last letter *(e)* in the name of the hydrocarbon chain, adding the suffix *-oic,* and following it with the word *acid.* The structures of the two simplest carboxylic acids, methanoic (or formic) acid and ethanoic (or acetic) acid, are shown in the figure below.

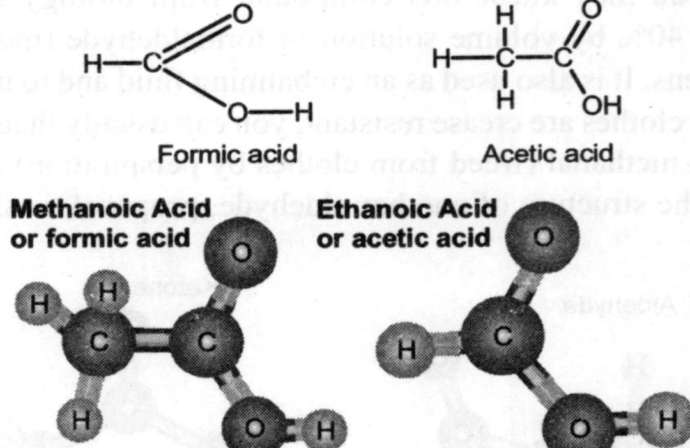

Carboxylic acids are weakly ionized acids, which means that in a water solution only a small fraction of their molecules react with water to produce 1 hydronium ions, H_3O^+

$$HC_2H_3O_2 + H_2O \rightleftharpoons H_3O^+ + C_2H_3O_{2\text{-}}$$

or

$$H-\underset{\displaystyle H}{\overset{\displaystyle H}{C}}-\overset{\displaystyle O}{\overset{\|}{C}}-O-H + H_2O \rightleftharpoons H_3O^+ + H-\underset{\displaystyle H}{\overset{\displaystyle H}{C}}-\overset{\displaystyle O}{\overset{\|}{C}}-O^-$$

ethanoic (acetic acid) water Hydronium ion Acetate ion

Vinegar is a solution of 4% to 5% by volume ethanoic acid in water.

Methanoic acid is used in the textile, rubber, and leather industries. It is also one of the substances that causes itching and swelling after a bite from stinging nettles and insects such as red ants and bees. Some of the worst odours, such as those from rancid butter, vomit, and limburger cheese, come from carboxylic acids with 4 to 10 carbon atoms. Acids with even longer chains (usually 12 or more carbon atoms) are called *fatty acids.*

Esters

An ester is an organic compound represented by the general formula. The IUPAC name for an ester is given by first naming the alkyl chain (R′) attached to the oxygen atom (methyl, ethyl, and so on) and then naming the carboxylic acid derived from the remainder of the molecule but replacing the *-oie* suffix with *-oate*. Thus, the two simplest esters have the following structural formulas and IUPAC names.

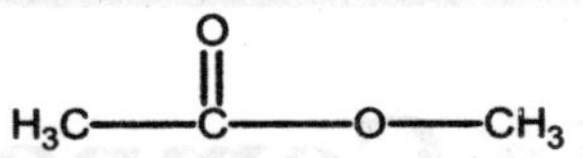

$$H_3C—\overset{\overset{\Large O}{||}}{C}—O—CH_3 \quad \text{and} \quad H_3C—\overset{\overset{\Large O}{||}}{C}—O—CH_2—CH_3$$

Methyl acetate
Methyl Ethanoate

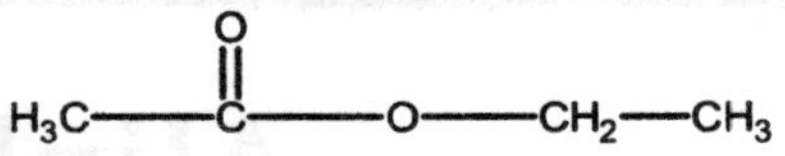

Ethyl acetate
Ethyl ethylethanoate

Ethyl ethanoate is used as a solvent for lacquers and plastics and is often used as a fingernail polish remover. Some of the odours of common fruits are due to the presence of esters. For example, ethyl butyrate, is responsible for the odour of pineapples. In contrast, esters with high molar masses often have an unpleasant odour.

This chapter has illustrated two important chemical ideas: (I) that carbon and only a few other elements can be used to build millions of organic compounds, and (2) that the key to the behavior of complex organic molecules is the location of key functional groups or reaction sites.

CHAPTER

5

Aliphatic Compounds
Saturated Hydrocarbons - Paraffins or Alkenes

HYDROCARBONS

The hydrocarbons are those compounds which contain hydrogen and carbon and sometimes oxygen (combines in the water molecule). They are mainly of two types. (1) Saturated hydrocarbon and (2) the unsaturated hydrocarbon.

In saturated hydrocarbon all the carbon valences are fully satisfied. They are stable, less reactive compounds and are known as paraffins (*Parva* - little *affinita* - affinity Latin).

Unsaturated hydrocarbons are farther differentiated into two classes **(*a*) olefines and (*b*) acetylenes.** In the case of unsaturated, giving rise to double or triple bonds.

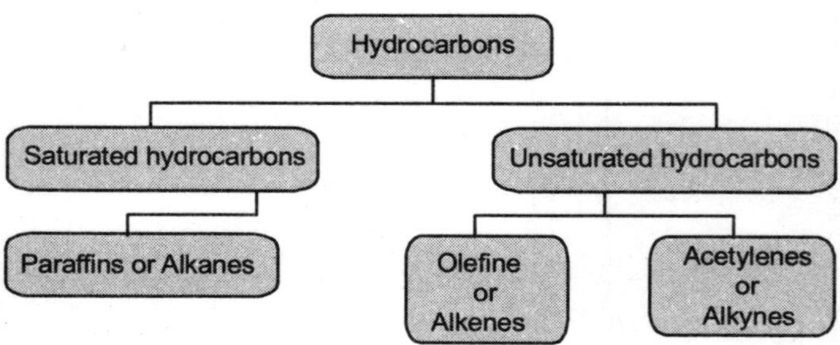

These different types of hydrocarbons are related to one another as shown in Table 5.1 on next page:

Table 5.1 : Different types of Hydrocarbons

Parameters	Paraffins	Olifines	Acetylenes
General	C_nH_{2n+2}	C_nH_{2n}	C_nH_{2n-2}
Formula	CH_4(methane)	CH_2 (Methylene not found)	
Members	C_2H_6(ethene)	C_2H_4(Ethylene or ethane)	C_2H_2(acetylene)
	C_3H_8(propane)	C_3H_6(propylene or propane)	C_3H_7(propyne)
	C_4H_{10}(butane)	C_3H_8(butylene or butane)	C_4H_6(butane)
	C_5H_{12}(pentane)	C_3H_{10}(pentylene or pentene)	C_5H_8(pentyne)

Saturated Hydrocarbons

Paraffins or Alkanes

The saturated hydrocarbons or paraffins form the homologous series of the formula C_nH_{2n+2} (RH)

The various homologous series are:

CH_4 .... Methane
C_2H_6 .... Ethane
C_3H_8 .... Propane
C_4H_{10} .... Butane
C_5H_{12} .... Pentane
C_6H_{14} .... Hexane
C_7H_{16} .... Heptane
C_8H_{18} .... Octane
C_9H_{20} .... Nonane
$C_{10}H_{22}$ .... Decane

Occurrence

(a) Petroleum
(b) Natural Gas accompanying petroleum
(c) Coal Gas, produced by destructive distillation of coal
(d) Coal mines
(e) Gases produced by the decay of vegetable matter –leaves grasses etc.
(f) Ozokerite or earth-wax a mineral found in the petroleum districts. This contains a mixture of higher solid hydrocarbons

General Methods of Preparation

The following are the general methods preparation of the paraffin

(I) By heating the sodium acetate soda lime [mixture of sodium hydroxide and calcium oxide]

Sodium hydroxide cannot be used alone because it is delequescent and having a low melting point

$$\underset{\text{Sodium Acetate}}{CH_3COONa} + \underset{\text{Soda-lime}}{NaOH(CaO)} \rightarrow \underset{\text{Methane}}{CH_4} + \underset{\text{Sodium Carbonate}}{Na_2CO_3}$$

The active component of soda-lime is sodium hydroxide.

The method also be regarded as decarbooxydation of fatty acids.

(II) By the reactions of Alkyl halides (preferably the iodides)

$$\underset{\text{Methyl Iodide}}{CH_3I} + \underset{\text{Nescent Hydrogen}}{2H} \rightarrow \underset{\text{Methane}}{CH_4} + \underset{\text{Hydrogen Iodide}}{HI}$$

Some of the important reacting agents for this preparations are, are sodium amalgam and water, zinc and a dilute acid , zinc and water 160°C, zinc copper couple and water or alcohol aluminium or magnesium mercury couple and alcohol and conc. Hydrogen Iodide a high temperature especially at the presence of red phosphorous.

(III) By the reduction of hydroxyl compounds with hydroiodic acid at high temperature in the presence of red phosphorous.

Acids are reduced with diffculty.

$$\underset{\text{Methyl Alcohol}}{CH_3-OH} + \underset{\text{Hydrogen Iodide}}{2HI} \rightarrow \underset{\text{Methane}}{CH_4} + \underset{\text{Hydrogen}}{H_2} + \underset{\text{Iodine}}{I_2}$$

Wurtz's Reaction

A method of synthesizing hydrocarbons by treating alkyl halides with sodium (Wurtz's reaction) that was adapted by the German chemist Rudolf Fittig to the preparation of mixed aliphatic and aromatic hydrocarbons (Wurtz-Fittig reaction).

Wurtz-Fittig Reaction

$$\underset{\text{1,4 bromobenzene}}{C_6H_4Br_2} + \underset{\text{Methyl iodide}}{2CH_3I} \xrightarrow[\text{Sodium ether}]{Na,\ Et_2O} \underset{\text{p-toluene}}{CH_3C_6H_4CH_3}$$

$$\underset{\text{ethyl iodide}}{C_2H_5I} + 2Na + \underset{\text{Methyl iodide}}{ICH_3} \longrightarrow \underset{\text{Propane}}{C_2H_5CH_3} + 2NaI$$

$$\underset{\text{Ethyl Bromide}}{C_2H_5Br} + 2Na + \underset{\text{Methyl Iodide}}{ICH_3} \xrightarrow{\text{Ether}} \underset{\text{Propane}}{C_2H_5CH_3} + NaI + NaBr$$

The reaction is not very satisfactory for the preparation of hydrocarbon with odd number of carbon atoms why?

In Wurtz's reaction the alkyl bromides give best results. Zinc magnesium silver or copper can also used in place of sodium in the above reaction.

Wurtz's reaction can be used for stopping up the homologous series:

Kolbe's Reaction

Kolbe believed that organic compounds could be derived from inorganic ones, directly or indirectly, by substitution processes. He validated his theory by converting carbon disulfide, in several steps, to acetic acid (1843-45). Introducing a modified idea of structural radicals, he contributed to the establishment of structural theory. He also predicted the existence of secondary and tertiary alcohols. He worked on the electrolysis of the salts of fatty and other acids and prepared ***salicylic acid***, a building block of aspirin. With Sir Edward Frankland he found that nitriles can be hydrolyzed to the corresponding acids. As editor of the Journal *für praktische Chemie* (from 1869), he was sometimes severely critical of the work of others.

Kolbe's Reaction Equation (for salicylic acid)

$$C_6H_5OH + NaOH \longrightarrow C_6H_5ONa \xrightarrow[4\text{–}7\ \text{atm}]{CO_2,\ 400K} C_6H_4(OH)COONa \xrightarrow{H_3\overset{+}{O}} C_6H_4(OH)COOH$$

2-Hydroxybenzoic acid
(Salicylic acid)

Methane cannot be obtained from this reaction, why?

Ethane is obtained by the electrolysis of potassium acetate solution

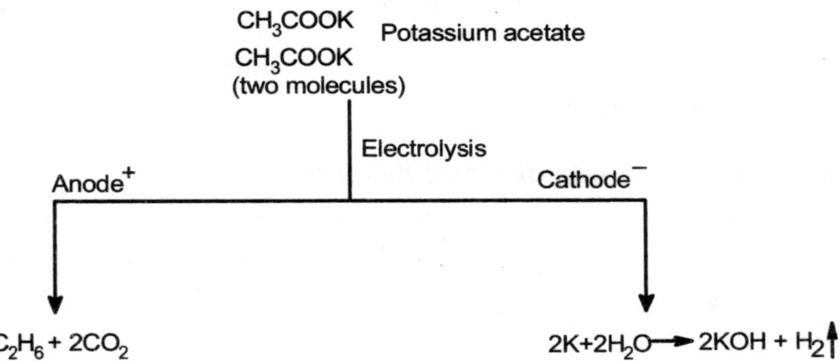

A mixture of the ethane and carbon dioxide obtained at the anode is passed through a strong solution of potassium hydroxide to remove carbon dioxide

CH_3COOK potassium acetate
C_2H_5COOK Potassium Propionate

Anode$^+$ → $C_3H_8 + 2CO_2$ (Propane)

Cathode$^-$ → $2K + 2H_2O \longrightarrow 2KOH + H_2$

Propane is obtained by the electrolysis of mixed solution of potassium acetate and potassium propionate.

Sabatier and Senderen's Reaction

$$C_nH_{2n} + 2H \xrightarrow{Ni} C_nH_{2n+2}$$

$$C_nH_{2n-2} \xrightarrow[Ni\ 200°C]{+2H} C_nH_{2n} \xrightarrow[Ni\ 200°C]{+2H} C_nH_{2n+2}$$

$$C_2H_2 \xrightarrow[Ni\ 200°C]{+2H} C_2H_4 \xrightarrow[Ni\ 200°C]{+2H} C_2H_5$$

By the hydrolysis of Grignard's Reagent

Grignard's reagents are named after Victor Grignard, a French chemist, who received a Nobel Prize (1912) for their discovery. Grignard's Reagent is the magnesium alkyl halide or magnesium ethyl chloride or iodide.

$$Mg(C_2H_5)(I) + H_2O \longrightarrow Mg(OH)(I) + CH_4$$

Grignard's Reagent → Magnesium Oxy Iodide + Methane

$$Mg(C_2H_5)(Cl) + H_2O \longrightarrow Mg(OH)(Cl) + C_2H_6$$

Grignard's Reagent → Magnesium Oxychloride + Ethane

Grignard Reactions with Aldehydes and Ketones

All reactions of Grignard reagents are like the carbon dioxide case: the R adds to the C of the carbonyl group and eventually - H to the O. So from an aldehyde or ketone we can make an alcohol - either primary, secondary or tertiary. The only difference between these three reactions is the number and nature of the group(s) attached to the C=O.

$$R{-}Br \longrightarrow \overset{\delta-\ \ \delta+}{R{-}MgBr} + H_2C{=}O \longrightarrow R{-}CH_2{-}OMgBr \xrightarrow{H^+} R{-}CH_2{-}OH$$

Making a primary alcohol from an alkyl bromide and formaldehyde

R—Br ⟶ R—MgBr (δ— δ+) + R'(H)C=O ⟶ R—C(R')(H)—OMgBr —H^+⟶ R—C(R')(H)—OH

Making a secondary alcohol from an alkyl bromide and an aldehyde

R—Br ⟶ R—MgBr (δ— δ+) + R'(R'')C=O ⟶ R—C(R')(R'')—OMgBr —H^+⟶ R—C(R')(R'')—OH

Making a tertiary alcohol from an alkyl bromide and a ketone.

And the great thing is that these reactions tend to all give excellent yields.

So let's see what we can do with our first tool so far.

(1) ⟵ (2) OH ⟵ acetone + BrMg-butyl (3)

How to make 2-methylhexane using a Grignard reagent?

Say we want to make a hydrocarbon, 2-methylhexane (1): if we can think of a way of converting the corresponding alcohol (2) to the hydrocarbon then we have an easy solution! Start with acetone (propanone) and n-butyl bromide! Not too difficult? So all we need to is convert the molecule we need to synthesise into one that we can make with a Grignard reagent.

Grignard Reactions - with Epoxides and Esters

If you think of a C=O as a 'two-membered ring' then you wouldn't be too surprised to discover that three-membered rings containing oxygen are also reactive with Grignard reagents. Compare the two reactions below:

H_2C—O (ring) —HMgBr⟶ RCH_2OMgBr

Formaldehyde depicted as a two-membered ring

H_2C—CH_2 (O bridge) —RMgBr⟶ RCH_2CH_2OMgBr

Ethylene oxide

We can now make primary, secondary and tertiary alcohols using Grignard reagents. If at least two of the groups on a tertiary alcohol are the same, there is a nice one-step method that we can use to make it. We react *two* moles of a Grignard reagent with one of an *ester*:

$CH_3C(=O)OEt$ + 2PhMgBr ⟶ Me—C(Ph)(Ph)—OH

$CH_3C(=O)OEt$ —PhMgBr⟶ Me—C(OMgBr)(Ph)—OEt —–EtOMgBr⟶ Me—C(=O)—Ph —PhMgBr⟶ Me—C(Ph)(Ph)—OMgBr —H^+⟶ Me—C(Ph)(Ph)—OH

A one-step route to tertiary alcohols via esters

Enolates

A method for making all kinds of ketones and acids and their derivatives. We are going to concentrate on two key molecules, but the principle can easily be extended to other related systems. The two compounds are ethyl acetoacetate and diethyl malonate. The reason they are so useful is because they both contain 'active' hydrogens - or in other words the red hydrogens shown below are surprisingly acidic.

MeO—C(=O)—CH₂—C(=O)—OEt   EtO—C(=O)—CH₂—C(=O)—OEt

Ethyl acetoacetate   Diethylmalonate

With a base such as sodium ethoxide (NaOEt) a proton is easily removed. But first, how are they made? Ethyl acetoacetate is made by use of the 'Claisen condensation', the interaction of two molecules of ethyl acetate. You'll notice below that the first step is the removal of the weakly acidic proton. This is followed by the attack of this carbanion on the C=O of another molecule of ethyl acetate and then loss of OEt. So the NaOEt is only needed as a catalyst.

$MeCO_2Et \xrightleftharpoons{NaOEt} :CH_2CO_2Et$ + Me—C(=O)—OEt ⟶ Me—C(O⁻)(OEt)—CH_2CO_2Et ⟶ Me—C(=O)—CH_2CO_2Et

The Claisen Condensation—The Simple Method to Make Ethyl Acetoacetate

Diethyl malonate is made from acetic (ethanoic) acid in a series of simple steps. As each product in the synthesis is useful in its own right, the rather lengthy synthesis is worthwhile.

$$CH_7CO_2H \xrightarrow[\text{phosphorus}]{Cl_2} ClCH_2CO^2H \xrightarrow{NaCN} NCCH_2CO_2H \xrightarrow{NAOH} HO_2COH_2CO_2H \xrightarrow{EtOH,\ H} EtO_2CCH_2CO_2Et$$

The Synthesis of Diethyl Malonate

These two molecules are the key 'spanners' that will undo many synthetic nuts. Let's focus first on ethyl acetoacetate (we'll refer to it as EAA from now on). Why enolate? An enol is exactly what the name says - an ene with an OH attached. The resonance that stabilizes the anion of ethyl acetoacetate involves two ions that derive from an enol as you can see below:

Me—C(=O)—CH₂—C(=O)—OEt $\xrightarrow{NaOEt}$ [Me—C(=O)—C⁻H—C(=O)—OEt ⟷ Me—C(O⁻)=CH—C(=O)—OEt ⟷ Me—C(=O)—CH=C(O⁻)—OEt]

The enolate ion

This ion reacts as if it were a carbanion (the first resonance structure) with most reagents and typical

reagents are alkyl halides. So for example if we take EAA and treat it firstly with a mole of NaOEt and then one of EtBr we ethylate the methylene carbon.

$$MeCOCH_2CO_2Et \xrightarrow{NaOEt} MeCO\bar{C}HCO_2Et \xrightarrow{EtBr} MeCOCH(Et)CO_2Et + NaBr$$

If this was all we could do, the chemistry would not be of much interest. But if we now hydrolyse the ester (with acid or base) a surprising thing happens:

$$MeCOCH_2CO_2Et \xrightarrow{H^+} MeCOCH(Et)CO_2H \xrightarrow{-CO_2} MeC(OH){=}CHEt \rightleftharpoons MeCOCH_2Et$$

Notice the nice 6-membered hydrogen-bonded ring. The shift of a proton in the last step is an example of **tautomerism**.

Acids with a beta-carbonyl group are very unstable and easily lose carbon dioxide. That's why we couldn't make EAA by esterifying the acid - the acid loses carbon dioxide as soon as you try to make it!

$$MeCOCH_2CO_2H \xrightarrow[CO_2]{\text{easily loses}} MeCOMe$$

So we have a method overall for converting ethanoic acid (acetic acid) to a methyl ketone.

$$MeCO_2H \xrightarrow[H^+]{EtOH} MeCO_2Et \xrightarrow{NaOEt} MeCOCH_2CO_2Et \xrightarrow[EtBr]{NaOEt} MeCOCH(CH_2Me)CO_2Et \longrightarrow MeCOCH_2Me$$

How to Convert Ethanoic Acid into Pentan-2-One

We can repeat the reaction making another enolate with NaOEt and reacting with the same or a different alkylating reagent.

Direct Systhesis

Methane and ethane along with ethylene and acetylene can also be prepared by passing the electric current between the carbon electrodes in an atmosphere of hydrogen.

$$C + 2H_2 \xrightarrow{\text{Electric Arc}(1200^\circ C)} CH_4$$

$$2C + 3H_2 \xrightarrow{\text{Electric Arc}} C_2H_6$$

The above reactions can however be brought about at a lower temperature 475 % 500°C in the presence of a catalyst of Nickel.

Properties of Hydrocarbons

1. The first four members of this family methane, ethane, propane and butane re colourless and odourless gasses. The next twelve members (from C_5H_{12} to $C_{13}H_{64}$) are colourless liquids. The higher members (from $C_{17}H_{36}$ onwards) are colourless solids.
2. They are almost insoluble in water but soluble in ether, benzene or carbon disulphide
3. Their boiling points melting points and specific gravities regularly increase as the series is ascended.

Typical Specifications of n-Paraffins

Properties	Units	Product				
		C5-C6	C7-C8	C9-C11	C12-C13	C14-C20
Flash point	°C	<-20	0	45	90	110
Distillation:						
Initial boiling point	°C	55	96	156	213	250
Dry point	°C	98	128	195	225	315
Appearance		Bright & clear	Bright & clear	Bright & clear	Bright & clear	Bright & clear
Colour	Saybolt	+25	+30	+30	+30	+30
Sulphur	mg/kg	2 max	2 max	2 max	5 max	1 max
Evaporation rate*	90%	6.75	1.97	0.199	0.01	<0.015
	100%	5.1	1.78	0.139	5n/a	n/a
Aromatics	mass %	0.1 max	0.1 max	0.1 max	0.5 max	0. 5 max
KB value		27	27	25	22	19
Analine point	°C	72	74	80	89	102
Pour point	°C	<-40	<-40	-30	-10	6

*relative to butyl acetate = 1

Physical and Chemical Properties of Alkenes and Paraffines

Property	Unit	C10-13	C11-12	C11-14	C12-14	C14-17	C17-20
Apprearance	–	clear liquid	clear liquid	clear liquid	clear liquid	clear liquid	clear liquid
Odour	–	none	none	none	none	none	none
Colour	Saybolt	+ 30	+ 30	+ 30	+ 30	+ 5	+ 30
Average Molecular	g/mole	185	193	174	186	214	256
Carbon Distribution	–	C10~ C13	C11~ C12	C11~ C14	C12~ C14	C14~ C17	C17~ C20
Density	kg/l at 15 C	0.749	0.749	0.756	0.762	0.773	0.789
Flash Point	Deg C	73	76	84	92	120	160
Total n-Paraffins	%	98.9	98.9	98.7	98.6	99.4	99.0
Aromatics	%	0.16	0.16	0.2	0.3	0.2	1.0
Sulfur	ppm	1 max	1 max	1 max	1 max	1 max	1 max
Bromine index	mg/100g	10	10	12	15	10	40
Aniline point	Deg C	83	84	85	88	93.5	102
Destillation Range (at 1.013 bar)							
IBP	Deg C	185	193	200	220	251	300
FBP	Deg C	230	213	239	237	391	330
Kinematic viscosity	cST	1.83	1.90	2.21	2.61	3.9	4.7
Pour Point	Deg C	-19	-22	-10	-6	9	28

Chemical Properties

Paraffins are very stable and do not react with chemical reagents such as fuming nitric acid, sulphuric acid sodium alkalies potassium permanganate and chromic acid. It also reacts with **magic acid (the Fluoro sulphonic acid, antimony pentachloride) FSO_3H-SbF_5**

Fuming nitric acid and sulphuric acid under pressure and high temperature may react slowly, ***higher paraffins*** in the gaseous phase to form sulphuric acids and nitro compounds.

Paraffins however, show the following reactions:

1. ***Halogenation of hydrocarbons:*** Paraffins react with chlorine in diffused sunlight to form chlorine substitution products. Bromine acts similar to chlorine but the reaction is slow. Iodine has no reaction on the hydrocarbons. Iodine reacts in the presence of an oxidizing agent

Carbon Number	Name	Formula (mol Wt.)	Melting Point °C	Boiling Point °C	CAS RN
C(5)	n-Pentane	C_5H_{12} (72.15)	-130	36	109-66-0
C(6)	n-Hexane	C_6H_{14} (86.18)	-95	69	110-54-3
C(7)	n-Heptane	C_7H_{16} (100.20)	-90.5	98.5	142-82-5
C(8)	n-Octane	C_8H_{18} (114.23)	-57	126	111-65-9
C(9)	n-Nonane	C_9H_{20} (128.26)	-53	158	111-84-2
C(10)	n-Decane	$C_{10}H_{22}$ (142.28)	-30	174	124-18-5
C(11)	n-Undecane	$C_{11}H_{24}$ (156.31)	-25.5	196	1120-21-4
C(12)	n-Dodecane	$C_{12}H_{26}$ (170.34)	-9.5	216	112-40-3
C(13)	n-Tridecane	$C_{13}H_{28}$ (184.36)	5.5	235	629-50-5
C(14)	n-Tetradecane	$C_{14}H_{30}$ (198.39)	6	254	629-59-4
C(15)	n-Pentadecane	$C_{15}H_{32}$ (212.42)	10	271	629-62-9
C(16)	n-Hexadecane	$C_{16}H_{34}$ (226.44)	18	287	544-76-3
C(17)	n-Heptadecane	$C_{17}H_{36}$ (240.47)	22	302	629-78-7
C(18)	n-Octadecane	$C_{18}H_{38}$ (254.50)	28	316	593-45-3
C(19)	n-Nonadecane	$C_{19}H_{40}$ (268.53)	32	329	629-92-5
C(20)	n-Eicosane	$C_{20}H_{42}$ (282.55)	37	343	112-95-8
C(21)	n-Henicosane	$C_{21}H_{44}$ (296.58)	40.5	356.5	629-94-7
C(22)	n-Docosane	$C_{22}H_{46}$ (310.61)	44.5	368.5	629-97-0
C(23)	n-Tricosane	$C_{23}H_{48}$ (324.63)	47.5	380	638-67-5
C(24)	n-Tetracosane	$C_{24}H_{50}$ (338.66)	52	391	646-31-1
C(25)	n-Pentacosane	$C_{25}H_{52}$ (352.69)	54	402	629-99-2

Carbon Number	Name	Formula (mol Wt.)	Melting Point °C	Boiling Point °C	CAS RN
C(26)	n-Hexacosane	$C_{26}H_{54}$ (366.71)	56.5	412	630-01-3
C(27)	n-Heptacosane	$C_{27}H_{56}$ (380.74)	59		593-49-7
C(28)	n-Octacosane	$C_{28}H_{58}$ (394.77)	64.5	431.5	630-02-4
C(29)	n-Nonacosane	$C_{29}H_{60}$ (408.79)	65		630-03-5
C(30)	n-Triacontane	$C_{30}H_{62}$ (422.82)	66	450	638-68-6
C(31)	n-Hentriacontane	$C_{31}H_{64}$ (436.85)	67		630-04-6
C(32)	n-Dotriacontane	$C_{32}H_{66}$ (450.87)	69	467	544-85-4
C(33)	n-Triatriacontane	$C_{33}H_{68}$ (464.90)	71		630-05-7
C(34)	n-Tetratriacontane	$C_{34}H_{70}$ (478.93)	72		14167-59-0
C(35)	n-Pentatriacontane	$C_{35}H_{72}$ (492.95)	73		630-07-9
C(36)	n-Hexatriacontane	$C_{36}H_{74}$ (506.98)	74		630-06-8
C(37)	n-Heptatriacontane	$C_{37}H_{76}$ (521.01)	76		7194-84-5
C(38)	n-Octatriacontane	$C_{38}H_{78}$ (535.03)	78		7194-85-6
C(39)	n-Nonatriacontane	$C_{39}H_{80}$ (549.05)	80		7194-86-7
C(40)	n-Tetracontane	$C_{40}H_{82}$ (563.09)	82		4181-95-7
C(42)	n-Dotetracontane	$C_{42}H_{86}$ (591.13)	84		7098-20-6
C(44)	n-Tetratetracontane	$C_{44}H_{90}$ (619.20)	85		7098-22-8
C(46)	n-Hexatetracontane	$C_{46}H_{94}$ (647.24)	86		7098-24-0
C(48)	n-Octatetracontane	$C_{48}H_{98}$ (675.30)	90		7098-26-2
C(50)	n-Pentacontane	$C_{50}H_{102}$ (703.36)	91		6596-40-3
C(52)	n-Dopentacontane	$C_{52}H_{106}$ (731.40)	92		7719-79-1

$$CH_4 + Cl_2 \xrightarrow{\text{Sunlight}} CH_3Cl + HCl$$

Methyl chloride
Monochloromethane

$$CH_3Cl + Cl_2 \xrightarrow{\text{Sunlight}} CH_2Cl_2 + HCl$$

Methylene Chloride
Dichloromethane

$$CH_2Cl_2 + Cl_2 \xrightarrow{\text{Sunlight}} CHCl_3 + HCl$$

Trichloromethane
Chloroform

$$CHCl_3 + Cl_2 \xrightarrow{\text{Sunlight}} CCl_4 + HCl$$

Carbon tetrachloride
Tetrachloromethane

The halogenation may be catalyzed by the presence of some suitable halogen carrier such as iron filling phosphorous etc.

Control of Chlorination

Chlorination of methane may yield anyone of four organic products, depending upon the stage to which the reaction is carried. Can we control this reaction so that methyl chloride is the principal organic product? That is, can we limit the reaction to the first stage, mono-chlorination?

We might at first expect-naively, as it turns out-to accomplish this by providing only one mole of chlorine for each mole of methane. But let us see what happens if we do so. At the beginning of the reaction there is only methane for the chlorine to react with, and consequently only the first stage of chlorination takes place. This reaction, however, yields methyl chloride, so that as the reaction proceeds methane disappears and methyl chloride takes its place.

As the proportion of methyl chloride grows, it completes with the methane for the available chlorine. By the time the concentration of methyl chloride exceeds that of methane, chlorine is more likely to attack methyl chloride than methane, and the second stage of chlorination becomes more important than the first. A large amount of methylene chloride is formed, which in a similar way is chlorinated to chloroform and this, in turn, is chlorinated to carbon tetrachloride. When we finally work up the reaction product, we find that it is a mixture of all four chlorinated methanes together with some unreacted methane.

The reaction may, however, be limited almost entirely to monochlorination if we use a large excess of methane. In this case, even at, the very end of the reaction unreacted methane greatly exceeds methyl chloride. Chlorine is more likely to attack methane than methyl chloride, and thus the first stage of chlorination is the principal reaction.

Because of the great difference in their boiling points, it is easy to separate the excess methane (b.p. -161.5 °C) from the methyl chloride (b.p. -24°C) so that the methane can be mixed with more chlorine and put through the process again. While there is a low conversion of methane into methyl chloride in each cycle, the yield of methyl chloride based on the chlorine consumed is quite high.

The use of a large excess of one reactant is a common device of the organic chemist who wishes to limit reaction to only one of a number of reactive sites in the molecule of that reactant.

Reaction with other halogens: halogenation

Methane reacts with bromine, again at high temperatures or under the influence of ultraviolet light, to yield the corresponding bromomethanes methyl bromide, methylene bromide, bromoform, and carbon tetrabromide.

$$CH_4 + Br_2 \xrightarrow{\text{Sunlight}} CH_3Br + HBr$$

Methyl Bromide
Monobromomethane

$$CH_3Br + Br_2 \xrightarrow{\text{Sunlight}} CH_2Br_2 + HBr$$

Methylene Bromide
Dibromomethane

$$CH_2Br_2 + Br_2 \xrightarrow{\text{Sunlight}} CHBr_3 + HBr$$

Tribromomethane
Bromoform

$$CHBr_3 + Br_2 \xrightarrow{\text{Sunlight}} CBr_4 + HBr$$

Carbon tetrabromide
Tetrabromomethane

Bromination takes place somewhat less readily than chlorination.

Methane does not react with iodine at all. With fluorine it reacts so vigorously that, even in the dark and at room temperature, the reaction must be carefully controlled: the reactants, diluted with an inert gas, are mixed at low pressure.

We can, therefore, arrange the halogens in order of reactivity.

Reactivity of halogens:

$F_2 > Cl_2 > Br_2 > I_2$.

This same order of reactivity holds for the reaction of the halogens with other alkanes and, indeed, with most other organic compounds. The spread of reactivities is so great that only chlorination and bromination proceed at such rates as to be generally useful.

Relative reactivity

Throughout our study of organic chemistry, we shall constantly be interested in *relative reactivities*. We shall compare the reactivities of various reagents toward the same organic compound, the reactivities of different organic compounds toward the same reagent, and even the reactivities of different sites in an organic molecule toward the same reagent.

It should be understood that when we compare reactivities we compare rates of reaction. When we say that chlorine is *more reactive* than bromine toward methane, we mean that under the same conditions (same concentration, same temperature, etc.) chlorine reacts with methane *faster* than does bromine. From another point of view, we mean that the bromine reaction must be carried out under more vigorous conditions (higher concentration or higher temperature) if it is to take place as fast as the chlorine reaction. When we say that methane and iodine do not react at all, we mean that the reaction is too slow to be significant.

We shall want to know not only what these relative reactivities are but also, whenever possible, how to account for them. To see what factors cause one reaction to be [aster than another, we shall take up in more detail this matter of the different reactivities of the halogens toward methane. Before we can do this, however, we must understand a little more about the reaction itself.

Reaction mechanisms

It is important for us to know not only *what* happens in a chemical reaction but also *how* it happens, that is, to know not only the *facts* but also *the theory.*

For example, we know that methane and chlorine under the influence of heat or light form methyl chloride and hydrogen chloride. Just how is a molecule of methane converted into a molecule of methyl chloride? Does this transformation involve more than one step, and, if so, what are these steps? Just what is the function of heat or light?

Bromination takes place somewhat less readily than chlorination. Methane does not react with iodine at all. With fluorine it reacts so vigorously that, even in the dark and at room temperature, the reaction must be carefully controlled the reactants, diluted with an inert gas, are mixed at low pressure.

We can, therefore, arrange the halogens in order of reactivity.

Reactivity of halogens $F_2 > Cl_2 > Be_2 > I_2$.

This same order of reactivity holds for the reaction of the halogens with other alkanes and, indeed, with most other organic compounds. The spread of reactivities is so great that only chlorination and bromination proceed at such rates as to be generally useful.

Mechanism of chlorination Free radicals

It will be worthwhile to examine the mechanism of chlorination of methane in some detail. The same mechanism holds for bromination as well as chlorination, and for other alkanes as well as methane; it even holds for many compounds that, while not alkanes, contain alkane-like portions in their molecules. Closely related mechanisms are involved in oxidation (combustion) and other reactions of alkanes. More important, this mechanism illustrates certain general principles that can be carried over to a wide range of chemical reactions. Finally, by studying the evidence that supports the mechanism, we can learn something of how a chemist finds out what goes on during a chemical reaction.

Among the facts that must be accounted for are these:

(a) Methane and chlorine do not react in the dark at room temperature.

(b) Reaction takes place readily, however, in the dark at temperatures over 250°C, or

(c) Under the influence of ultraviolet light at room temperature.

(d) The wavelength of light that induces chlorination is that known independently because dissociation of chlorine molecules.

(e) In the light-induced reaction, many (several thousand) molecules of methyl chloride are obtained for each photon of light that is absorbed by the system.

(f) The presence of a small amount of oxygen slows down the reaction for a period of time after which the reaction proceeds normally; the length, of this period depends upon how much oxygen is present.

(1) $Cl_2 \xrightarrow{\text{heat or light}} 2Cl^-$

(2) $Cl^- + CH_4 \xrightarrow{\text{heat or light}} HCl + CH_3^-$

(3) $CH_3^- + Cl_2 \xrightarrow{\text{heat or light}} CH_3Cl + Cl^-$

The first step is the breaking of a chlorine molecule into two chlorine atoms. Like the breaking of any bond, this requires energy, the *bond dissociation energy* The energy is supplied as either heat or light.

$$:\ddot{Cl}:\ddot{Cl}: \longrightarrow :\ddot{Cl}\cdot + \cdot\ddot{Cl}:$$

The chlorine molecule undergoes *homolysis* that is, cleavage of the chlorine-chlorine bond takes place in a symmetrical way, so that each atom retains one electron of the pair that formed the covalent bond. This **odd electron** is not *paired* as are all the other electrons of the chlorine atom; that is, it does not have a partner of opposite spin. *An atom or group of atoms possessing an odd (unpaired) electron is called a* **free radical.** In writing the symbol for a free radical, we generally include a dot to represent the odd electron just as we include a plus or, minus sign in the symbol of an ion.

Once formed, what is a chlorine atom most likely to do? Like most free radicals, it is extremely reactive because of its tendency to gain an additional electron and thus have a complete octet; from another point of view, energy was supplied to each chlorine atom during the cleavage of the chlorine molecule, and this energy-rich particle tends strongly to lose energy by the formation of a new chemical bond.

To form a new chemical bond, that is, to react, the chlorine atom must collide with some other molecule or atom. What is it most likely to collide with? Obviously, it is most likely to collide with the particles that are present in the highest concentration: chlorine molecules and methane molecules. Collision with another chlorine atom is quite unlikely simply because there are very few of these reactive, short-lived particles around at any time. Of the likely collisions, that with a chlorine molecule causes no net change; reaction may occur, but it can result only in the exchange of one chlorine atom for another:

$$:\ddot{Cl}\cdot + :\ddot{Cl}:\ddot{Cl}: \longrightarrow :\ddot{Cl}:\ddot{Cl}: + \cdot\ddot{Cl}:$$ Collision probable but not productive

Collision of a chlorine atom with a methane molecule is both *probable* and *productive*. The chlorine atom abstracts a hydrogen atom, with one electron, to form a molecule of hydrogen chloride:

$$:\!\ddot{Cl}\cdot + :\!\ddot{Cl}\!:\!\ddot{Cl}\!: \longrightarrow :\!\ddot{Cl}\!:\!\ddot{Cl}\!: + \cdot\ddot{Cl}\!:$$ Collision probable but not productive

$$H\!:\!\underset{H}{\overset{H}{\ddot{C}}}\!:\!H + \cdot\ddot{Cl}\!: \longrightarrow H\!:\!\ddot{Cl}\!: + H\!:\!\underset{H}{\overset{H}{\ddot{C}}}\cdot$$ Collision probable and productive

Methane — Methyl radical

Now the methyl group is left with an odd, unpaired electron; the carbon atom has only seven electrons in its valence shell. One free radical, the chlorine tom, has been consumed, and a new one, the methyl radical, CH_3 has been for ed in its place. This is step (2) in the mechanism.

Now, what is this methyl radical most likely to do? Like the chlorine atom, It is extremely reactive, and for the same reason: the tendency to complete its octet, to lose energy by forming a new bond. Again, collisions with chlorine molecules or methane molecules are the probable ones, not collisions with the relatively scarce chlorine atoms or methyl radicals. But collision with a methane molecule could at most result only in the exchange of one methyl radical for another:

$$H\!:\!\underset{H}{\overset{H}{\ddot{C}}}\!:\!H + \cdot\underset{H}{\overset{H}{\ddot{C}}}\!:\!H \longrightarrow H\!:\!\underset{H}{\overset{H}{\ddot{C}}}\cdot + H\!:\!\underset{H}{\overset{H}{\ddot{C}}}\!:\!H$$ Collision probable but not productive

The collision of a methyl radical with a chlorine molecule is, then, the important one. The methyl radical abstracts a chlorine atom, with one of the bonding electrons, to form a molecule of methyl chloride:

$$H\!:\!\underset{H}{\overset{H}{\ddot{C}}}\cdot + :\!\ddot{Cl}\!:\!\ddot{Cl}\!: \longrightarrow H\!:\!\underset{H}{\overset{H}{\ddot{C}}}\!:\!\ddot{Cl}\!: + :\!\ddot{Cl}\cdot$$ Collision probable and productive

The other product is a chlorine atom. This is step (3) in the mechanism.

Here again the consumption of one reactive particle has been accompanied by the formation of another. The new chlorine atom attacks methane to form a methyl radical, which attacks a chlorine molecule to form a chlorine atom, and so the sequence is repeated over and over. Each step produces not only a new reactive particle but also a molecule of product: methyl chloride or hydrogen chloride.

This process cannot, however, go on forever. As we saw earlier, union of two short-lived, relatively scarce particles is not likely; but every so often it does happen, and when it does, this particular sequence of reactions stops. Reactive particles are consumed but not generated.

$$:\!\ddot{Cl}\cdot + \cdot\ddot{Cl}\!: \longrightarrow :\!\ddot{Cl}\!:\!\ddot{Cl}\!:$$

$$CH_3 + CH_3 \longrightarrow CH_3\!:\!CH_3$$

$$CH_3 + \cdot\ddot{Cl}\!: \longrightarrow CH_3\!:\!\ddot{Cl}\!:$$

Chain reactions

The chlorination of methane is an example of a chain reaction, *a reaction that involves a series of steps, each of which generates a reactive substance that brings about the next step.* While chain reactions may vary widely in their details, they all have certain fundamental characteristics in common.

First in the chain of reactions is a chain-initiating step, in which energy is absorbed and a reactive particle generated, in the present reaction it is the cleavage of chlorine into atoms (step 1).

There are one or more chain-propagating steps, each of which consumes a reactive particle and generates another; here they are the reaction of chlorine atoms with methane (step 2), and of methyl radicals with chlorine (step 3).

Finally, there are chain-terminating steps, in which reactive particles are consumed but not generated; in the chlorination of methane these would involve the union of two of the reactive particles, or the capture of one of them by the walls of the reaction vessel.

Under one set of conditions, about 10,000 molecules of methyl chloride are formed for every quantum (photon) of light absorbed. Each photon cleaves one chlorine molecule to form two chlorine atoms, each of which starts a chain. On the average, each chain consists of 5000 repetitions of the chain-propagating cycle before it is finally stopped.

Inhibitors

Finally, how does the mechanism of chlorination account for fact (t), that a small amount of oxygen slows down the reaction for a period of time, which depends upon the amount of oxygen, after which the reaction proceeds normally?

Oxygen is believed to react with a methyl radical to form a new free radical:

$$CH_3 + O_2 \longrightarrow CH_3 - O - O$$

The CH_3OO radical is much less reactive than the CH_3. Radical, and can do little to continue the chain. By combining with a methyl radical, one oxygen molecule breaks a chain, and thus prevents the formation of thousands of molecules of methyl chloride; this, of course, slows down the reaction tremendously. After all the oxygen molecules present have combined with methyl radicals, the reaction is free to proceed at its normal rate.

A substance that slows down or stops a reaction even though present in small amount is called an **inhibitor**. *The period of time during which inhibition lasts, and after which the reaction proceeds normally, is called the inhibition period.* Inhibition by a relatively small amount of an added material is quite characteristic of chain reactions of any type, and is often one of the clues that first leads us to suspect that we are dealing with a chain reaction. It is hard to see how else a few molecules could prevent the reaction of so many. (We shall frequently enter the use of oxygen to inhibit free-radical reactions.)

Heat of reaction

In our consideration of the chlorination of methane, we have so far been concerned chiefly with the particles involved - molecules and atoms - and the changes that they undergo. As with any reaction, however, it is important to consider also the energy changes involved, since these changes determine to a large extent how fast the reaction will go, and, in fact, whether it will take place at all.

By using the values of homolytic bond dissociation energies given in the appendix we can calculate the energy changes that take place in a great number of reactions. In the conversion of methane into methyl chloride, two bonds are broken, CH_3–H and Cl–Cl, consuming 104 + 58, or a total of 162 kcal/mol. At the same time two new bonds are formed, CH_3–Cl and H–Cl, liberating 84 + 103, or a total of 187 kcal/mol. The result is the liberation of 25 kcal of heat for every mole of methane that is converted into methyl chloride; this is, then, an exothermic reaction. (This calculation, we note, does not depend on our knowing the mechanism of the reaction.)

$$CH_3 - H + Cl - Cl \rightarrow CH_3 - Cl + H - Cl$$

$$\underbrace{104 \quad 58}_{162} \qquad \underbrace{84 \quad 103}_{187} \qquad \Delta H = -25\ \text{Kcal}$$

When heat is liberated, the heat content (enthalpy), *H,* of the molecule themselves must decrease; the change in heat content, *ΔH,* is therefore given negative sign. (In the case of an endothermic reaction, where heat is absorbed, the increase in heat content of the molecules is indicated by a positive *ΔH.)*

Prob: 5.1 Calculate ΔH for the corresponding reaction of methane with: (a) bromine (b) iodine (c) chlorine.

The value of -25 kcal that we have just calculated is the *net* AH for the overall relation. A more useful picture of the reaction is given by the ΔH values of 1 individual steps. These are calculated below:

(1) $$Cl{-}Cl \;(58) \longrightarrow 2Cl \qquad \Delta H = +58\ \text{Kcal}$$

(2) $$Cl + H_3C{-}H \longrightarrow CH_3 + H{-}Cl\;(103) \qquad \Delta H = +1$$

(3) $$CH_3 + Cl{-}Cl\;(58) \longrightarrow H_3C{-}Cl\;(84) + Cl \qquad \Delta H = -26$$

It is clear why this reaction, even though exothermic, occurs only at temperature {in the absence of light}. The chain-initiating step, without which reaction cannot occur is highly **endothermic,** and takes place (at a significant rate) only at a high temperature. Once the chlorine atoms are formed, the two chain - propagating steps - one only slightly endothermic, and the other exothermic - occur readily many times before the chain is broken. The difficult cleavage of chlorine is the barrier that must be surmounted before the subsequent easy steps can be taken.

We have assumed so far that exothermic reactions proceed readily, that is, are reasonably fast at ordinary temperatures, whereas endothermic reactions proceed with difficulty, that is, are slow except at very high temperatures. This assumed relationship between *ΔH* and rate of reaction is a useful rule of thumb when other information is not available; it is *not,* however, a *necessary* relationship, and there are many exceptions to the rule. We shall go on, then, tQ a discussion of mother energy quantity, the *energy of activation,* which is related in a more exact lay to rate of reaction.

Energy of activation

To see what actually happens during a chemical reaction, let us look more loosely at a specific example, the attack of chlorine atoms on methane:

$$\underset{}{Cl} + \underset{(104)}{CH_3 - H} \rightarrow \underset{(103)}{H - Cl} + CH_3 \qquad \Delta H = +1 \text{ kcal} \quad E_{act} = 4\text{cal}$$

This reaction is comparatively simple: it occurs in the gas phase, and is thus not complicated by the presence of a solvent; it involves the interaction of a single atom and the simplest of organic molecules. Yet from it we can learn certain principles that apply to any reaction.

Just what must happen if this reaction is to occur? First of all, a chlorine atom and a methane molecule must collide. Since chemical forces are of extremely short range, a hydrogen-chlorine bond can form only when the atoms are in close contact.

Next, to be *effective,* the collision must provide a certain *minimum amount of energy.* Formation of the H-CI bond liberates 103 kcal/mol; breaking the CH_3–H bond requires 104 kcal/mol. We might have expected that only 1 kcal/mol additional energy would be needed for reaction to occur; however, this is not so. Bond-breaking and bond-making evidently are not perfectly synchronized, and the energy liberated by the one process is not completely available for the other. Experiment has shown that, if reaction is to occur, an additional 4 kcal/mol of energy must be supplied.

The minimum amount of energy that must be provided by a collision for reaction to occur is called the energy of activation, E_{act} . Its source is the kinetic energy of the moving particles. Most collisions provide less than this minimum quantity and are fruitless, the original particles simply bouncing apart. Only solid collisions between particles one or both of which are moving unusually fast are energetic enough to bring about reaction. In, the present example, at 275°C only about one collision in 40 is sufficiently energetic.

Finally, in addition to being sufficiently energetic, the collisions must occur when the particles are properly oriented. At the instant of collision, the methane molecule must be turned in such a way as. to present a hydrogen atom to the full force of the impact. In the present example, only about one collision in eight is properly oriented.

In general, then, *a chemical reaction requires collisions of sufficient energy (*E_{act} *and of proper orientation.* There is an energy of activation for nearly every reaction where bonds are broken, even for exothermic reactions, in which bond-making liberates more energy than is consumed by bond-breaking.

The attack of bromine atoms on methane is more highly endothermic, with a ΔH of +16 kcal.

$$Br + CH_3 - H \rightarrow H - Br + CH_3 \qquad \Delta H = +\,16 \text{ kcal} \qquad E_{act} = 18 \text{ kal}$$

Breaking the CH_3-H bond, as before, requires 104 kcal/mol, of which only 88 kcal is provided by formation of the H-Br bond. It is evident that, even if this 88 kcal were completely available for bond-breaking, at least an additional 16 kcal/mol would have to be supplied by the collision. In other words, the E_{act} of an endothermic reaction must be at least as large as the ΔH. As is generally true, the E_{act} of the present reaction (18 kcal) is actually somewhat larger than the ΔH.

Progress of Reaction: Energy Changes

These energy relationships can be seen more clearly in diagrams like Figs. 5.1. and 5.2. Progress of reaction is represented by horizontal movement from reactants on the left to products on the right. Potential energy (that is, all energy except kinetic) at any stage of reaction is indicated by the height of the curve.

Let us follow the course of reaction in Fig. 5.2. We start in a potential energy valley with a methane molecule and a chlorine atom. These particles are moving, and hence possess kinetic energy in addition to the potential energy shown. The exact amount of kinetic energy varies with the particular pair of particles, since some move faster than others.

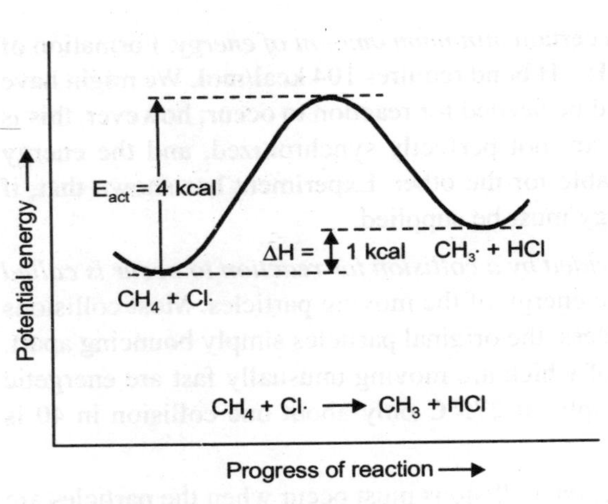

Fig. 5.1. Potential energy changes during the progress of reaction the methane–chlorine atom reaction

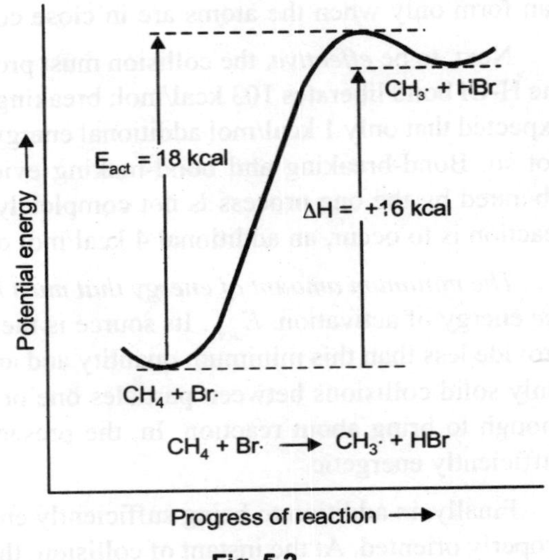

Fig. 5.2.

They collide, and kinetic energy is converted into potential energy. With this increase in potential energy, reaction begins, and we move up the energy hill. If enough kinetic energy is converted, we reach the top of the hill and start down the far side.

During the descent, potential energy is converted back into kinetic energy, until we reach the level of the products. The products contain a little more potential energy than did the reactants, and we find ourselves in a slightly higher valley than the one we left. With this net increase in potential energy there must be a corresponding decrease in kinetic energy. The new particles move-apart, and since they are moving more slowly than the particles from which they were formed, we observe a drop in temperature. Heat will be *taken up* from the surroundings.

In the bromine reaction, shown in Fig. 5.1, we climb a much higher hill and end up in a much higher valley. The increase in potential energy-and the corresponding decrease in kinetic energy - is much larger than in the chlorine reaction; more heat has to be taken up from the surroundings.

An exothermic reaction follows much the same course. (Take, for example, the reverse of the bromine reaction; that is, read from right to left in Fig. 5.1. In this case, however, the products contain less potential energy than did the reactants so that we end up in a lower valley than the one we left. Since this time the new particles contain more kinetic energy than the particles from which they were formed, and hence move faster, we observe a rise in temperature. Heat will be *given off* to the surroundings.

In any reaction there are many collisions that provide too little energy for us to reach the top of the hill. These collisions are fruitless, and we slide back to our original valley. Many collisions provide sufficient energy, but take place when the molecules are improperly oriented. We then climb an energy hill, but we are off the road; we may climb very high without finding the pass that leads over into the next valley.

The difference in level between the two valleys is, of course, the *AH;* the difference in level between the reactant valley and the top of the hill is the E_{act} We are concerned only with these differences, and not with the absolute height at any stage of the reaction. We are not even concerned with the relative levels of the reactant valleys in the chlorine and bromine reactions. We need only to know that in the chlorine reaction we climb a hill 4 kcal high and end up in a valley 1 kcal higher than our starting point; and that in the bromine reaction we climb a hill 18 kcal high and end up in a valley 16 kcal higher than our starting point.

As we shall see, it is the height of the hill, the E_{act}, that determines the rate of reaction, and not the difference in level of, the two valleys, ΔH. In going to a lower valley, the hill might be very high, but could be very low-or even non-existent. In climbing to a higher valley, however, the hill can be no lower than the valley to which we are going; that is to say, *in an endothermic reaction the* E_{act} *must be at least as large as the* ΔH.

An energy diagram of the sort shown in Figs. 5.1 and 5.2 is particularly useful because it tells us not only about the reaction we are considering, but also about the reverse reaction. Let us move from right to left in Fig. 5.1, for example.

We see that the reaction

$$CH_3 + H - Cl \rightarrow CH_3 - H + Cl^- \qquad \Delta H = -1, \ E_{act} = 3$$

(103) (104)

has an energy of activation of 3 kcal, since in this case we climb the hill from the higher valley. This is, of course, an exothermic reaction with a ΔH of - 1 kcal

In the same way we can see from Fig. 5.2 that the reaction

$$CH_3 + H - Br \rightarrow CH_3 - H + Br^- \qquad \Delta H = -16 \quad E_{act} = 2$$

(88) (104)

has an energy of activation of 2 kcal, and is exothermic with a ΔH of -16 kcal. (We notice that, even though exothermic, these last two reactions have energies of activation.)

Reactions like the cleavage of chlorine into atoms

$$\underset{(58)}{Cl-Cl} \rightarrow Cl^{-} + Cl \qquad \Delta H = +58 \qquad E_{act} = 58$$

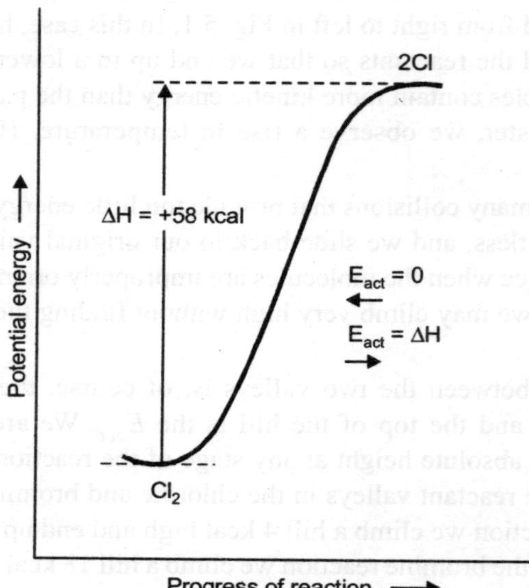

Fig. 5.3 : The potential energy changes during the progress of reaction simple.dissociation.

fall into a special category: a bond is broken but no bonds are formed. The reverse of this reaction, the union of chlorine atoms, involves no bond-breaking and hence

$$Cl^{-} + Cl \rightarrow \underset{(58)}{Cl-Cl} \qquad \Delta H = -58 \qquad E_{act} = 0$$

would be expected to take place very easily in fact, with no energy of activation at all. This is considered to be generally true for reactions involving the union of two free radicals.

If there is no hill to climb in going from chlorine atoms to a chlorine molecule, but simply a slope to descend, the cleavage of a chlorine molecule must involve simply the ascent of a slope as shown in the Fig. 5.3. The E_{act} for the cleavage of a chlorine molecule, then, must equal the ΔH, that is, 58 kcal. This equality of E_{act} and ΔH is believed to hold generally for reactions in which molecules dissociate into radicals.

Relative Reactivities of Halogens toward Methane

With this background let us return the reaction between methane and the various halogen, and we see if we can account for the order of reactivity given before, $F_2>Cl_2>Br_2>I_2$ and in particular for the fact that iodine does not react at all. From the table of the bond dissocioation energies we can calculate for each of the four halogens the ΔH for each of these reactions, let us see what tentative conclusions we can reach using only ΔH

		X =	F	Cl	Br	I
(1)	$X_2 \rightarrow 2X$	ΔH =	+38	+58	+46	+36
(2)	$X + CH_4 \rightarrow HX + CH_3$		–32	+1	+16	+33
(3)	$CH_3 + X_2 \rightarrow CH_3X + X$		–70	–26	–24	–20

Since step (1) involves simply dissociation of molecules into atoms, we may quite confidently assume that ΔH in this case is equal to E_{act}. Chlorine has the largest E_{act} and should dissociate most slowly, iodine has the smallest E_{act} and should dissociate more rapidly. Yet this does not agree with the observed order of the reactivities.

Step (3) attack of methyl radicals on halogen, is exothermic for all four halogens, and for chlorine, bromine, and iodine it has very nearly the same ΔH. For these reactions, E_{act} could be very small, and does indeed seem to be so; probably only a fraction of a kilocalorie. Even iodine has been found to react readily with methyl radicals generated in another way, for example, by the heating of tetra methyl lead. In fact, iodine is sometimes employed as a free-radical "trap" or "scavenger" in the study of reaction mechanisms. The third step, then, cannot be the cause of the observed relative reactivities.

This leaves step (2), abstraction of hydrogen from methane by a halogen atom. Here we see a wide spread of ΔH values, from the highly exothermic reaction with the fluorine atom to the highly endothermic reaction with the iodine atom. The endothermic bromine atom reaction must have an E_{act} of at least 16 kcal; as we have seen, it is actually 18 kcal. The slightly endothermic chlorine atom reaction could have a very small E_{act}; it is actually 4 kcal. At a given temperature, then, the fraction of collisions of sufficient energy is much larger for methane and chlorine atoms than for methane and bromine atoms. To be specific, at 275°C the fraction is about 1 in 40 for chlorine and 1 in 15 million for bromine.

A bromine atom, on the average, collides with many methane molecules before it succeeds in abstracting hydrogen; a chlorine atom collides with relatively few. During its longer search for the proper methane molecule, a bromine atom is more likely to encounter another scarce particle - a second halogen atom or a methyl radical - or be captured by the vessel wall; the chains should therefore be much shorter than in chlorination. Experiment has shown this to be so: where the average chain length is several thousand for chlorination, it is less than 100 for bromination. Even though bromine atoms are formed more rapidly than chlorine atoms at a given temperature because of the lower Eact of step (1), overall bromination is slower than chlorination because of the shorter chain length.

For the endothermic reaction of an iodine atom with methane, Eact can be no less than 33 kcal, and is probably somewhat larger. Even for this minimum value of 33 kcal, an iodine atom must collide with an enormous number of methane molecules (1013 or ten million million at 275°C) before reaction is likely to occur. Virtually no iodine atoms last this long, but instead they recombine to form iodine molecules; the reaction therefore proceeds at a negligible rate. Iodine atoms are easy to form; it is their inability to abstract hydrogen from methane that prevents iodination from occurring.

We cannot predict the E_{act} for the highly exothermic attack of fluorine atoms on methane, but we would certainly not expect it to be any larger than for the attack of chlorine atoms on methane. It appears actually to be smaller (about 1 kcal), thus permitting even longer chains. Because of the surprising weakness of the fluorine-fluorine bond, fluorine atoms should be formed faster than chlorine atoms; thus there should be not only longer chains in fluorination but also *more* chains. The overall reaction is extremely exothermic, with a ΔH of - 102 kcal, and the difficulty of removing this heat is one cause of the difficulty of control of fluorination.

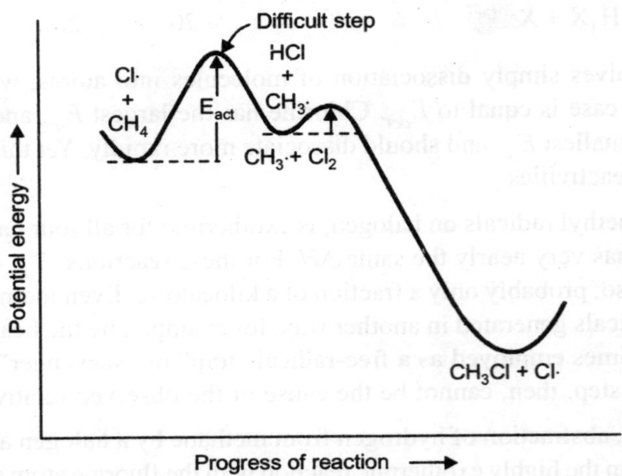

Fig. 5.4. The Potential energy changes during the progress of reaction: chlorination of methane. The formation of the radical is the difficult step

Of the two chain-propagating steps, then, step (2) is more difficult than step (3) (see Fig. 5.4). Once formed, methyl radicals react easily with any of the halogens; it is how fast methyl radicals are formed that limits the rate of overall reaction. Fluorination is fast because fluorine atoms rapidly abstract hydrogen atoms from methane; Eact is only 1 kcal. Iodination does not take place because iodine atoms find it virtually impossible to abstract hydrogen from methane; Eact is more than 33 kcal.

Values of E_{act} for step (2), we notice, parallel the values of ΔH. Since the same bond, CH_3-H, is being broken in every case, the differences in ΔH reflect differences in bond dissociation energy among the various hydrogen-halogen bonds. Ultimately, it appears, the reactivity of a halogen toward methane depends upon the strength of the bond which that halogen forms with hydrogen.

One further point requires clarification. We have said that an E_{act} of 33 kcal is too gıeat for the reaction between iodine atoms and methane to proceed at a significant rate; yet the initial step in each of these halogenations requires an even greater E_{act}. The difference is this: since halogenation is a chain reaction, dissociation of each molecule of halogen gives rise ultimately to many molecules of methyl halide; hence, even though dissociation is very slow, the overall reaction can be fast. The attack of iodine atoms on methane, however, is a chain-propagating step and if it is slow the entire reaction must be slow; under these circumstances chain-terminating steps (e.g., union of two iodine atoms) become so important that effectively there is no chain.

An Alternative Mechanism for Halogenation

In the preceding section we were concerned with the relative reactivities of the various halogens toward methane. In the next chapter we shall change our viewpoint, and .look at the relative reactivities of various alkanes - or various positions in one alkane - toward a given halogen. All this helps make up an important part of our study of organic chemistry: how variations in structure lead to variations in reactivity. But there is an even more fundamental point to consider: how a particular (type of structure leads to a particular type of reaction in the first place. How is it, not that one halogen or one alkane reacts faster or slower than another, but that any halogen and any alkane react together *in the way they do?*

To answer this question, let us take the chlorination of methane as an example, and examine it closely. The chain-propagating steps in our mechanism are (2a) and (3a)

$$Cl + CH_4 \rightarrow HCl + CH_3 \quad (2a)$$

$$CH_3 + Cl_2 \rightarrow CH_3Cl + Cl \quad (3a)$$

But consider, instead, the sequence (2b) and (3b), which represent an alternative mechanism.

$$Cl^- + CH_4 \rightarrow HCl + Cl^- \quad (2b)$$

$$H + Cl_2 \rightarrow HCl + Cl \quad (3b)$$

On the face of it, this mechanism would certainly seem to be worth considering; indeed, until 1940 it was just as consistent with the evidence as the now-accepted mechanism was. But this alternative mechanism is *not* the one by which chlorination takes place, and we shall see direct evidence against it.

Our question, then, becomes this: how is it that chlorination goes by steps (2a) and (3a) rather than by (2b) and (3b)? The crux of the matter lies in step (2). It is here that the two reaction paths divide; what happens in (2) determines the entire course of reaction. *If(2b)* were to occur, (3b) would inevitably follow; (3b) is a known reaction that takes place readily in a different system. But (2b) does *not* occur.

So we have narrowed our inquiry still further. We now ask: how is it that (2a) takes place rather than (2b)? In both reactions the chlorine atom attacks methane molecule.

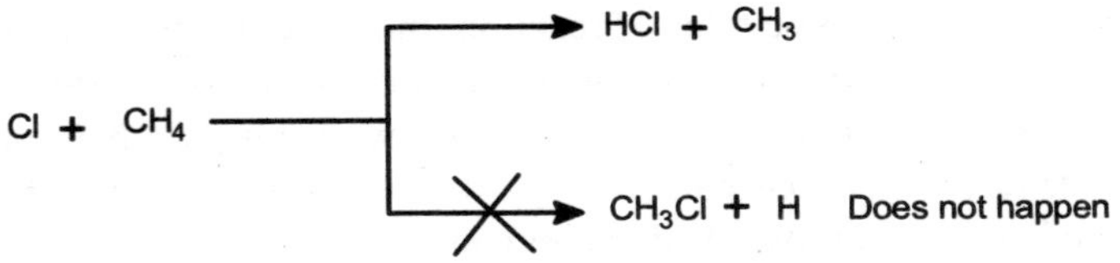

$$Cl + CH_4 \rightarrow HCl + CH_3$$

$$Cl + CH_4 \not\rightarrow CH_3Cl + H \quad \text{Does not happen}$$

It can either attach itself to hydrogen and expel a methyl radical, or attach itself to carbon and expel a hydrogen atom. There is thus *competition* between the two reactions, and the faster reaction wins. If step (2a) predominates, it can only mean that (2a) *goes faster than* (2b).

How do we account for this? Ideally we would like to know the E_{act} values for the competing reactions. But E_{act} cannot, of course, be measured for (2b), since this reaction does not take place. For (2a) ΔH is + 1 kcal; E_{act} could therefore be as small as 1 kcal and, as we have seen, is actually 4 kcal.

$$(2a) \quad Cl + CH_3 - H \rightarrow H - Cl + CH_3 \qquad \Delta H = +1 \text{ kcal} \quad E_{act} = 4 \text{ kcal}$$

For (2b), *DH* is +20 kcal; E_{act} must therefore be *at least* 20 kcal-and is probably considerably larger.

(2b) $Cl + CH_3 - H \rightarrow Cl + H$ $\Delta H = +20$ kcal E_{act} = at leat 20 kcal

The fraction of collisions providing 4 kcal or more is enormously larger than the fraction providing 20 kcal: at 275°C, for example, 2:5 *million times larger!* Just on the basis of this *minimum* estimate of the difference in E_{act}, we see that (2a) must proceed so much faster than (2b) that, in effect, (2a) is the *only* reaction that takes place.

The point is *not* that 20 kcal is in itself too high a barrier for reaction to occur; after all, the attack of Br on methane has an E_{act} of 18 kcal, and it occurs. 'The point here is that a reaction with an E_{act} of 20 kcal *cannot compete successfully* with a reaction whose E_{act} is only 4 kcal. When a chlorine atom collides with a methane molecule, the collision is overwhelmingly more likely to provide enough energy for (2a) than for (2b). And so (2a) is what happens.

Finally, let us see what structural feature makes (2a) the easier of the two reactions. Both reactions involve breaking of a carbon-hydrogen bond. The difference lies in which bond is being formed, hydrogen-chlorine or carbon chlorine. Breaking the carbon-hydrogen bond requires 104 kcal/mol - a great deal of energy. A small fraction of this is supplied by collisions. But most of it comes from the concerted formation of another bond: hydrogen-chlorine in the case of (2a), or carbon-chlorine in the case of (2b). The hydrogen-chlorine bond is a strong one (103 kcal) and its formation can supply nearly all the needed energy. But the carbon-chlorine bond is weaker (only 84 kcal) and, even if all this were available to help break the carbon-hydrogen bond, 20 kcal more would have to be supplied by collisions. The course of this reaction is, then, ultimately determined by the fact that the hydrogen-chlorine bond is stronger than the carbon-chlorine bond.

Examination of the bond dissociation energies shows that what we have just described is part of a pattern: each halogen forms a stronger bond to hydrogen than to carbon - not only carbon in methane but carbon in other alkanes as well. The result is that, whatever the halogen and whatever the alkane, halogenation follows a mechanism that is analogous to (2a) and (3a) and *not* to (2b) and (3b).

Again we have encountered relative rates of reaction, this time determining the most fundamental aspect of chemical behaviour: *what type of reaction takes place.* Whenever different kinds of molecules are mixed together, there will, in principle, be more than one way in which they can react. There will be a competition between different reactions paths - very often, as we shall find, a closer competition than the one we have just used as our example. And, by and large, what the molecules actually do is *what is easiest for them.* As we encounter such cases of competition, we shall try to see what factors tend to favor one path or the other; we shall even try to see what we can do to make the path that we prefer be the easier one for the reaction to follow.

Structure of the methyl radical Sp2 Hybridization

We have spent a good part of this chapter discussing the formation and reactions of the methyl free radical CH_3. Just what is this molecule like? What is its shape? How are the electrons distributed and, in particular, where is the odd electron ?

These are important questions, for the answers apply not only to this simple radical but to any free radical, however complicated, that we shall encounter. The *shape,* naturally, underlies the three-dimensional chemistry-the stereochemistry -of free radicals. The *location of the odd electron* is intimately involved with the stabilization of free radicals by substituent groups.

As we did when we "made" methane, let us start with the electronic configuration of carbon,

	1s	2s	2p		
C	↑↓	↑↓	↑	↑	

and, to provide more than two unpaired electrons for bonding, promote a 2s electron to the empty *2p* orbital

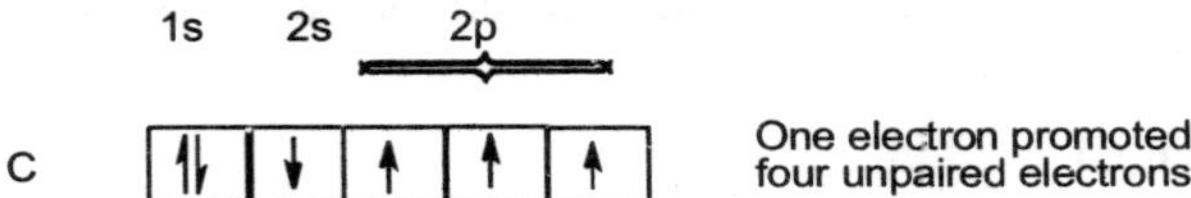

Like boron in boron trifluoride carbon here is bonded to three other atoms. Hybridization of the *2s* orbital and two of the *p* orbitals provide the necessary orbitals: three strongly directed *Sp2* orbitals which, as we saw before, lie in a plane that includes the carbon nucleus, and are directed to the corners of an equilateral triangle.

If we arrange the carbon and three hydrogen of a methyl radical to permit maximum overlap of orbitals, we obtain the structure shown in Fig. 5.5. It is flat, with the carbon atom at the centre of a triangle and the three and the three hydrogen atoms at the corners. Every bond angle is 120°C.

Fig. 5.5. Methyl radical (a) Only a bonds shown (b) the odd electron is the *p* orbital above and below the plane of the bonds.

Now where is the odd electron? In forming the Sp_2 orbitals, the carbon atom has used only two of its three *p* orbitals. The remaining *p* orbital consists of two equal lobes, one lying above and the other lying below the plane of the three Sp_2 orbitals; it is occupied by the odd electron.

This is not the only conceivable electronic configuration for the methyl radical: an alternative treatment would lead to a pyramidal molecule like that of ammonia, except that the fourth Sp_3 orbital contains the odd electron instead of an electron pair. Quantum mechanical calculations do not offer a clear-cut decision between the two configurations. Spectroscopic studies indicate that the methyl radical is actually flat, or nearly so. Carbon is trigonal, or not far from it; the odd electron occupies a *p* orbital, or at least an orbital with much *p* character.

Compare the shapes of three molecules in which the central atom is bonded to three other atoms: (a) boron trifluoride, with no unshared electrons, trigonal; (b) ammonia, with an unshared *pair,* tetrahedral; and (c) the methyl radical, with a *single* unshared electron, trigonal or intermediate between trigonal and tetrahedral.

There is stereo chemical evidence that most other free radicals are either flat or, if pyramidal, undergo rapid *inversion* like that of the ammonia molecule.

Chlorofluorocarbons and the Ozone Shield

Once upon a time, not so long ago, a project was set up to design and synthesize gases for use in refrigerators, air-conditioners, and aerosol spray cans. These gases were to be easily compressible and, unlike their predecessors, inert and non-toxic. And the project succeeded: all these aims were met. The products of the project were the fully halogenated methanes known as *chlorofluorocarbons* or CFCs.

CF_3Cl	CF_2Cl_2
Chlorotrifluoromethane	Dichlorodifluoromethane
CFC 11	CFC 12

For decades these CFCs have been found to work admirably and when old refrigerators and air-conditioners are abandoned and when coiffures and armpits are sprayed, the CFCs liberated drift slowly upward in the atmosphere. The very inertness that they were designed to have ensures that they survive, unchanged, as they climb higher and higher. Until, finally, they reach the stratosphere and are exposed to ultraviolet light from the sun.

And now, it turns out, the CFCs, so carefully created with the best of intentions, are actually Frankenstein's monsters that threaten their creators and, in fact, all life on earth.

To see what is involved, let us start at the beginning. Earth is bathed in electromagnetic radiation from the sun. This warms earth directly and provides energy for the production of carbohydrates in plants - carbohydrates which, directly or indirectly, feed all forms of animal life. The radiation is made up of light, visible and invisible. Light of wavelength between about 400 nm and 750 nm is visible. (A *nanometer,* nm, is 10^{-7} cm.) Just beyond the red end of the visible spectrum there is infrared light of longer wavelength (greater than 750 nm) and lower energy. Just beyond the violet end there is ultraviolet light, of shorter wavelength (less than 400 nm) and higher energy. Now, the more energetic of the ultraviolet light can cause photolysis ("Cleavage by light") and thus bring about chemical reactions-break a chlorine molecule into two atoms, for example, as we have seen. Even wavelengths of somewhat less energy, up to 300 nm, can damage the molecules involved in biological processes: the vital *biomolecules.* Chief among these is DNA, which controls heredity on the molecular level, and directs the synthesis of other biomolecules

The major components of the atmosphere, O_2 and N_2, absorb wavelength up to 250 nm and thus shield earth from much of this radiation - but *not all of it* they are transparent to wavelengths between 250 and 300 nm, and allow this light' to pass through.

Fortunately, in - the higher atmosphere - in the stratosphere and beyond - there is another, minor component: *ozone,* O_3, This ozone "layer" is extremely diffuse, extending from an altitude of about 90 km down to about 30 km. (If compressed to sea-level pressure, it would make a layer only a few centimetres thick!) Ozone absorbs wavelengths between 25d and 300 nm, and provides earth's only protection against this biologically harmful radiation. Without this shield wispy and, as we shall see, fragile life could never have crawled out of the protecting sea and established itself on land. The high altitude, "good" ozone must not be confused with the low level, "bad" ozone that is a serious component of smog. Ozone is much too reactive to survive a climb from ground level to the stratosphere. The high altitude ozone is formed by the action of ultraviolet light on O_2, as shown in equations (1) – (5).

$$O_2 + h\nu \rightarrow O^- + O \qquad (1)$$

$$O + O_2 \rightarrow O_3 + M \qquad (2)$$

$$O_3 + h\nu \rightarrow O_2 + O \qquad (3)$$

Then (2), (3) (2) (3) etc. until finally

$$O + O + M \rightarrow O_2 + M \qquad (4)$$

$$O + O_3 \rightarrow O_2 + O_3 \qquad (5)$$

We recognize the familiar components of a free-radical chain reaction. First, in the *chain-initiating step* (1), an oxygen molecule, O_2, is broken into two oxygen atoms, O′. The energy for this dissociation (represented as *hv*) is provided by absorption of ultraviolet light of wavelengths less than 250 nm. (It is through this absorptioh that O_2 shields earth from these wavelengths.)

Next, there occur the *chain-propagating steps,* that is, (2) and (3), repeated again and again.

In (2), O collides with the abundant species O_2, and combines with it to form ozone, O_3, And here we encounter something we have not discussed before. An O – O_2 bond is formed, releasing energy. No bond is broken, and we might (naively) expect reaction to proceed with no trouble. but things are not so simple. What is to become of the energy released by the bond-formation? If it remains long in the newly formed ozone molecule, it would break the molecule (the *excited* ozone) apart. If a stable product is to result there must be-very soon after the initial bond-formation-a collision with a *third body,* M, that is, some other particle (O_2 or N_2, say) that can, in effect, carry away the excess energy as kinetic energy. (In other words, heat is generated.)

The energy cannot simply be converted into kinetic energy of O_3 since this would violate the law of conservation of momentum. After the collision with M, however, O_3 and M can move apart, each moving faster than before but in such directions as to preserve the net momentum.

In step (3), ozone is broken apart to regenerate O_2 and O The energy for this is provided by ultraviolet light-but of a different wavelength - from that required in step (1). Here, absorption is of wavelengths between 250 and 300 nm; it is through this absorption that ozone shields earth from this damaging radiation.

Finally, every so often, there occurs a *chain-terminating step,* like (4) or (5), in which reactive particles are consumed but not generated.

Thus, in step (2) O_2 is converted into ozone, and in step (3) ozone is converted back into O_2. It is this interconversion that has maintained the ozone concentration at a constant level; the balance between O_2 and O_3 has for millennia kept the ozone shield intact.

Until now.

In the 1970s, laboratory studies showed that chlorine atoms could break the (2), (3) chain and reduce the rate of formation of ozone. Soon afterwards, it was realized that there was a source of chlorine atoms in the upper atmosphere: photolysis of CFCs by ultraviolet light. The CFCs break the ozone-producing chain, as shown in equations (6)-(8), with CF_3Cl used as an example.

$$F_3C:Cl + h\nu \rightarrow F_3C + Cl \quad (6)$$

$$Cl + O_3 \rightarrow ClO + O_2 \quad (7)$$

$$ClO + O \rightarrow Cl + O_2 \quad (8)$$

Here we encounter another free-radical chain reaction. In the chain-initiating step (6), the chlorofluorocarbon is cleaved by ultraviolet light to yield a chlorine atom.

There then follow two chain-propagating steps. In (7), a chlorine atom attacks ozone to give O_2 and ClO. In (8), the ClO attacks O to give O_2 and regenerate the chlorine atom. Each of these reactions breaks the ozone-producing chain, (2), (3). Reaction (7) destroys the ozone needed for step (3); reaction (8) destroys the O: needed for step (2).

This effect has a familiar look. We discussed the action of an inhibitor: by breaking a chain, a relatively few molecules can prevent the formation of many molecules of product - until the inhibitor is all consumed. But here the chain-breaker is *not* consumed. *Steps* (7) *and* (8) *regenerate the original Cl.,* which is now ready to break another ozone-producing chain, and be regenerated, and so on, over and over. The chain-breaking factor is multiplied by the Cl. regeneration, and the net effect is enormous: each molecule of CFC that is photolyzed in step (6) eventually prevents the formation of some 100000 molecules of ozone.

Although CFCs are to be phased out by international agreement, their concentration in the atmosphere continues to rise, and the concentration of ozone continues to drop. The hole in the ozone shield over the Antarctic continues to grow, and there have been indications that severe depletion is occurring over the Arctic as well.

Perhaps the most frightening aspect of the effect of CFCs is the time lag. CFCs rise slowly to the height where they are photolyzed; once there, they are destroyed very slowly because of their low concentration. The ozone depletion observed so far is due to only a small fraction of the CFCs already in the atmosphere. Even when - and if - production of CFCs is completely stopped, their maximum effect will be felt decades in the future. They are waiting up above us like a time-bomb.

Molecular formula: its fundamental importance

In this section we have been concerned with the structure of methane: the way in which atoms are put together to form a molecule of methane. But first we had to know what kinds of atoms these are and how many of them make up the molecule; we had to know that methane is CH_4 Before we can assign a structural formula to a compound, we must, first know its molecular formula.

Much of the section has been spent in discussing the substitution of chlorine for the hydrogen of methane. But first we had to know that there *is* substitution, that each step of the reaction yields a product that contains one less hydrogen atom and one more chlorine atom than the reactant; we had to know that CH_4 is converted successively into CH_3Cl, CH_2Cl_2, CHCl), and CCI4. Before we can discuss the reactions of an organic compound, we must first know the molecular formulas of the products.

Let us review a little of what we know about the assigning of a molecular formula to a compound. We must carry out:

(*a*) A *qualitative elemental analysis,* to find out what kinds of atoms are present in the molecule;

(*b*) A *quantitative elemental analysis,* to find out the relative numbers of the different kinds of atoms, that is, to determine the *empirical formula;*

(c) A *molecular weight determination,* which (combined with the empirical formula) shows the actual numbers of the different kinds of atoms, that is, gives us the *molecular formula.*

Most of this should be familiar to the student from previous courses in chemistry. What we shall concentrate on here will be the application of these principles to organic analysis.

Combustibility or Oxidation: Paraffins burn in air forming carbon dioxide and water.

$$CH_4 + 2O_2 \rightarrow CO_2 + 2H_2O$$

$$2C_2H_5 + 7O_2 \rightarrow 4CO_2 + 6H_2O$$

The oxidation of the gaseous hydrocarbon in a eudiometer tube is used for the determination of their molecular formula.

When hydrocarbons are heated with oxygen at high temperature, a low combustion of hydrocarbon takes place by successive stages of dehydroxydation.

$$\underset{\text{ethane}}{H_3C{-}CH_3} \xrightarrow{O} \underset{\text{ethanol}}{H_3C{-}CH_2{-}OH} \xrightarrow{O} \underset{\text{ethane-1,1-diol}}{H_3C{-}HC(OH)_2} \xrightarrow{-H_2O}$$

$$\underset{\text{acetaldehyde}}{H_3C{-}CH{=}O} \xrightarrow{O} \underset{\text{acetic acid}}{H_3C{-}C({=}O)OH} \xrightarrow{O} \underset{\text{Carbon Dioxide}}{CO_2} + \underset{\text{Hydrogen}}{2H_2}$$

$$\underset{\text{methane}}{CH_4} \xrightarrow{O} \underset{\text{methanol}}{H_3C{-}OH} \xrightarrow{O} \underset{\text{methanediol}}{H_2C(OH)_2} \xrightarrow{O} \underset{\text{formaldehyde}}{H_2C{=}O} + \underset{\text{water}}{H_2O}$$

$$\underset{\text{formaldehyde}}{H_2C{=}O} \xrightarrow{O} \underset{\text{formic acid}}{H{-}C({=}O)OH} \rightarrow \underset{\text{Carbon Dioxide}}{CO_2} + \underset{\text{Hydrogen}}{2H_2}$$

Thermal decomposition: When vapours of higher paraffins through a metal tube at 500 - 700°C thermal decomposition cracking or pyrolysis takes place and lower hydrocarbons are foermed.

(a) $CH_4 \xrightarrow{\text{Pyrolysis}} 2H_2 + C$

(b) $C_2H_6 \xrightarrow{\text{Pyrolysis}} C_2H_4 + H_2$

(c) $C_8H_{18} \xrightarrow{\text{Pyrolysis}} C_4H_{10} + CH_4 + C_2H_4 + C$

The process is extremely useful for the production of super motofuels.

Rate of reaction

A chemical reaction is the result of collisions of sufficient energy and proper orientation. The rate of reaction, therefore, must be the rate at which these effective collisions occur, the number of effective collisions, let us say, that occur during each second within each liter of reaction space. We can then express the rate as the product of three factors. (The number expressing the probability that a collision will have the proper orientation is commonly called the **probability factor.**) Anything that affects anyone of these factors affects the rate of reaction.

The **collision frequency** depends upon:

(*a*) how closely the particles are crowded together, that is, concentration or pressure;
(*b*) how large they are; and

(*c*) how fast they are moving, which in turn depends upon their weight and the temperature.

We can change the concentration and temperature, and thus change the rate. We are familiar with the fact that an increase in concentration causes an increase in rate; it does so, of course, by increasing the collision frequency. A rise in temperature increases the collision frequency; as we shall see, it also increases the energy factor, and the latter effect is so great that the effect of temperature on collision frequency is by comparison unimportant.

The size and weight of the particles are characteristic of each reaction and cannot be changed. Although they vary widely from reaction to reaction, this variation does not affect the collision frequency greatly. A heavier weight makes the particle move more slowly at a given temperature, and hence tends to decrease the collision frequency. A heavier particle is, however, generally a larger particle, and the larger size tends to increase the collision frequency. These two factors thus tend to cancel out.

The probability factor depends upon the geometry of the particles and the kind of reaction that is taking place. For closely related reactions it does not vary widely.

Kinetic energy of the moving molecules is not the only source of the energy needed for reaction; energy can also be provided, for example, from vibrations among the various atoms within the molecule. Thus the probability factor has to do not only with what atoms in the molecule suffer the collision, but also with the alignment of the other atoms in the molecule at the time of collision.

By far the most important factor determining rate is the **energy factor**: the fraction of collisions that are sufficiently energetic. This factor depends upon the temperature, which we can control, and upon the energy of activation, which is characteristic of each reaction.

At a given temperature the molecules of a particular compound have an average velocity and hence an average kinetic energy that is characteristic of this system; in fact, the temperature is a measure of this average kinetic energy. But the individual molecules do not all travel with the same velocity, some moving faster than the average and some slower. The number of molecules with a particular kinetic energy is greatest for an energy hear the average and decreases as the energy becomes larger or smaller than the average.

The distribution of collision energies, as we might expect, is described by a similar curve, Fig. 5.6. Let us indicate collisions of a particular energy, E_{act}, by a vertical line. The number of collisions with energy equal to or greater than Eact is indicated by the shaded area under the curve to the right of the vertical line. The fraction of the total number of collisions that have this minimum energy, E_{act}, is then the fraction of the total area that is shaded. It is evident that *the greater the value of* E_{act}, *the smaller the fraction of collisions that possess that energy.*

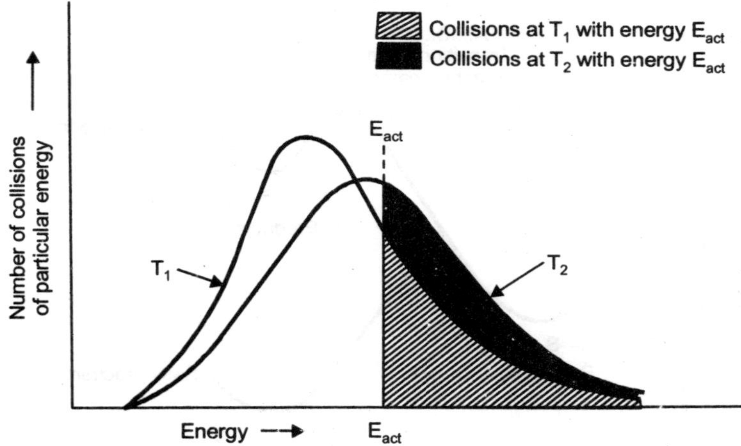

Fig. 5.6. The Change in collision energy with change in temperature

The exact relationship between energy of activation and fraction of collisions with that energy is:

$$e^{-E_{act}/RT} = \text{Fraction of collisions with energy greater than } E_{act}$$

E_{act} = energy of activation in cal (not Kcal)

e = 2.718 (base of natural log)

R = 1.986 (gas constant)

T = absolute temperature.

Using P for the probability factor and Z for the collision frequency, we arrive at the rate equation:

$$\text{rate} = \text{PZe}^{-E_{act}/RT}$$

This exponential relationship is important to us in that it indicates that a small difference in E_{act} has a large effect on the fraction of sufficiently energetic collisions, and hence on the rate of reaction. For example, at 275°C, out of every million collisions, 10 000 provide sufficient energy if E_{act} = 5 kcal, 100 provide sufficient energy if E_{act} = 10 kcal, and only one provides sufficient energy if E_{act} = 15 kcal. This means that (all other things being equal) a reaction with E_{act} = 5 kcal will go 100 times as fast as one with E_{act} = 10 kcal, and 10 000 times as fast as one with E_{act} = 15 kcal.

We have so far considered a system held at a given temperature. A rise in temperature, of course, increases the average kinetic energy and average velocities, and hence shifts the entire curve to the right, as shown in Fig. 5.7. For a given energy of activation, then, a rise in temperature increases the fraction of sufficiently energetic collisions, and hence increases the rate, as we already know.

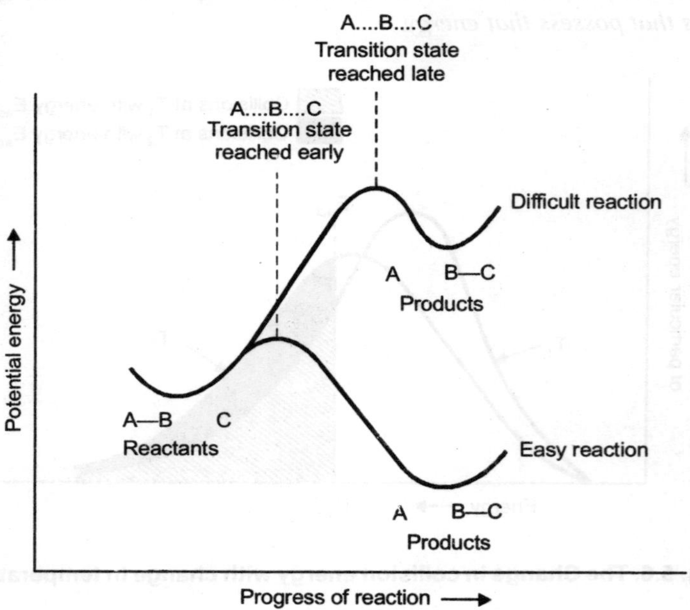

Fig. 5.7. Potential energy changes during the progress of reaction: reactivity and development of the transition state. Difficult reaction: transition state reached late, resembles products. Easy reaction: transition state reached early, resembles reactants.

The exponential relationship again leads to a large change in rate, this time for a small change in temperature. For example, a rise from 250 to 300°C, which is only a 10% increase in absolute temperature, increases the rate by 50% if E_{act} = 5 kcal, doubles the rate if E_{act} = 10kcal, and trebles the rate if E_{act} = 15 kcal. As this example shows, the greater the E_{act}, the greater the effect of a given change in temperature; this follows from the $e^{-E_{act}/RT}$ relationship. Indeed, it is from the relationship between rate and temperature that the E_{act} of a reaction is determined: the rate i~ measured at different temperatures, and from the results E_{act} is calculated.

We have examined the factors that determine rate of reaction. What we have learned may be used in many ways. To speed up a particular reaction, for example, we know that we might raise the temperature, or increase the concentration of reactants, or even (in ways that we shall take up later) lower the E_{act}. Of immediate interest, however, is the matter of relative reactivities. Let us see, therefore, how our knowledge of reaction rates can he]p us to account for the fact that one reaction proceeds faster than another, even though conditions for the two reactions are identical.

Relative rates of reaction

We have seen that the rate of a reaction can be expressed as a product of three factors: rate = collision frequency X energy factor X probability factor.

Two reactions could proceed at different rates because of differences in any or all of these factors. To account for a difference in rate, we must first see in which of these factors the difference lies.

As an example, let us compare the reactivities of chlorine and bromine atoms toward methane; that is, let us compare the rates, under the same conditions, of the two reactions:

$$Cl + CH_3 - H \rightarrow H - Cl + CH_3 \qquad \Delta H = +1 \quad E_{act} = 4$$

$$Br + CH_3 - H \rightarrow H - Br + CH_3 \qquad \Delta H = +16 \quad E_{act} = 18$$

Since temperature and concentration must be the same for the two reactions if we are to compare them under the same conditions, any difference in collision frequency would have to arise from differences in particle weight or size. A bromine atom is heavier than a chlorine atom, and it is also larger; as we have seen, the effects of these two properties tend to cancel out. In actuality, the collision frequencies differ by only a few percent. It is generally true that, for the same temperature and concentration, two closely related reactions differ but little in collision frequency - A difference in collision frequency therefore cannot be the cause of a large difference in reactivity.

The nature of the probability factor is very poorly understood. Since our two reactions are quite similar, however, we might expect them to have similar probability factors. Experiment has shown this to be true: whether chlorine or bromine atoms are involved, about one in every eight collisions with methane has the proper orientation for reaction. In general, where closely related reactions are concerned, we may assume that a difference in probability factor is *not likely* to be the cause of a large difference in reactivity.

We are left with a consideration of the energy factor. At a given temperature, the fraction of collisions that possess the amount of energy required for reaction depends upon how large that amount is, that is, depends upon the E_{act}. In our example E_{act} is 4 kcal for the chlorine reaction, 18 kcal for the bromine reaction. As we have seen, a difference of this size in the E_{act} causes an enormous difference in the energy factor, and hence in the rate. At 275°C, of every 15 million collisions, 375000 are sufficiently energetic when chlorine atoms are involved and only *one* when bromine atoms are involved. Because of the difference in E_{act} alone, then, chlorine atoms are 375000 times as reactive as bromine atoms toward methane.

As we encounter, again and again, differences in reactivity, we shall in general attribute them to differences in E_{act}. In many cases we shall be able to account for these differences in E_{act} on the basis of differences in molecular structure; *it must be understood that we are justified in doing this only when the reactions being compared are so closely related that differences in collision frequency and in probability factor are comparatively insignificant.*

Transition state

Clearly, the concept of E_{act} is to be our key to the understanding of chemical reactivity. To make it *useful,* we need a further concept: *transition state.*

A chemical reaction is presumably a continuous process involving a gradual transition from reactants to products. It has been found extremely helpful, however, to consider the arrangement of atoms a1 an intermediate stage of reaction as though it were an actual molecule. This intermediate structure is called the **transition** state; its energy content corresponds to the top of the energy hill (Fig. 5.6)

Just as ΔH is the difference in energy content between reactants and products, so *difference in energy content between reactants and transition state* transition state concept is useful for this reason: we can analyze the : of tile transition state very much as though it were a molecule, and to estimate its stability. Any factor that stabilizes the transition state to the reactants tends to lower the energy of activation; that is to say, any at lowers the top of the energy hill more than it lowers the reactant valley the net height we must climb during reaction. Transition state stability the basis whether explicit or implicit - of almost every discussion of reactivity in this book the transition state is only a fleeting arrangement of atoms which, by its very nature-lying at the top of an energy hill-cannot be isolated and examined. We possibly know anything about its structure? Well, let us take as an the transition state for the abstraction of hydrogen from methane by a atom, and see where a little thinking will lead us.

To start with, we can certainly say this: the carbon-hydrogen bond is stretched entirely broken, and the hydrogen-halogen bond has started to form but is complete. This condition could be represented as

Reactants — Transition state — Products

$$CH_4 + X \cdot \longrightarrow \left[H_3C \overset{\delta}{\cdots} H \cdots \overset{\delta}{X} \right] \longrightarrow CH_3 \cdot + H{-}X$$

where the dashed lines indicate partly broken or partly formed bonds.

Now, what can we say about the shape of the methyl group in this transition In the reactant, where methyl holds the hydrogen, carbon is tetrahedral (sp^3-hybridised); in the product, where methyl has lost the hydrogen, carbon is trigonal (sp^2-hybridized). In the transition state, where the carbon-hydrogen bond is partly broken, hybridization of carbon is somewhere between *Sp3* and *Sp2,* The methyl group is partly but not completely flattened; bond angles are greater than 109.5° but less than 120°.

C—H + X → C—H—X → C + H—X

Reactent Tetrahedral | Transition Rate Becomming trigonal | Product Trigonal

Finally, where is the odd electron? It is on chlorine in the reactants, on the methyl group in the products, and divided between the two in the transition state. (Each atom's share is represented by the symbol δ) The methyl group *partly* carries the odd electron it will have in the product, and to this extent has taken on some of the character of the free radical it will become. .

Thus, in a straightforward way, we have drawn a picture of the transition state that shows the bond-making and bond-breaking, the spatial arrangement of the atoms, and the distribution of the electrons.

(This particular transition state is intermediate between reactants and products not only in the time sequence but also in structure. Not *all* transition states are intermediate in structure, reactant and product in SN2 reactions are tetrahedral, whereas the transition state contains pentavalent carbon.)

In rate of reaction, we looked at the matter of reaction rates from the standpoint of the *collision theory.* An alternative, more generally useful approach is the *transition state* (or *thermodynamic) theory* of reaction rates. An equilibrium is considered to exist between the reactants and the transition state, and this is handled in the same way as true equilibria of reversible reactions Energy of activation E_{act} and probability factor are replaced by, respectively, *heat (enthalpy) of activation* and *entropy of activation (ΔH),* which together make up *free energy of activation (ΔG*).*

$$\Delta G = \Delta H - T\Delta S$$

The smaller (the less positive) the ΔH and the larger (the more positive) the ΔS, the smaller ΔG will be, and the faster the reaction.

Entropy corresponds roughly, to the randomness of a system; equilibrium tends to favor the side in which fewer restrictions are placed on the atoms and molecules. Entropy of activation, then, is a measure of the relative randomness of reactants and transition state; the fewer the restrictions that are placed on the arrangement of atoms in the transition state-relative to the reactants - the faster the reaction will go. We can see, in a general way, how probability factor and entropy of activation measure much the same thing. A low probability factor means that a rather special orientation of atoms is required on collision. In the other language, an unfavorable (low) entropy of activation means that rather severe restrictions are placed on the positions of atoms in the transition state.

Reactivity and Development of the Transition State

For the abstraction of hydrogen from methane by a halogen atom, we have just seen that the transition state differs from the reactants - and this difference is, of course, "that we are looking for - chiefly in being like the products. This is generally true for reactions in which free radicals (or, for that matter, carbocations or carbanions) are formed.

But just *how much* does this particular transition state resemble the products? How far have bond-breaking and bond-making gone? How flat has the methyl group become, and to what extent does it carry the odd electron?

Surprisingly, we can answer even questions like these, at least in a relative way. *In a set of similar reactions, the higher the* E_{act} *the later the transition state is reached in the reaction process.* Of the theoretical considerations underlying this postulate, we shall mention only this: the difference in electronic distribution that we call a difference in structure corresponds to a difference in energy; the greater the difference in structure, the greater the difference in energy. If E_{act} is high, the transition state differs greatly from the reactants in energy and, presumably, also in electronic structure; if E_{act} is low, the transition state differs little from the reactants in energy and, presumably, also in electronic structure (see Fig. 8).

Practically, this postulate has been found extremely useful in the interpretation of experimental results; among other things, as we shall see, it enables us to account for the relationship between reactivity and selectivity

Abstraction of hydrogen by the highly reactive chlorine atom has a low E_{act}. According to tile postulate, then, the transition state is reached before the reaction has proceeded very far and when the carbon-hydrogen bond is only slightly stretched. Atoms and electrons are still distributed much as they were in the reactants; carbon is still nearly tetrahedral. The methyl group has developed little free-radical character.

Abstraction of hydrogen by the less reactive bromine atom, in contrast, has a very high E_{act}. The transition state is reached only after reaction is well along toward completion and when the carbon-hydrogen bond is more nearly broken. The geometry and electron distribution have begun to approach those of the products, and carbon may well be almost trigonal. The methyl group has developed much free-radical character.

Thus, *in the attack by a reagent of high reactivity, the transition state tends to resemble the reactants; in the attack by a reagent of low reactivity, the transition state tends to resemble the products.*

Use of Paraffins

The chief uses of the paraffins are:

(a) ***As Illumines:*** Methane ethane, natural gas coal, paraffin wax (candle) and kerosene oil are used for the purpose

(b) ***As fuel:*** Petrol and kerosene oil are used as fuel. The gas obtained during the distillation of petroleum is also used for the heating purpose

(c) ***As Dry cleaning agent:*** Petrol benzene etc. obtained during refining of natural petroleum are used as dry cleaning agents.

(d) ***As solvents for fats and oils:*** Petrol benzene ligroin are used as solvents for fats and oils

(e) ***As lubricant:*** Vaseline is an important lubricant for machinery.

(f) ***As a source for commercial hydrogen:*** The source is methane CH_4

METHANE

Formula: - CH_4

Occurrence

Distribution of Methane gas in the different places of earth (shown in a pie graph) Fig. 5.8.

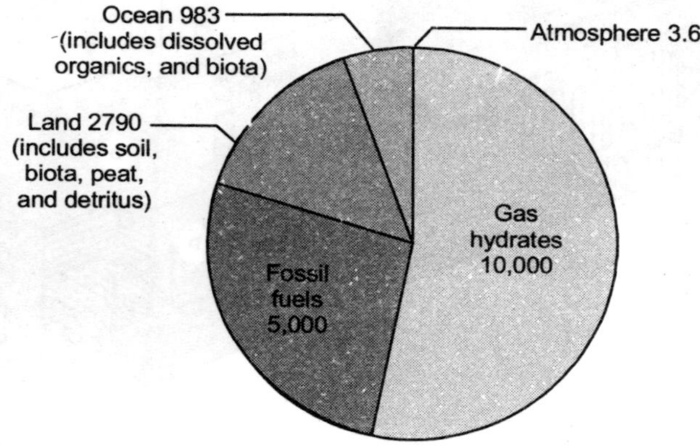

Fig. 5.8. The Methane Distribution.

The Methane Cycle and Green House Effect

During the 1980s the world became aware of the threat of global warming brought about by the *greenhouse effect.* And here the chief culprit is, not a noxious side product, but an unavoidable principal product of the burning of hydro-carbons: *carbon dioxide.*

The atmosphere contains a number of gases that act like the glass in a greenhouse. They are transparent to visible light from the sun, and allow it to pass through to the earth below; but they absorb and trap infrared light radiated outward from the earth and convert it into heat. The principal "greenhouse gases" are CFCs, methane, and, most important of all, carbon dioxide. The concentration of greenhouse gases is rising today because of two factors, both stemming from our technology: the manufacture of CFCs, never before present on our planet; and the burning of fossil fuels at a tremendous - and still increasing - rate. If this trend is not reversed, it is predicted, there will be a steady rise in the temperature of the earth. This global warming could cause drastic climatic changes with devastating results: shifting of seasonal winds, with the creation and spreading of deserts; alterations in crop patterns, with 'social upheaval and starvation; melting of the polar ice caps, with a rise in the level of the sea and the inundation of large areas of land and many of the biggest and most important cities.

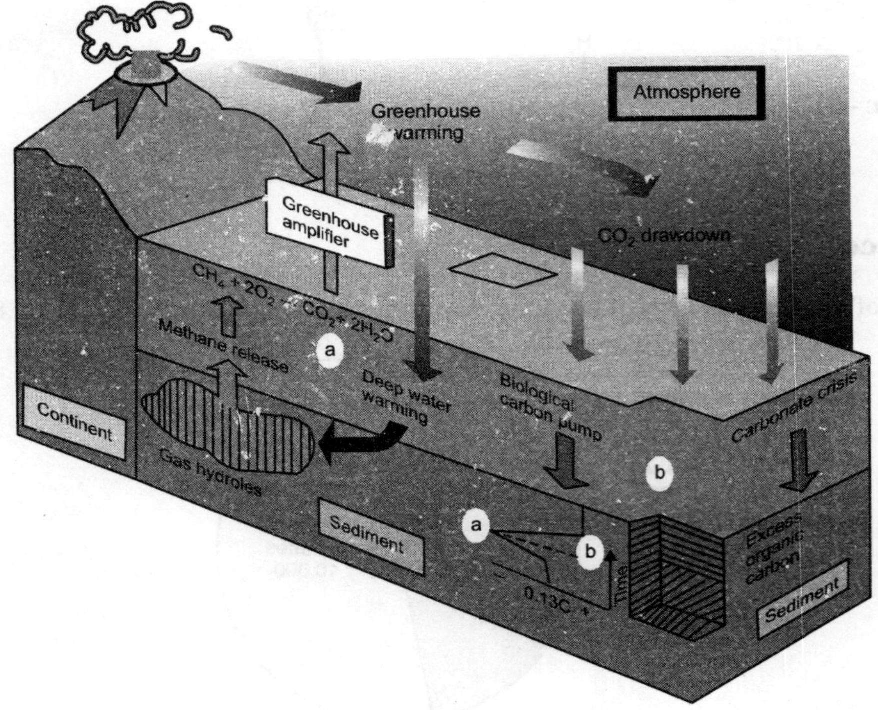

Fig. 5.9. The Methane Cycle in Nature

The solution to the problem of CFCs - in principle, though not in practice - is simple: stop making them. (There is an even more pressing reason for doing that: the unrelated matter of depletion of the ozone shield.) But one cannot burn carbon compounds without producing carbon dioxide-and our way of life is centered on such burning. Clearly, other sources of energy must be developed.

This is the simplest hydrocarbon, methane occurs in the following sources:-

(*a*) In natural gases, issuing from the petroleum oil mills

(*b*) In the gases issuing from coal mines. Methane combines with air in the mine and forms an explosive mixture called free damp by the coal-mines. All can be detected in mines by Marsh gas.

(*c*) In the gases rising from strongest water where it is produced by the decomposing of vegetable matter by bacteria in the absence of air. For the same reason, the gas is found in marshy places and therefore called 'marsh gas'.

(*d*) In coal gas produced by the dry distillation of coal.

(*e*) In wood gas, produced by the distillation of wood.

Preparation

Methane can be prepared by the following methods:

1. **Lab preparation of Methane**: In the lab methane is prepared by strongly heating a mixture of dry sodium acetate and soda lime in a glass tube fitted with a delivery tube (See Fig. 5.10).

$$\underset{\text{Sodium Aceate}}{CH_3COONa} + NaOH \rightarrow \underset{\text{Metahne}}{CH_4} + Na_2CO_3$$

The gas can be collected over water. The gas obtained by this method is not pure and contain traces of hydrogen and ethylene as impurities. The gas can be purified by passing it through a cold alkaline solution of potassium permanganate in which the impurities are obtained while methane remains unaffected.

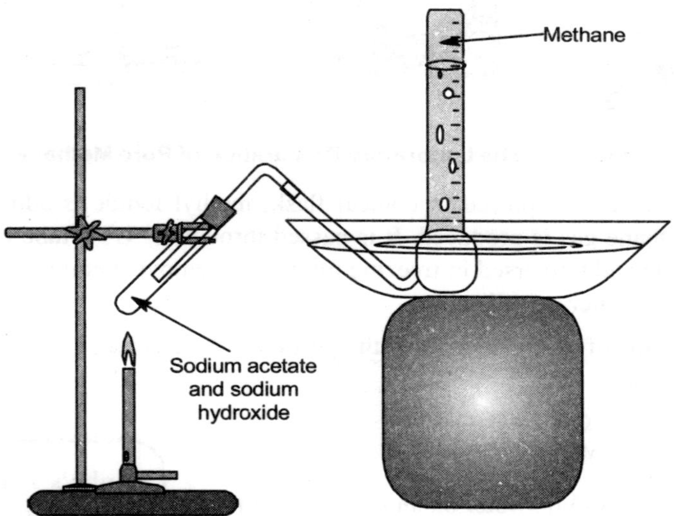

Fig. 5.10. The Laboratory preparation of Methane.

2. **Pure methane**: It is best prepared in the lab by the reduction of methyl iodide with zinc-copper couple* and methyl alcohol. (See Fig. 5.11 on next page).

Equation:

$$2CH_3 - I + Zn + [2H] \rightarrow 2CH_3 - H + Zn$$

*Zinc copper couple is prepared by dipping pieces of zinc in a solution of copper sulphate zinc gets coated with copper. It is filtered, washed and used to produce nascent hydrogen.

$$Zn \rightarrow Zn^{++} + 2e$$

$$CH_3I + e \quad I^- + CH_3$$

$$CH_3 \rightarrow :\bar{C}H_3$$

$$:CH_3 + H^+ \xrightarrow{C_2H_2OH} CH_3 - H + (C_2H_5O)^-$$

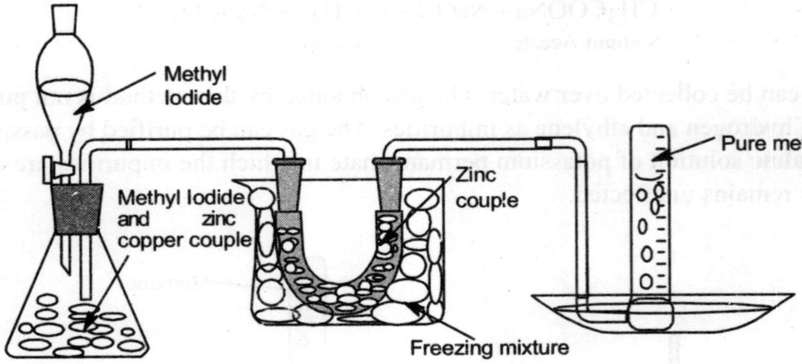

Fig. 5.11. The Laboratory Preparation of Pure Methane

Zinc copper couple is placed in conical flask; methyl iodide is added drop wise from top formed. Methane gas is produced. It is passed through a "U" – tube containing some zinc – copper couple and immersed in freezing mixture to remove methyl iodide vapours. The gas is collected over water.

3. By the reduction of methyl alcohol with hydroiodic Acid at high temperature in the presence of red phosphorous.

$$\underset{\text{Methyl Alcohol}}{CH_3OH} + 2HI \rightarrow \underset{\text{Methane}}{CH_4} + H_2O + I_2$$

4. By the action of boiling water on aluminium carbide

$$\underset{\text{Alu min ium Carbide}}{Al_4C_3} + 12H_2O \rightarrow 4Al(OH)_3 + 3CH_4$$

Koibe's reaction, Senderen's reaction and Wurtz's reaction cannot produce methane

This reaction can be also used in the lab for the preparation of methane. Methane cannot be obtained by Wirtz reaction. Kolbe's reaction or Sabatier Senderen's reaction because they involve the union of at least two carbon atoms but methane has only one carbon atom.

Synthesis and Industrial Preparation

1. By passing hydrogen gas through a mixture of carbon and finely dividednickel heated to 4475°C

$$C + 2H_2 \xrightarrow{Ni(475°C)} CH_4$$

2. By passing a mixture of carbon monoxide or carbon dioxide and hydrogen over finely divided nickel heated to about 300°C.

$$CO + 3H_2 \xrightarrow{Ni(300^\circ C)} CH_4 + H_2O$$

$$CO_2 + 4H_2 \xrightarrow{Ni(300^\circ C)} CH_4 + 2H_2O$$

3. By passing a mixture of carbon disulphide vapour and H_2S

(*a*) $\underset{\text{Carbon Disulphide}}{2CS_2} + 4H_2S + 8Cu \rightarrow \underset{\text{Cuprous Sulphide}}{4Cu_2S} + 2CH_4$

(*b*) $2CS_2 + 4H_2O + 8Cu \rightarrow 2CuS + 4CuO + 2CH_4$

Landfill Riser Shafts

The landfill riser system is composed of three segments: the **Base**, the **Riser-Segment** and the **Cap**. All these pipe components are fabricated from ruggedized HDPE pipe with at least 46 psi pipe stiffness. Each segment hand-screws into the other using two men laborers. As the garbage level rises and the riser needs to be vertically extended, the debris cap is unscrewed and removed by hand. The new Riser-Segment is hand threaded into place and the debris cap is reinstalled by hand threading it back into position. The short-length and lightweight HDPE pipe combine to make this a very practical system (Fig. 5.12).

Landfill Riser Shafts

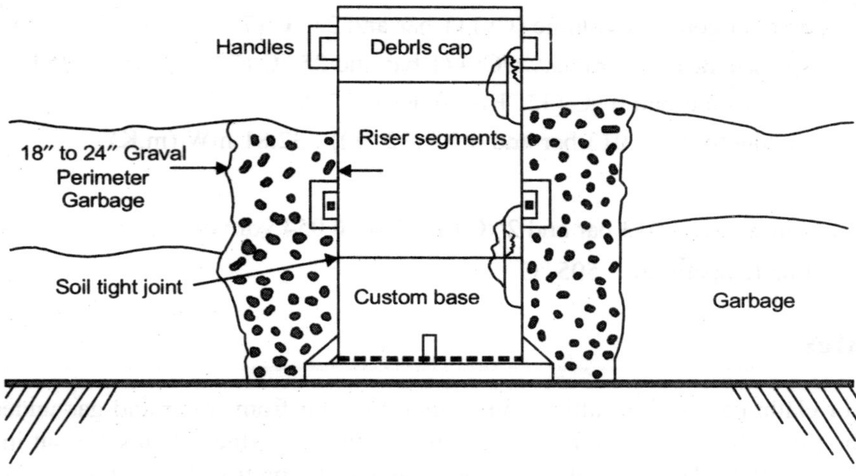

Properties (Physical)

Molecular Weight

- Molecular weight : 16.043 g/mol

Solid phase

- Melting point : –182.5 °C
- Latent heat of fusion (1,013 bar, at triple point) : 58.68 kJ/kg

Liquid phase

- Liquid density (1.013 bar at boiling point) : 422.62 kg/m^3
- Liquid/gas equivalent (1.013 bar and 15 °C (59 °F)) : 630 vol/vol
- Boiling point (1.013 bar) : -161.6 °C
- Latent heat of vaporization (1.013 bar at boiling point) : 510 kJ/kg

Critical point

- Critical temperature : -82.7 °C
- Critical pressure : 45.96 bar

Gaseous phase

- Gas density (1.013 bar at boiling point) : 1.819 kg/m^3
- Gas density (1.013 bar and 15 °C (59 °F)) : 0.68 kg/m^3
- Compressibility Factor (Z) (1.013 bar and 15 °C (59 °F)) : 0.998
- Specific gravity (air = 1) (1.013 bar and 21 °C (70 °F)) : 0.55
- Specific volume (1.013 bar and 21 °C (70 °F)) : 1.48 m^3/kg
- Heat capacity at constant pressure (Cp) (1 bar and 25 °C (77 °F)) : 0.035 kJ/(mol.K)
- Heat capacity at constant volume (Cv) (1 bar and 25 °C (77 °F)) : 0.027 kJ/(mol.K)
- Ratio of specific heats (Gamma:Cp/Cv) (1 bar and 25 °C (77 °F)) : 1.305454
- Viscosity (1.013 bar and 0 °C (32 °F)) : 0.0001027 Poise
- Thermal conductivity (1.013 bar and 0 °C (32 °F)) : 32.81 mW/(m.K)

Miscellaneous

- Solubility in water (1.013 bar and 2 °C (37 °F)) : 0.054 vol/vol
- Autoignition temperature : 595 °C

Gas Hydrates

Gas hydrates or clathrates are crystalline compounds forming from water and gas under certain P/T conditions. The title "clathrates" coming from Latin 'clathratus' which means "to put in a cage" was given by Pawll in 1948. These compounds are non-stoichiometric that is they are of a variable composition. Gas hydrates (of sulphur dioxide and chlorine gases) were observed for the first time in the end of XVIII century by G. Pristly, B. Pelete and V. Karsten. Water molecules form a framework (i.e. host lattice) with the cages. These cages can be occupied with gas molecules ("guest" molecules). Gas molecules are bonded with water molecules by Van der Waals forces. Gas hydrates are described by general formula $M*nH_2O$, where *M* is a molecule of hydrate-forming gas, and *n* is the ratio that indicative the quantity of water molecules to 1 molecule of gas (*n* varies from 5.75 to 17).

At present three crystal modifications of gas hydrates are known:

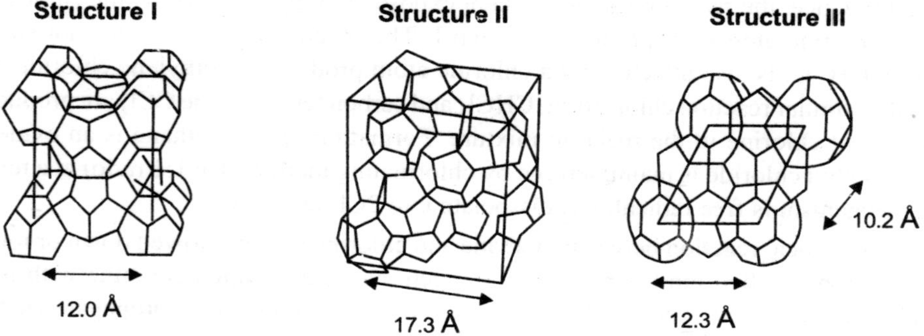

Chemical Properties

1. Halogenation: - Fluorine decomposes violently with methane to carbon and hydrogen fluoride. Chlorine does not react with it in the dark but in strong sunlight, the reaction is explosive. Carbon and HCl are thus produced. In diffuse sunlight, a chain reaction occurs in quick succession; methyl chloride, methylene chloride, chloroform and finally carbon tetrachloride, are thus formed, depending on the proportion of methane and chlorine. These are difficult to separate and in practice generally carbon tetrachloride in this mode. The mechanism has been discussed below

(i) Initiation: Chlorine molecules absorb light energy and dissociate into atoms by homolythic fission:

$$Cl:Cl \xrightarrow{h\nu} Cl\bullet + Cl\bullet$$

(ii) Propagation: A chlorine atom then abstracts a hydrogen atom H^- from methane and forms H:Cl; A free methyl radical CH_3 abstracts chlorine atom from a chlorine molecule forming methyl chloride and a chlorine atom

(*i*) $Cl^- + H:CH_3 \rightarrow H:Cl + CH_3$

(*ii*) $CH_3 + Cl:Cl \rightarrow CH_3:Cl + Cl$

The reactions take place because it is energetically favorable; H–Cl (bonds energy 103 kcal/mole) is more stable than $H–CH_3$ (bond energy 99kcal/mole) 103-99=4 kcal energy is liberated during the reaction. Reaction (*i*) and (*ii*) constitute the propagation step steps I and II could go on indefinitely so long as there are enough methane and chlorine molecules in the vessel.

(iii) Termination: (by combination or coupling)

(*i*) $CH_3 + CH_3 \rightarrow CH_3 : CH_3$
Ethane

(*ii*) $CH_3 + Cl \rightarrow CH : Cl$
Methyl Chloride (Majotr Product)

In actual practice, the reaction mixture is kept exposed to light through out the chlorination period so that fresh chlorine atoms are produced by step-I. The reaction however, does not stop at CH_3:Cl stage because CH_3:Cl is also attached by a chlorine atom producing methylene chloride, CH_2Cl_2 and HCl. By further similar reaction chloroform, $CHCl_3$ and carbon tetrachloride CCl_4 are produced provided there is sufficient chlorine in the reaction mixture. Formation of little ethane is an evidence for this mechanism. Methyl chloride is manufactured by chlorinating methane (in 1:1 ratio) at a temperature of 375°C the mechanism is free radical; Cl_2 dissociate into $2Cl^-$ at 300°C.

The photochemical reaction between methane and chlorine may be slowed down or totally stopped by some compounds called inhibitors such as oxygen. No I_2 etc. which combine with the alkyl free radical to produce reality non-reactive (as stable) free radicals which do not propagate the kinetic chain e.g.

$$CH_3 + O : O \rightarrow CH_3 - O - O$$

Bioradical Peroxymethyl Free Radical

Free radical fluorination of alkenes is violent 104kcal/mole heat is evolved which cleaves both C–H and C–C bonds (bond energy 99 kcal/mole and 80 kcal/mol respectively) Iodination is energetically unfavourable; the reaction is endothermic by 14 kcal/mole. Bromination proceeds slowly heat evolved by 6/kcal/mole. Where as for chlorination it is 24kcal/mole. The order of stability of alkyl free radicals is the some as that of alkyl carbonium ions.

Tertiary>secondary>primary> methyl

Now regarding this photochemical reaction we see that reaction equation follows exactly the mechanism.

$$Cl_2 \xrightarrow{hv} 2Cl$$

$$CH_4 + Cl \rightarrow HCl + CH_3$$

$$CH_3 + Cl - Cl \rightarrow CH_3 - Cl + Cl^-$$

methyl chloirde

Similarly

$$CH_3 - Cl + Cl_2 \rightarrow CH_2Cl_2 + HCl$$

Methyl Chloride

$$CH_2Cl_2 + Cl_2 \rightarrow CHCl_3 + HCl$$

Chloroform

$$CHCl_3 + Cl_2 \rightarrow CCl_4 + HCl$$

Carbon tetrachloride

$$CH_4 + I_2 \; f \quad \text{Methyl Iodide}$$

2. *Nitration*: Methane may be nitrated to nitro methane in vapour phase with HNO_3 at 400 - 450°C the simplified reaction is

$$CH_4 + HNO_3 \rightarrow CH_3 - NO_2 + H_2O$$

Initialization: $HO\text{-}NO_2 \xrightarrow{400^\circ\text{-}500^\circ C} HO + NO_2$

Propagation: $\underset{\text{alkene}}{R - H} + HO \rightarrow R + HOH$

Termination: $R + NO_2 \rightarrow \underset{\text{Nitroaniline}}{R - NO_2}$

3. *Sulphonation*: An alkane does not react with concentrated sulphuric acid even at elevated temperature, but higher alkenes in presence of ultra-violet light or free radical generator SO_2Cl_2 and hydrolysis

$$R - H + SO_2Cl_2 \rightarrow R - SO_2Cl + HCl$$
$$\downarrow H_2O$$
$$R - SO_3H + HCl$$

4. *Oxidation*: Methane can be oxidised with steam to a mixture of O and H_2 (or CO_2 and H_2) almost completely at about 900°C, the reaction can be controlled. An excess of steam produces CO_2 at about 500°C. Hydrogen thus produced is converted into NH_3 with atmospheric nitrogen and finally into ammonium sulphide a common fertilizer. CO and H_2 are raw materials for synthesis of alcohols aldehydes acids etc. The mixture is known as synthesis gas.

$$CH_4 + H_2O \xrightarrow{850^\circ-900^\circ C} CO + 3H_2$$

$$CO + H_2O \xrightarrow{500^\circ C\ Fe_2CO_3} CO_2 + N_2$$

5. *Combustibility or Oxidation*: Methane burns in air or Oxygen with a non-luminous flame forming CO_2 and H_2O.

$$CH_4 + 2O_2 \rightarrow CO_2 + 2H_2O$$

A mixture of methane and air explodes violently on ignition

6. *Action of Ozone*: On treatment with ozone methane produces formaldehyde

$$\underset{\text{methane}}{CH_4} + \underset{\text{Ozone}}{2O_3} \longrightarrow \underset{\text{formaldehyde}}{H-\overset{O}{\overset{\|}{C}}-H} + \underset{\text{water}}{H_2O} + \underset{\text{Oxygen}}{2O_2}$$

As formaldehyde can be identified by its characteristic smell and colour reactions. It serves as a delicate test for methane.

Methane contains carbon and hydrogen

A mixture of pure methane and excess of pure oxygen is exploded by sparking in a closed tube. As a result of the combustion CO_2 and H_2O are formed.

$$CH_4 + 2O_2 \rightarrow CO_2 + 2H_2O$$

The exploded mixture is cooled when water vapours colourless to a colourless liquid CO_2 is identified by testing with clear lime water (milk ness appears) and water is tested with anhydrous copper sulphate which turns blue in the presence of water. As CO_2 contains carbon and H_2O hydrogen, the positive results prove methane contains carbon and hydrogen (as no other substance in the experiment contains hydrogen and carbon.

Methane is a saturated hydrocarbon

Methane is a saturated hydrocarbon because it does not appear in the addition reaction but always in the substitution reaction. It does not decolorize bromine water.

The four hydrogen atoms in the methane cannot be distinguished from one another's in other words, they are equivalent. This fact is amply proved by the reaction of methane. For example, in what ever procedure dichloromethane (CH_2Cl_2) is prepared; we get the same compound in every effort. The three dimensional regular tetrahedral structure of CH_4 molecule with one hydrogen atom situated at each vertex points to the equivalence of four atoms of methane

Uses of Methane

Methane is used:

(a) As an illuminant

(b) For preparing commercial hydrogen

(c) For preparing formaldehyde, methyl chloride and carbon tetrachloride which are technically very important.

(d) For preparing "carbon black" burning methane in the insufficient supply of air carbon black is used in making printing inks.

(e) Another major use of the methane is the use of bio-gas

Bio-gas Plants

Biogas is a mixture of methane and carbon dioxide. It also has traces of hydrogen sulphide, ammonia, oxygen, hydrogen, water vapour etc., depending upon feed materials and other conditions. Biogas is generated by fermentation of cellulose rich organic matter under anaerobic conditions. In anaerobic conditions, the methane-producing bacteria become and more active. Thus, the gas produced becomes rich in methane. Biogas is one of the renewable sources of energy receiving popularity in rural areas and has successfully met the cooking energy needs of families. The optimum utilization depends upon the successful physical installations, which in turn depend upon plant design and its selection (Fig. 5.13).

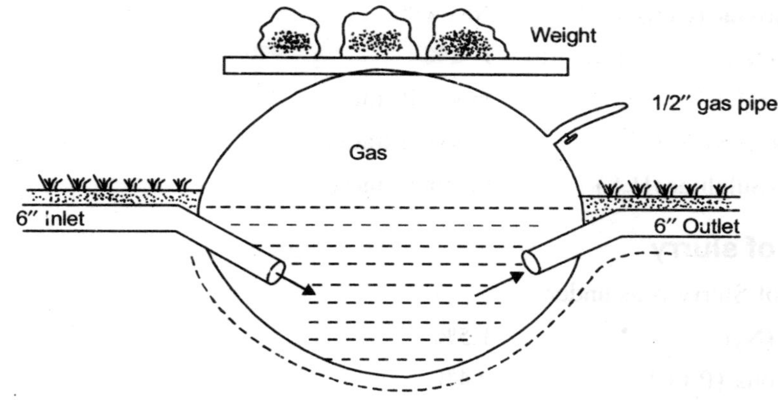

Fig. 5.13. Biogas plant

Potential Gas Production from Different Feed Stocks

Sl. No.	Type of feed stock	Gas yield per kg. (Cu.m)	Normal feed availability per animal per day kg (wet weight)	Gas yield per day (Cu.m)
1.	Cow dung	0.036	10.0	0.36
2.	Buffalo dung	0.036	15.0	0.54
3.	Camel dung	0.056	6.0	0.336
4.	Horse dung	0.045	10.0	0.45
5.	Sheep dug	0.04	21.0	0.042
6.	Pig (Approx. 50 kg)	0.08	2.25	0.18
7.	Chicken (Approx 2 kg)	0.062	0.18	0.011
8.	Human excreta	0.07	0.40	0.28

Biogas is a zero waste technology. The products of biogas plants, like biogas and digested slurry can be utilized economically for cooking and as manure for agriculture and horticulture. Biogas is a non-poisonous and non-toxic gas which when mixed with air burns with blue flame, without soot or any offensive smell. The slurry is rich in nitrogen, phosphorous, potash and humus material. It has good applications in agriculture and horticulture.

Composition of biogas and slurry composition of biogas

The composition of Biogas is as below:

(i) Methane (CH_4) : 55-70%

(ii) Carbon dioxide (CO_2) : 30 - 45%
(iii) Nitrogen (N_2) : 1 -2%
(iv) Hydrogen (H_2) : In small traces
(v) Carbon Monoxide (CO) : In small traces
(vi) Hydrogen Sulphide (H_2S) : In small traces

Composition of slurry

The composition of Slurry is as under:

(i) Nitrogen (N_2) : 1.5%
(ii) Phosphorous (P_2O_5) : 0.4%
(iii) Potash (K_20) : 2.2%

Number of Cattle Required for Different Capacity of Plants and Family sizes

Sl. No.	Size of plant (m^3)	Amount of wet dung required daily (kg)	Approximate no. of cattle required	Approximate family size
1.	1	25	2-4	1-4
2.	2	50	4-5	5-8
3.	3	75	6-8	9-12
4.	4	100	9-10	13-16
5.	6	150	11-15	17-22
6.	8	200	13-16	23-26
7.	10	250	17-20	27-32

Raw materials for Biogas Generation

In India, we have large population of cattle and animals. About 43 million families own more than four cattle, which is minimum number of cattle required to own a small biogas plant. For biogas, various crop residues, aquatic plants and animal dung can be used as feedstock. The potential gas production from different feed stock is given in the table :

Types of Biogas Plants

The common types of biogas plants are :

(*a*) KVIC floating drum type,
(*b*) Janata fixed dome type,
(*c*) Deenbandhu fixed dome type, and
(*d*) Shramik Bandhu biogas plant.

KVIC Floating Drum Type

This is one of the most commonly used biogas plant in India. In this type of plant, the gas is stored in a metallic drum (which is inverted upside down and which moves on a guided frame according to quantity of gas stored in it).

The gasholder rests on a guide frame which is fixed in digester walls. The KVIC type plant consists of inlet tank, outlet tank, gasholder, digester, inlet-outlet pipes and partition wall. The cow dung and water is mixed in 4:5 proportion respectively and agitated thoroughly in the inlet tank. Pipe cover is open so the mixture can go into digester. In the digester, the reaction takes place and biogas is produced.

The produced biogas gets stored in inverted drum. The slurry goes out through the outlet pipe.

For the drum, new alternative materials like ferrocement, HDPC or fibreglass are also being used. In this type of biogas plant, pressure remains constant, which is equivalent to weight of gasholder over unit area. The pressure can be increased by putting additional weight stones/ bricks on the gas holder.

The side of gasholder remains dipped in the slurry to seal the leakage of gas to the atmosphere. The design is available in various sizes for 1-140 Cu.M gas per day. The designs for various capacity plants are readily available with Khadi and Village Industries Commission, Vile Parle, Mumbai.

Heating Value of Various Fuel used for Cooking and Their Replacement

Sl. No.	Name of fuel	Heating value	Replacement value
1.	Biogas	4713 K Cal/cu. M	1.0 m^3
2.	Electricity	860 K Cal/cu. M	4.968 kWh
3.	Fuel wood	4700 K Cal/cu. M	3.414 kg
4.	Coal	6930 K Cal/cu. M	1.458 kg
5.	L P Gas	10882 K Cal/cu. M	0.433 kg
6.	Dung Cake	2092 K Cal/cu. M	12.296 kg
7.	Kerosene oil	9122 K Cal/cu. M	0.620 litre

In KVIC design, the cost is more as the cost of gasholder is high. The plant also requires regular check-up and maintenance.

Janata Biogas Plant

It is a fixed dome type plant. It has got an inlet tank, outlet tank, digester and fixed dome.

The digester of biogas plant is of cylindrical shape constructed with bricks and cement. It holds the dung slurry, and the biogas produced is stored in gas portion, which is an integral part of plant between dome and digester. The height of gas portion is above the inlet and outlet openings as the gas volume increases slurry is displaced. In Janata type plant, the pressure of biogas is not constant. The cost of this plant is comparatively less in comparison to KVIC plant.

Deenbandhu Biogas Plant

This is also a faxed dome plant developed by Action for Food Production, New Delhi. It is a low cost biogas plant. The principle of working of plant is same as that of Janata type, except for the configuration of inlet entrance and digester.

The digester of this type of plant is spherical. In this digester, surface area is reduced for same volume of material that reduces the cost of plant.

At the top of the foundation, a window opening is provided (outlet portion) for outward movement of the digested slurry. In the inlet, asbestos cement pipe of 15 mm is used.

The pipe is embedded in the digester at a fixed position, just opposite to outlet opening, to avoid short-circuiting of fresh material and digested slurry. The volume of the outlet is increased to produce requisite gas pressure through the weight of the displaced slurry. At the top of dome, a gas outlet pipe is fixed.

The cost of construction of Deenbandhu fixed dome type biogas plant is lower compared to other types. The Deenbandhu model was designed by making effective use of the principles of structural engineering as well as the long term experience of one of the NGOs with the fixed dome biogas plant. Initially, studies on performance aspects were conducted by scientists, engineers and specialists.

The design was then finalized on the basis of feedback from various NGOs who had set up Deenbandhu plants on an experimental basis in different parts of the country. The model was finally approved by the Government of India in 1986 for propagation under NPBD. Since then, it has become the most popular model in the country.

Quantities of Biogas Required for Different Applications

Sl. No.	Use	Specification of Appliances	Quantity of Gas Required
1.	Cooking	2" Burner, 0.33 cu. M/hr 4" Burner, 0.47cu. M/hr 6" Burner, 0.64 cu. M/hr (@ 55-60% eficiency)	0.3 cu. M. per day per person
2.	Lighting	100 Candle power lamp	0.15 cu.M. per hr.
3.	Motive Power	75-80% replacement of diesel in dual fuel engine	0.50 cu.M. per bhp hour

Shramik Bandhu Biogas Plant

The Shramik Bandhu Biogas Plant was developed by one of the NGOs during 1995-96. The objective was to reduce the construction cost by using locally available low cost material. The plant design and working principle is similar to that of Deenbandhu biogas plant. However, the construction material used for both the plants differ. Bamboo is used in SBBP and bricks are used in Deenbandhu. This can be advantageous in areas where there is difficulty in procuring bricks. The technology is being tested and the cost of the plant works out to be approximately Rs 1000 less than the Deenbandhu model. In areas where bamboo is abundantly available, the cost of the plant would be still lower. As the mode] is labour intensive it would also create more employment opportunities, especially for rural women, since weaving of bamboo is primarily undertaken by them.

Selection of Size of Biogas Plants

For selection of size of biogas plants, there are two important factors, namely:

(*a*) Total availability of biodegra-dable materials for biogas production and

(*b*) Consumption of Biogas.

(*a*) Availability of total bio-degradable material: The biogas is generated mainly from cow dung. Other animal's dung and human excreta, agricultural and forest wastes could also be used as biogas feed. Total quantity of dung available per day should be determined and then size of biogas plant can be decided. The table gives the cattle required for different capacity of plants, and family size for which they are suitable.

(*b*) Consumption of biogas: When availability of biogas is more and requirement is less, then consumption biogas is calculated and size is decided. The consumption of biogas can be calculated based upon the purpose for which it is required. The quantities of biogas required are given in the table.

Selection of Type of Biogas Plant

The biogas plants available in India fall in two categories, namely, floating gas holder type and fixed dome type. The selection of type of plant to be installed should be based upon their merits and demerits and with reference to physical condit-ions. A comparison of floating gas holder and fixed dome type biogas plant is give in the table below.

Comparison of floating gas holder and fixed dome type biogas plant design

Floating Gas Holder Design	Fixed Dome Design
• Initial cost of installation is more	Capital investment for the corresponding size of plant is less
• Repair and Maintenance cost during operation is high	As there are no moving parts, the maintenance cost is minimum
• Steel gas holder needs to be replaced due to corrosion effect	Steel gas holder is not required
• The life of digester is expected to be 30 years and that of gas holder is 5-8 years	The life of the plant is expected to be comparatively more
• Drum, which is exposed over the ground, does not allow the use of space over the plant for other purposes	As the unit is underground, the space above the plant can be used for other purposes
• Because of metallic gas holder the low temperature during winter effects gas production	Effect of low temperature is less

Floating Gas Holder Design	Fixed Dome Design
• It is most suitable for dung the night soil, other organic materials may clog the inlet pipe	It could be easily modified and adopted for use of other types of organic wastes
• The construction of digester is simple, and therefore no specially trained person is required but the fabrication of gas holder requires workshop facility	Construction of dome portion of the unit is a skilled job and only trained mason can do it
• Requires relatively less excavation work during installation	Requires more excavation work
• The location of defects in the gas holder and its repair is easy and can be done in workshop	Location of defects in the dome and repairing are difficult
• Release of gas is at constant pressure because of fixed weight of steel gas holder and hence the efficiency of plant is higher	Release of gas is at variable pressure and it may cause reduction in the efficiency of gas appliances
• In areas having a high water table, horizontal plants could be installed	Construction of the plant is difficult in high water table areas

Factors to be Considered

The type of biogas plant to be installed is decided upon following factors:

- Technical.
- Climatological.
- Geographical.
- Economical.

(*i*) **Technical considerations:** The level of skill required, quality of building material available locally, suitability of site and gas consumption pattern influences the selection of particular design of plant. The fixed dome type plants require skilled and trained mason and good quality building material whereas KVIC floating drum type plant can be constructed by a mason doing building works.

(*ii*) **Consideration of climatological factors:** The climate, specially temperature, affects the process of fermentation, and hence, the working of plant. In the area where mean temperature in winter season is more than 20°C, any plant can be chosen. In case the mean temperature is less than 20°C, then fixed dome type underground plants should be preferred.

(*iii*) **Consideration of geographical factors:** The fixed dome type plants should be constructed underground completely. It should be constructed in raised platform to achieve automatic flow of slurry.

(*iv*) **Economic considerations:** The selection should be made based upon economic standard of beneficiary. The cost of floating drum biogas plants are more in comparison to fixed dome biogas plant, as it involves cost of drum and more earth work.

Utilization of Biogas and Slurry

(*a*) **Utilization of biogas:** The biogas is a mixture of methane (59%), Carbon dioxide (38%), Oxygen (2%) and traces of other gases. It is non-toxic and non-poisonous. One of its most common uses is in cooking. In rural areas one of the major energy needs is energy for cooking. The heating value of various alternative fuels used for cooking and their replacement by Cu-meter of biogas is given in the following table.

(*b*) **Biogas burners: Single and Double Burners:** Biogas burns after getting mixed with air. The biogas burners are designed for 7-10 cm of water column pressure. Small burner consumes 225 litre per hour of gas whereas big burner consumes 450 litre per hour of gas. Depending upon use, type and size of burners can be chosen. The commercially available burners are: Single Burner Double Burner. For controlling the gas and regulating the flame, a spring loaded knob is provided. The air control opening is also provided near gas outlet which is adjusted once only. The efficiency of these burners ranges between 55-60%.

(*c*) **Chapati burner:** In certain parts of our country, baked chapatti is a must in diet of the people. For chapatti baking, special burners are used. The types of available burners used for chapatti making are Sevagram and Spiral burner.

(*d*) **Biogas lamps:** One of the common applications of biogas in rural area is lighting where electricity is not available, Biogas lamp is similar to kerosene lamps. It provides 100 candle (60 W) light illumination.

In the top of burner, an “on and off” locking lever is provided to supply gas for lighting. The intensity of lighting is regulated by gas regulator. The mantle is fixed in clay nozzle and covered with glass to check entry of insects etc.

(*e*) **Utilization of slurry as manure:** The chemical fertilizers which are used by farmers have undesirable effects in long term. For long period sustained agriculture, organic manure is very important. Biogas slurry is rich in nutrients compared to farm yard manure. The use of biogas slurry as manure has following advantages.

(*i*) Its decomposition time is less

(*ii*) Weed seeds of cow dung are destroyed completely

(*iii*) It is rich in nutrients compared to Farm Yard Manure (F.Y.M.)

The biogas slurry could be applied directly in wet condition or it could be dried and used afterwards when required.

Comparative Availability of Nutrients in Biogas Slurry and Farm Yard Manure (F.Y.M.)

Sl. No.	Type of Manure	Time of decompo-sition	Loss in quantity during process	Available		
				N	P	K
1.	Biogas slurry	1-2 months	7-10%	1.175%	1.10%	1.0%
2.	Farm yard manure (composted under cover)	3-4 months	20-25%	0.9%	0.75%	0.75%
3.	Farm yard manure composting	4-6 months	45%	0.75%	0.60%	0.60%

Wet slurry

Biogas plant effluents can be directly mixed with irrigation water. Wet slurry is rich in nutrients. Slurry should not be applied to the plants directly. It should always be diluted in water. It gives better yield compared to dry application.

Dried slurry

If the farmer's field is away from the biogas plant, then it is advisable to dry the slurry and apply whenever necessary. The slurry that comes from biogas plant contains 90% For drying, two to three manure pits of 2' × 2' × 1' are dug. After filling the pit with it, slurry is covered by agricultural waste. In 3-4 months, it gets dried and decomposed. Then it is taken out and applied in field during ploughing before sowing.

Other uses of slurry

The decomposed slurry is rich in nutrition and Ins number of other uses :

(*i*) **In Pisciculture :** The biogas slurry can be applied in fish pond which results in rich growth of fish. As slurry is free from harmful bacteria and pathogens, it reduces fish diseases. The slurry helps in quick growth of algae which is liked by fishes.

Annexure 1. Cost Estimate of 2 cu.m Modified Deenbandhu Biogas plant (for 30 days HRT)

Item	Unit	Rate (approx) Rs	Quantity	Cost (Rs)
Cement	ba	160	13	2080
Sand	cft	4	80	320
Bricks	nos	1	750	750
Ms rod (6mm dia)	kg	14	8	112
Stone chips	cft	4	35	140
Ac pipe 6 dia				70
GI pipe 3/4 dia with sockets				30
Paint	litr	130	0.5	65
Labour charges for pit digging				350
Mason charges	man-days		7	500
Labour charges for construction	man-days		7	260
Miscellaneous like transport etc.				195
Total				4872
				Say Rs. 5,000

Annexure 2. Cost estimate of 2 cu.m Deenbandhu biogas plant (for 40 days HRT)

Item	Unit	Rate (approx) Rs	Quantity	Cost (Rs)
Bricks (I Class)	nos	1.20	1000	1200
Cement	bag	140	14	1960
Stone chips of 1/2 size	cft	5	40	480
Sand	cft	5	40	200
Coarse sand	cft	12	40	480
GI Pipe 3/4 diameter with sockets	30 cm	no	25	1.25
AC pipe 6 diameter	ft	10	6	60
!ron bars (6mm diameter for outlet tank cover	kg	15	7	105
Paint	litre	100	1	100
Labour charges for pit digging	man-days	40	10	400
Labour charges for construction	man-days	40	22	880
Masons charges	90	11	990	
Miscellaneous like				
Local transport etc				250
Total				**7130**

Annexure 3. Cost estimate of 2 cu.m. Shramik Bandhu Biogas Plant (for 40 days HRT)

Item	Unit	Rate (approx) Rs	Quantity	Cost (Rs)
Bamboo	Piece	10	100	1000
Cement	Bag	140	9	1260
Coarse sand	cft	12	30	360
Fine sand	cft	5	30	150
GI	Pipe			25
AC	pipe			40
DPC powder	kg	20	3	60
Binding wire	kg	25	1	25
Copper sulphate	kg	60	0.250	15
Paint	litre	100	1	100
Bamboo weaving charges				1000
Labour charges for pit diggin	gman-days	40	10	400
Mason's charges	man-days	90	8	720
Labour charges for construction	man-days	40	10	400
Miscellaneous like local transport etc				250
Labour for mould digging				300
Total				**6105**

***(ii)* In mushroom production:** The biogas slurry which is free of bacteria and pathogens is a goodmedia for culture of edible fungi.

It improves quality and output of mushrooms.

Advantages

- Highly acceptable and most convenient domestic fuel for rural areas
- Availability of smokeless fuel for cooking and lighting
- Digested manure is available for production in agriculture system. The slurry contains more nitrogen (1.5%) compared to FYM (0.8%)
- Better sanitation due to safe disposal of waste

Limitations

- Shortage of feed stock for the plant
- Fixed dome type biogas plant are severely affected by the local soil condition resulting into construction faiiure
- Pressure variation in the biogas plants affects the distribution of the gas to the utility point
- Low production of gas in winter season
- Slurry handling : Wet slurry is to be dried before its transportation to the field
- Choking of gas distribution line due to water vapor
- Shortage of water for mixing the cow dung for feeding into the plant
- Low methane content in the gas
- High retention time

Structure

(1) Qualitative and quantitative analysis show that the empirical formula for methane is CH_4

(2) Vapour density is 8

(3) Therefore the molecular weight is 16

(4) Hence the molecular formula is CH_4

(5) Since carbon is tetra valent and hydrogen monovalent, the only possible structure is all the hydrogen atom if methane are similarly placed within the molecule.

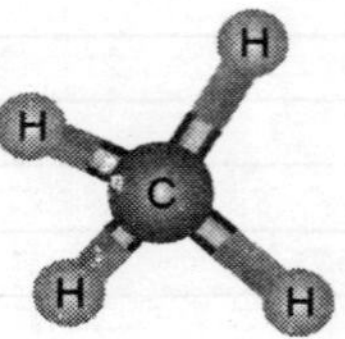

3D Model of Methane Molecule

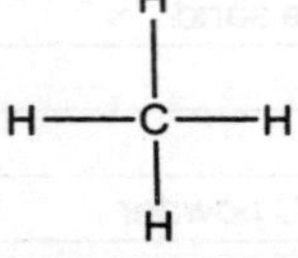

The Structure of Methane Molecule

(6) Methane forms only substitution products. This fact proves that methane is a saturated hydrocarbon

(7) The electronic formula of methane

```
   H
   •×
H ×• C •× H
   •×
   H
```

The carbon atom is sharing its four valency electrons with each of the four hydrogen atoms.

ETHANE

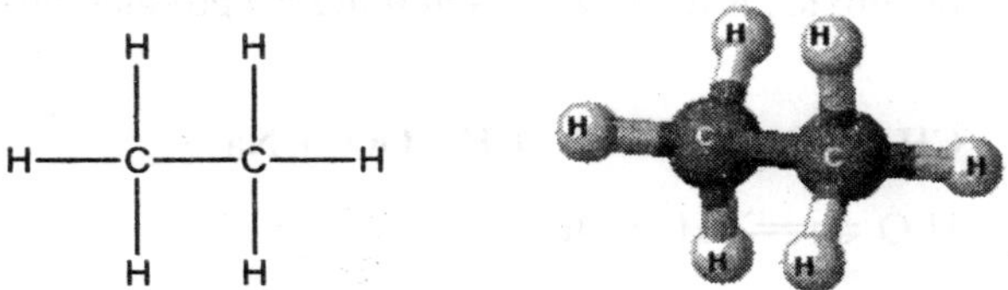

Occurrence

Ethane occurs in natural gas coal gas and petroleum

Preparation

Ethane can be easily prepared by the general method used for the preparation of paraffin

Laboratory Preparation of Ethane

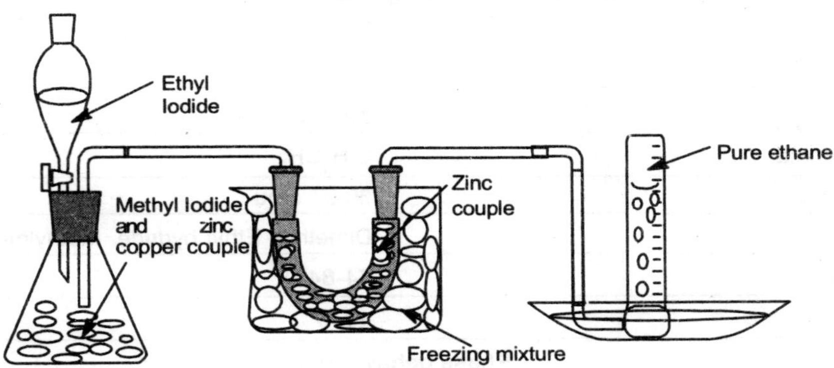

Fig. 5.14. Showing the Laboratory Preparation of Ethane

By reducing ethyl iodide with nascent hydrogen (from zinc-copper couple and aqueous alcohol)

$$C_2H_5I + 2H = C_2H_6\uparrow + HI$$

The other processes are also available

1. ***By heating a mixture of sodium propionate and soda lime***

$$C_2H_5COONa + NaOH(CaO) \rightarrow C_2H_6\uparrow + Na_2CO_3$$

2. **By Wurtz's synthesis** form metallic sodium and dry ethereal solution of methyl iodide

$$2CH_3I + 2Na \rightarrow C_2H_6 + 2NaI$$

This is a useful reaction and also a useful method for preparing paraffin containing as even number of carbon atoms in its molecule, the sodium serves the halogen atoms causing the residual alkyl radicals to combine together thus

$$CH_3 \boxed{I + Na} + \boxed{Na + I}\ CH_3 = CH_3 - CH_3 + 2NaI$$

3. ***By Kolbe's Method***: A concentrated solution of sodium acetate is electrolyzed by using platinum electrode. Hydrogen is evolved at the cathode and a mixture of ethane and carbon dioxide at the anode. The carbon dioxide is removed by bubbling the product through potassium hydroxide solution.

$$CH_3COONa \rightleftharpoons CH_3COO^- + Na^+$$

$$H_2O \rightleftharpoons H^+ + OH^-$$

$$\text{Cathode: } 2H^+ + 2e \rightarrow 2H \rightarrow H_2\uparrow$$

$$\text{Anode: } 2CH_3COO^- - 2e \rightarrow 2C_2H_6\uparrow + 2CO_2$$

5. ***From a mixture of ethylene acetylene and hydrogen:*** A mixture of ethylene acetylene and hydrogen is passed over finely grinded nickel powder about 200°C

$$C_2H_4 + H_2 \rightleftharpoons C_2H_6$$

Properties

Name	Ethane
Chemical formula	CH_3CH_3
Formula weight	30.07 amu
Synonyms	Dimethyl, Ethyl hydride, Methylmethane
CAS number	74-84-0
UN number	1035
Phase behaviour	
Melting point	90.34 K (-182.76 °C)
Boiling point	184.5 K (-88.6°C)
Triple point	91 K (-182°C) 0.000011 bar
Critical point	305 K (32°C) 49 bar
$\Delta_{fus}H$	0.583 kJ/mol
$\Delta_{fus}S$ The Entropy of fusion of a substance represents the increase in the degree of disorder involved in the transition from an organized crystalline solid to the disorganized structure of a liquid. It is denoted as ΔS and normally expressed in J / mol	6.46 J/mol·K
$\Delta_{vap}H$ The heat of vaporization is a physical property of substances. It is defined as the	10 kJ/mol

heat required to vaporize one mole of a substance at its boiling point under standard pressure (101325 Pa). The heat of vaporization is expressed in kJ/mol.	
Solubility	4.7 ml/100 ml water
	Liquid properties
$\Delta_f H^0_{liquid}$ The standard enthalpy of formation of a compound is the change of enthalpy that accompanies the formation of 1 mole of that substance from its component elements, at their standard states (the most stable form of the element or compound at 25 degrees Cels	? kJThe joule (symbol J also called newton metre or coulomb volt is the SI unit of energy and work. The unit is pronounced to rhyme with "tool", and is named in honour of the physicist James Prescott Joule (1818-1889). 1 joule 1 N· 1 m 1 newton · 1 metre 1 k/ mol. The mole (: symbol mol) is one of the seven SI base units and iscommonly used in chemistry. It measures the amount of substance of a system and is defined as the amount of substance that contains as many elementary entities as there are atoms in exactly 0
S^0_{liquid} In chemistry, the standard molar entropy is the entropy content of one mole of substance, under conditions of standard temperature and pressure. By comparing the entropies of products and reactants in a chemical reaction, we can determine whether the reac	126.7 J/mol·K
C_p Heat capacity (abbreviated C or just C also called thermal capacity is the ability of matter to store heat. The heat capacity of a certain amount of matter is the quantity of heat (measured in Joules) required to raise its temperature by one kelvin.	68 J/mol·K
Density	0.548×10^3 kg/ m^3
	Gas properties
$\Delta_f H^0_{gas}$	-83.8 kJ/ mol
S^0_{gas}	? J/mol·K
C_p	52.49 J/mol·K
Flash point	-135°C
Autoignition temperature	515°C
Explosive limits	3-12.5%

Chemical Properties

Combustion: It burns in air (or oxygen) forming CO_2 and H_2O and producing considerable amount of heat

$$2C_2H_6 + 7O_2 \rightarrow 4CO_2 + 6H_2O$$

Oxidation: Controlled oxidation of ethane in the presence of a catalyst, however produces in steps, ethyl alcohol, acetaldehyde and finally acetic acid.

$$C_2H_6 \xrightarrow{[O]} C_2H_5OH \xrightarrow{[O]} CH_3CHO \xrightarrow{[O]} CH_3COOH$$

Halogenation: Ethane reacts with halogen in a manner similar to that of methane to form substitution products the ultimate product being C_2X_6 (where X is a halogen atom). It should be noted that in this case a much greater number of products are obtained due to the presence of greater number (six) of hydrogen atoms in a molecule of ethane (compound to four in methane) and also due to the substitution products of ethane. [Thus, two dichlorothane viz CH_3CHCl_2(gen-dichloroethane) and $CH_2Cl.CH_2Cl$ (Vic dichloroethane are obtained]

$$\underset{\text{Ethene}}{C_2H_6} \xrightarrow[\text{-HCl}]{+Cl_2\,[\text{light}]} \underset{\text{Ethyl Chloride}}{C_2H_5Cl} \xrightarrow[\text{-HCl}]{+Cl_2\,[\text{light}]} \underset{\text{Dichloroethane (two isomers)}}{C_2H_4Cl_2}$$

$$\underset{\text{Dichloroethane}}{C_2H_4Cl_2} \xrightarrow[\text{-HCl}]{+Cl_2\,[\text{light}]} \underset{\text{Trichloroethane}}{C_2H_3Cl_3}$$

$$\underset{\text{Hexa Chloromethane}}{C_2Cl_6} \xleftarrow[\text{-HCl}]{+Cl_2\,[\text{light}]} \underset{\text{Penta Chloromethane}}{C_2HCl_5} \xleftarrow[\text{-HCl}]{+Cl_2\,[\text{light}]} \underset{\text{Tetra Chloromethane}}{C_2H_2Cl_4} \xleftarrow[\text{-HCl}]{+Cl_2\,[\text{light}]} \underset{\text{Trichloroethane}}{C_2H_3Cl_3}$$

Thermal decomposition: When heated with a catalyst in the absence of air at about 500°C ethane decomposes into ethane and hydrogen.

$$C_2H_6 \xrightarrow{\text{Catalyst}(500^\circ C)} CH_2 = CH_2 + H_2$$

Uses of Ethane

1. Ethane is used for making ethylene and epoxyethyle
2. Ethane is an important constituent of coal gas and natural gas which are used for illumination and heating purposes
3. Ethane is converted to C_2Cl_6 hexachloroethane and used as artificial camphor. It can be also used as refrigerant

Structure of Etahne

Next in size after the methane is the ethane, C_2H_6. If we cannot see the atoms of this molecule by co-valent bonds, following the rule of one bond one pair of electrons for each hydrogen and four bonds (four pairs of electrons) for each carbon, we arrive at the structure

```
  H  H            H       H
  ••  ••          |       |
H:C::C:H      H—C——————C—H
  ••  ••          |       |
  H  H            H       H
```

Each carbon is bonded to three hydrogen and to the other carbon. Since each carbon atom is bonded to these other atoms its bonding orbitals (sp^3 orbitals) are bonded towards the corners of the tetrahedron, as in the case of methane the carbon-hydrogen. The carbon-carbon bond arises from overlap of the sp^3 orbitals.

The carbon-hydrogen and carbon-carbon bonds have the same general electrons distribution, being cylindrically symmetrical about a line joining the atomic nuclei (See Fig. 5.15 a-b)

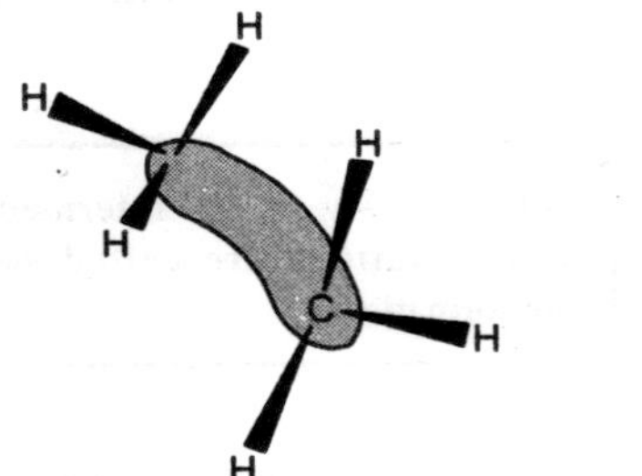

Fig. 5.15(*a*). Ethane molecule carbon—carbon single: σ-bond.

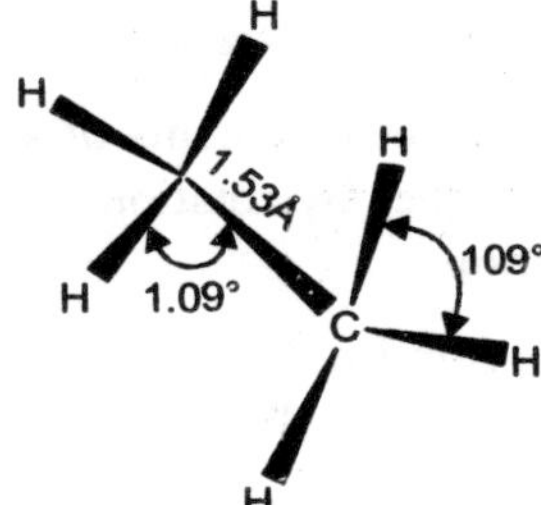

Fig. 5.15(*b*). Ethaner molecule shape and size.

In ethane, then, the bond angles and carbon hydrogen bonds lengths should be very much the same as in methane that is about 109.5°C and about 1.10Å respectively.

Electron diffraction and spectroscopic studies have verified this structure in all respects, giving the following measurements for the molecule, bond angles 109.5° - C – H length, 1.10Å C – C length 1.53Å. Similar studies have shown that, with only slight variations these values and quite characteristic of carbon – hydrogen and carbon – carbon bonds and of carbon bond angles in alkanes.

Free Rotation About The Carbon – Carbon Single Bond Conformation Torsinal Strain

This particular set of bond angles and bond lengths still does not limit us to a single arrangement of atoms for the ethane molecule, since the relationship between the hydrogen of one carbon and the hydrogen of the other carbon is not specified. If we examine models of ethane (see Fig. 5.16). We find that we could have an arrangement between one which hydrogen exactly oppose each other and infinitely of intermediate arrangement, which of these is the actual structure of ethane? The answer is all of them.

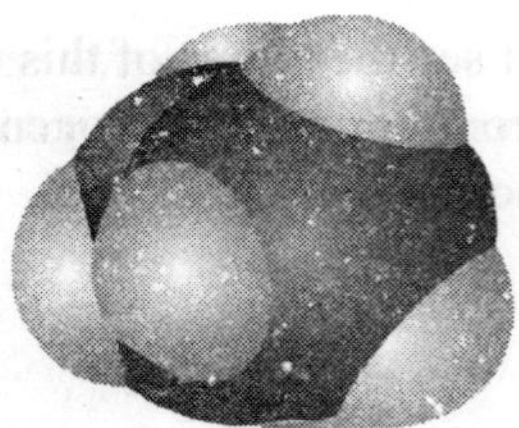

Fig. 5.16A. II Staggered Conformation

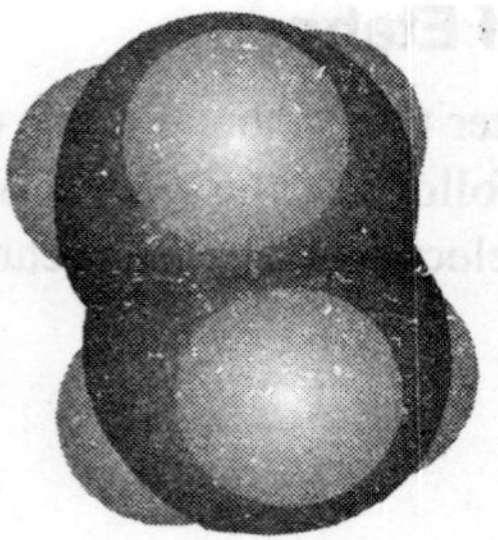

Fig. 5.16B. I Eclipsed Conformation

We have seen that the δ-bond forming the carbon atoms is cylindrically symmetrically about a line joining the two carbon nucleus overlap and hence bond strength should be the same for all these possible arrangements. If the various armaments do not differ in energy from them, the molecule is not restricted to nay of them but can change freely from one to another. Since the change from one to another involves rotation about the carbon - carbon bond. We describe this freedom to change by saying that *this is a free rotation about the carbon - carbon single bond.*

Different arrangements of atoms that can be converted into one another by rotation about single bonds are called conformation.

[The infinity of the intermediate conformations are called screw conformation]

Arrangement I called the eclipsed conformation arrangement II is called staggered conformation.

To represent such conformations we shall often use two kinds of three dimensional formulas

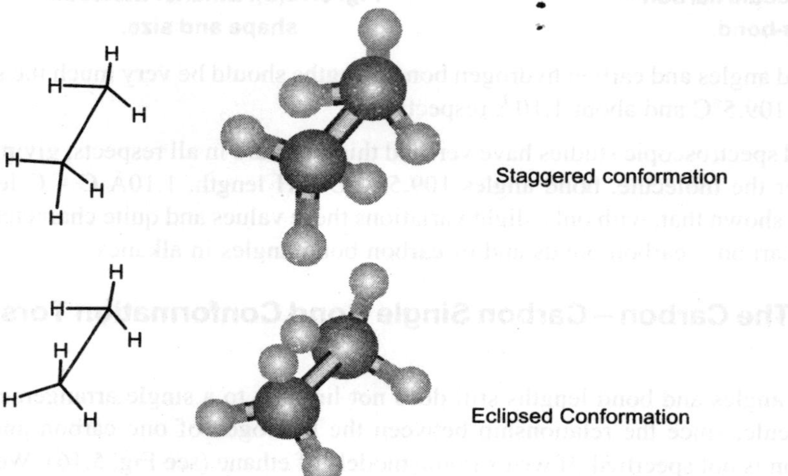

(and iron formulas) and Newman's projections named after **M.S Newman** from Ohio University who first proposed it.

Eclipsed Conformation

Staggered Conformations

The picture is not yet complete, certain physical properties show that rotation is not quite free, and there is an energy barrier of about 3 kcal/mol. The potential energy of molecule is at ethane molecules naturally exist in the most stable staggered conformation or but difficulty, any molecule spends most stable conformations.

How free are ethane molecules to rotate from one staggered arrangement to another? The 3 kcal barrier is not a very high one; even at room temperature the fraction of collision with sufficient energy is large enough that a rapid inter-conversion between staggered arrangements occurs. For most practical purposes we may still consider that the carbon - carbon single bond permits free rotation.

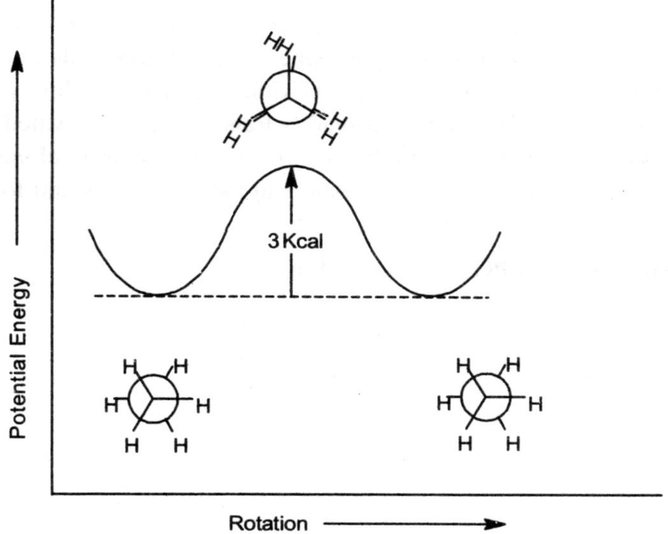

Fig. 5.17. Potential energy changes during roation about the carbon - carbon bond of ethane

The nature of the rotational barrier in ethane is not understood – what is not exactly the same thing – is not readily explained. It is too high to be due merely to Van der Waals forces. Although thrown closer together in the eclipsed conformation then in the staggered conformation, the hydrogen on opposite carbons are not big enough for this to cause appreciable crowding. The barrier is considered to arrive in some way from integration among the electron clouds of the carbon - hydrogen bonds. Quantum mechanical calculations show that the barrier should not exist and so perhaps "lack of understanding" amount of difficulty in paraphrasing the mathematics in physical terms like the bond orbitals in methane the two sets of orbitals in ethane tend to be staggered.

The energy required to rotate the ethane molecule about carbon – carbon bond is called ***torsinal energy***. We speak about the relative instability of the eclipsed conformation – any of the intermediate skew conformation – as being due to torsinal strain.

As the hydrocarbons of ethane are replaced by other atoms or groups of atoms other factors affecting the relative structure of conformations is accompanied by torsinal strain.

PROPANE AND BUTANE

Propane is a colourless, easily liquefied, gaseous hydrocarbon (compound of carbon and hydrogen), the third member of the paraffin series following methane and ethane. The chemical formula for propane is C_3H_8. It is separated in large quantities from natural gas, light crude oil, and oil-refinery gases and is commercially available as liquefied propane or as a major constituent of liquefied petroleum gas (LPG). As with ethane and other paraffin hydrocarbons; propane is an important raw material for the ethylene petrochemical industry. The decomposition of propane in hot tubes to form ethylene also yields another important product, propylene. From propylene such organic chemicals as acetone and propylene glycol are derived. The oxidation of propane to such compounds of carbon, hydrogen, and oxygen as acetaldehyde is also of commercial interest. Although a gas at ordinary atmospheric pressure, propane has a boiling point of -42.1°C and thus is readily liquefied under elevated pressures. It therefore is transported and handled as a liquid in cylinders and tanks. In this form, alone or mixed with liquid butane, it has great importance as a fuel for domestic and industrial uses and for internal-combustion engines.

The industrial preparation of LPG is shown in Fig. 5.18.

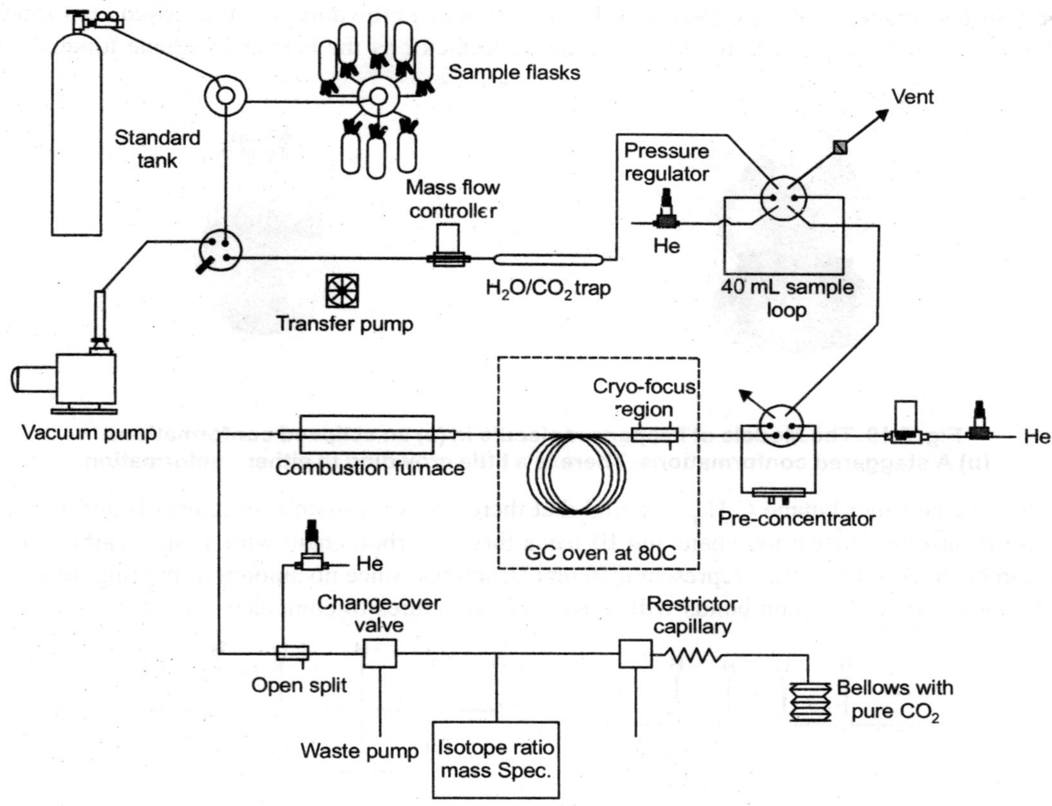

Fig. 5.18. The Industrial preparation of Liquid Petroleum Gas

Structure of Propane and Butane

The next member of alkane family is propane C_3H_8. Again the rule of one bond per hydrogen atom and four bonds for per carbon, we arrive at structure I.

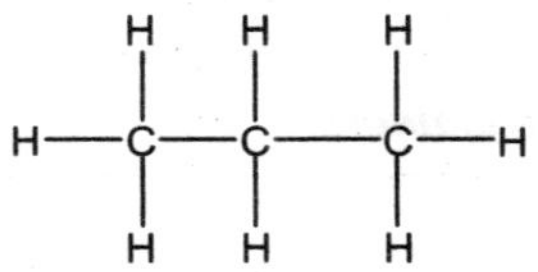

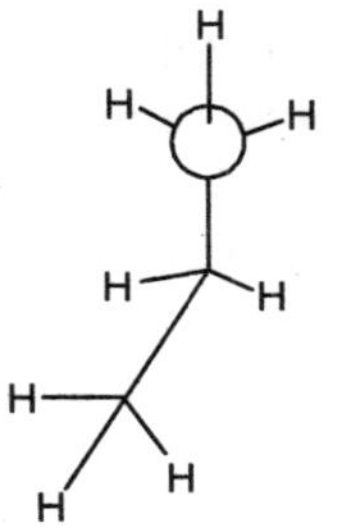

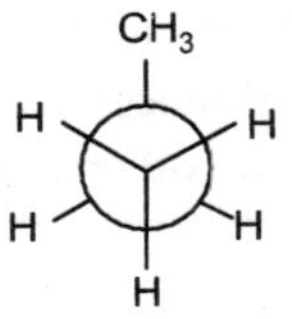

Here rotation can occur about two carbon-carbon bonds and again is essentially free. Although the methyl group is considerably larger than hydrogen, the rotational barrier (3.3 kcal/mol) is only a little higher than for ethane. Evidently there is still not very much crowding in the eclipsed conformation and the rotational barrier is due to chiefly to the same factor as the barrier in ethane torsinal strain (Fig. 5.19).

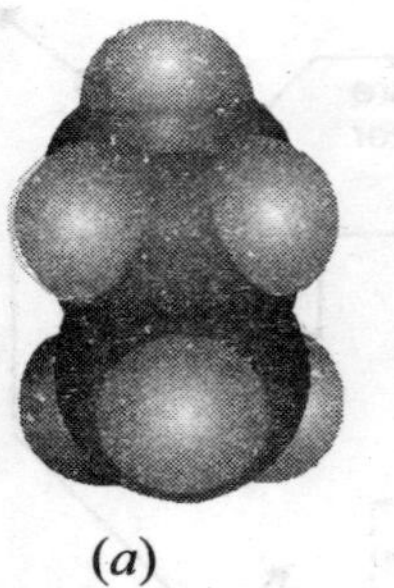

(*a*)

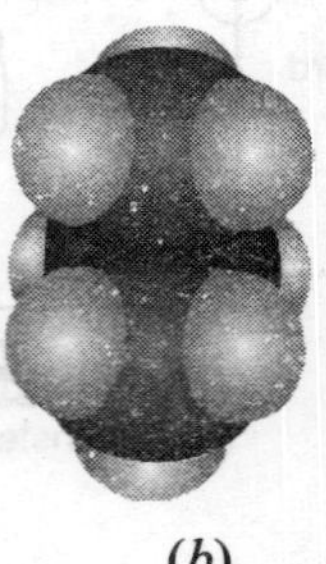

(*b*)

Fig. 5.19. The Models of Propane molecule in (a) an eclipsed conformations (b) A staggered conformations. There is a little crowding in either conformation.

When we consider butane C_4H_{10}, we find that there are two possible structures II and III below. Structure II has the four carbon chain and III has a three - carbon chain with a one - carbon branch. There can be no doubt that they represent different structures, since no amount of moving, twisting or rotating about carbon - carbon bonds will cause three structures to coincide.

```
    H   H   H   H                 H     H     H     H
    |   |   |   |                 |     |     |     |
H---C---C---C---C---H          H--C-----C-----C-----C--H
    |   |   |   |                 |     |     |     |
    H   H   H   H                 H     |     H     H
                                     H--C--H
                                        |
                                        H
```

butane
Structure II

Isobutane
Structure III

We can see that that in the straight chain structure (II) each carbon possesses at least two hydrogen where as in the branched chain structure (III) one carbon possesses only single hydrogen; or we may notice that in the branched chain structure (III) on carbon bond is bonded to more than two other atom.

In agreement with this prediction we find that two components of the same formula C_4H_{10} have been isolated. There can be no doubt that these two substances are different compounds, since they show definite differences in this physical and chemical properties (see table below) for example one boils at 0°C and the other one at – 12°C by definition ***they are isomers***

Boiling Point	0°C	–12°C
Melting Point	–138°C	–154°C
Relative Density	0.622	0.604
Solubility at 100ml of alcohol	1813 ml	1320 ml

Two compounds of formula C_4H_{10} are known as we have drawn two structures to represent them. The next question arises in front is that which formula represents which compound? For the answer we turn to the evidence of isomer number, like methane, the butanes can be obtained; the chlorination can be allowed to proceed until there are two chlorine atoms per molecule. Form the butane boiling point of B.P 0°C six isometric formulas $C_4H_8Cl_2$ are obtained form the butane with – 12°C b.p. only three. We find that we can allow just six dichlorobutanes containing a straight chain of carbon atoms and just three containing a branched chain. Therefore the butane of b.p. 0°C must have the straight chain and butane of b.p. – 12°C must have branched chain, to distinguish between two isomers the straight chain structures called n-butane (termed as normal butane) and the branched chain is called iso-butane

Conformations of n-butane Vander waals Repulsion

Let us look more closely at the n - butane molecule and the conformation in which it exists focusing over attention or the on the middle C — C bond, we see a molecule similar to that of ethane, but with a methyl group replacing one hydrogen on each carbon. As with ethane, staggered conformations have lower torsinal energies and hence are more stable than eclipsed conformations. But due to the presence of the methyl groups, two new points are explained here, first there are several different staggered conformations, and second a factor besides torsinal strain comes into play to affect conformational stabilities.

These is the conformation I, is which the methyl group are as far apart as they can be (dihedral angle 180°) there are two gauche conformations II and III in which methyl groups are only 60° apart (conformations II and III are mirror images) and of the same stability; nevertheless they are different. Make models and convince yourself that is so.

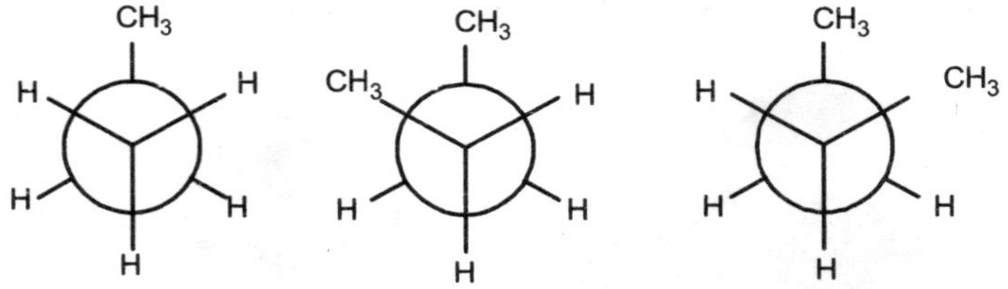

CH3 H H H H H — I Anticonformation

CH3 CH3 H H H H — II Gauche conformation

CH3 CH3 H H H H — III

n-butane

The anti-conformation, it has been is more stable (by 0.8 kcal/mol) then the gauche both are free of torsinal strain. But in a gauche conformation the methyl groups are crowded together, that is are known together closer than the sum of their Van der Waals radies under these conditions. Van der Waals forces are repulsive and raise the energy of the of the conformation. We say that there is Van der Waals repulsion (or steric repulsion) between the methyl groups and that molecule is less stable because of Van der Waals strain (or steric strain) we can see this crowding quite clearly in scale molecules.

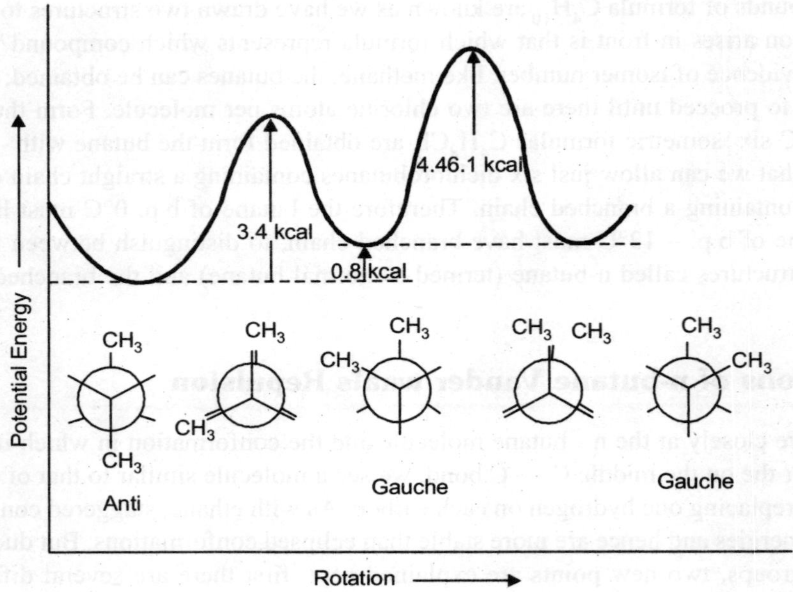

Fig. 5.20. Potential energy changes during rotation about C(2) – C(3) bond of n – butane

Van der Waals strain affect not only the relative stabilization of various staggered conformation but also the heights of the barriers between then – the energy maximum reached when two methyl groups saving past each other – rather than past halogens – is the highest rotational barrier of all, and has been essential at 4.4 - 6.1 kcal/mol. Even so, it is low enough that – at ordinary temperatures at least – the energy of molecular collisions causes rapid rotations; a given molecule exists now in a gauche conformations and the next instant in the anti - conformation.

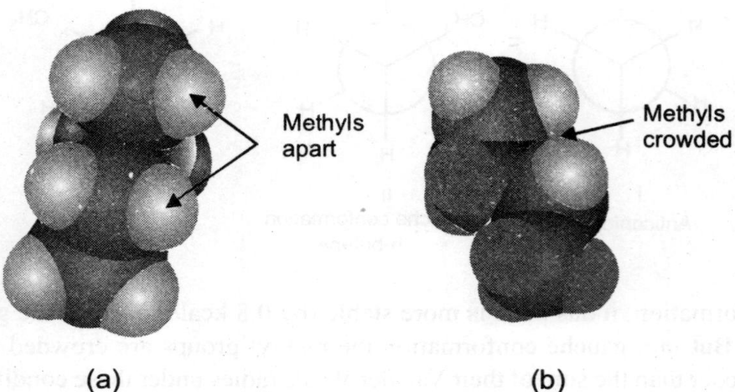

Fig. 5.21. The Models of n-butane in (a) an anti-conformation and (b) a gauche conformation. Note the crowding betwen the methyl groups in (b)

Higher Alkanes and the Homologous Series

If we examine the molecular formula of the alkanes we have been so far reading we see that butane contains one carbon and two hydrogens more than propane, and two halogens more than ethane, and so on.

A series of compounds in which each member differ from the next member by a constant amount and this is called ***homologous series***.

The family of alkanes forms such homologous series and they are called ***homologs***. The family of alkanes forms such a homologous series, the constant difference between successive members being CH_2 . . . . We also notice that in each of these alkanes the number of hydrogen atoms equals two more than twice the number of carbon atoms so that we may write as a general formula for members of this series C_nH_{2n+2}. Other homologous series has their own general formulas.

In agreement with this general formula, we find that the next alkane, pentane has the formula C_6H_{12} followed by hexane C_6H_{14} heptane C_7H_{16} and so on. We should expect that as the number of atoms increase so does the number of possible arrangements of these atoms. As we go up the series of alkanes, we find that this is true; the number of isomers of successive homologs increases at a surprising rate. There are three isomeric pentanes 5 hexanes 9 heptanes and 75 decanes (C_{10}) for the twenty – carbon icosane there are 366319 possible isometric structures.

The carbon skeletons of the isometric pentanes and hexanes are shown below: -

PENTANE

C—C—C—C—C
n - pentane
b.p 36°C

C—C—C—C (C branch on third carbon)
iso-pentane
b.p 28°C

C—C—C (C above and C below the central carbon)
neo-pentane
b.p 9.5°C

HEXANE

C—C—C—C—C—C
b.p 69°C

C—C—C—C—C—C (C branch on fifth carbon)
b.p 60°C

C—C—C—C—C (C branch on fourth carbon)
b.p 50°C

C—C—C—C (C branches on second and third carbons)
b.p 58°C

It is more important to practice drawing the possible isomeric structures that correspond to a single molecular formula. In doing this a set of molecular formula. In doing this a set of molecular modeling software called chem sketch (5.0V commercial) is especially helpful since it will show that many structures appear to be different when drawn on paper are actually identical.

Nomenclature of Higher Alkanes

We have seen that the names methane, propane, butane and pentane are used for alkanes containing respectively one two three four and five carbon atoms gives the names of many larger alkanes except for the first four members of the family, the name is simply derived from the Greek (or Latin) prefix for the particular number of carbons in the alkane this pentane for five, hexane for six heptane for seven and octane for eight and so on.

You should certainly understand the names of at least the first ten alkanes having done this; you will have at the same time essentially learned the names of compounds are closely related. Compare, for example the names propane, propene and propyne for the three carbon alkane, alkene and alkyne.

But nearly every alkane can have a number of isomeric structures and a number of isomeric structures and there are distinguished by the use of prefixes n – butane, iso – butane n – pentane, iso – pentane and neo – pentane. But there are 5 hexanes and 9 heptanes and 75 decanes it would be difficult to device and even more difficult to remember a difficult prefix of each of these isomers. It is obvious that some symmetric method of naming is need.

Preparations of Higher Alkanes

Each of the alkanes, from the methane through n - pentane and iso- pentane can be obtained by the fractional distillation of petroleum and natural gas; neo-pentane does not occur naturally above the pentanes the number of isomers in each homolog becomes so large and the boiling point differences become so small that it is no longer feasible to isolate individual pure compounds; these alkanes must be synthesized by one of the methods below.

In some of the reactions R is used as to represent any alkyl group. In writing these generalized equations; however, we must not loose sight of one important point. A equation involving RCl, to take a specific example, has meaning only in terms of reaction that we can carry out in the lab using as real compound, like methyl chloride. Although typical of alkyl halides a reaction may differ widely in rate or yield depending upon the particular alkyl group actually concerned. We may use quite different experimental condition for methyl chloride might go so slowly or give so many side products as to be completely useless for *tert* - butyl chloride.

Preparation

These Reactions are Extremly Important

1. ***Hydrogenation of Alkanes***

$$\underset{\text{Alkene}}{C_nH_{2n}} \xrightarrow{H_2+\text{Pt,Pd or Ni}} \underset{\text{Alkane}}{C_nH_{2n+2}}$$

2. ***Reduction of Alkyl Halides***

(a) Hydrolysis of Grignard's Reagent

$$R-x+Mg \rightarrow \underset{\substack{\text{Grignards}\\ \text{Reagent}}}{R-MgX} \xrightarrow{H_2O} R-H$$

Example:

$$\underset{\text{sec-butyl bromide}}{H_3CCH_2\underset{|\atop Br}{C}H\text{-}CH_3} \xrightarrow{Mg} \underset{\text{Sec-butyl magnesium bromide}}{H_3C\,CH_2\underset{|\atop Mg\,Br}{C}H\text{-}CH_3} \xrightarrow{H_2O} \underset{\text{n-butane}}{H_3C\,CH_2\underset{|\atop H}{C}HCH_3}$$

(b) Reduction by metal and acid

$$R - X + Zn + H^+ \longrightarrow Zn^{2+} + X^-$$

Example:

$$\underset{\text{Sec- butyl bromide}}{H_3CCH_2\underset{\substack{|\\ Br}}{CH}\cdot CH_3} \xrightarrow{Zn\ H^+} \underset{n\text{-butane}}{H_3C\cdot CH_2\cdot \underset{\substack{|\\ H}}{CH}\ CH_3}$$

3. *Coupling of alkyl halides with organo-metallic compounds*

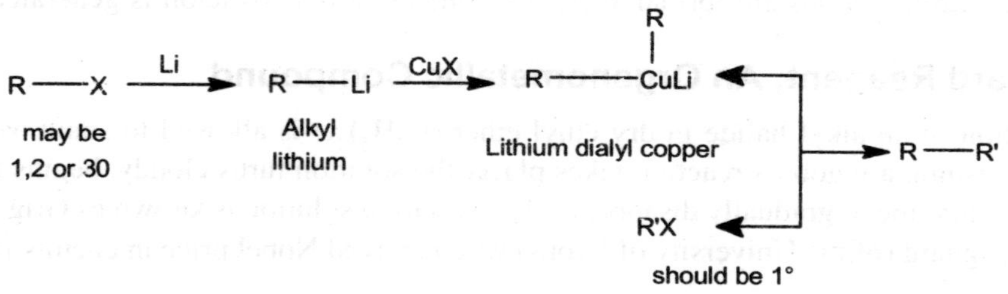

$$R-X \xrightarrow{Li} R-Li \xrightarrow{CuX} R-\underset{}{\overset{\substack{R\\|}}{Cu}}Li \xrightarrow{R'X} R-R'$$

R—X may be 1,2 or 30; R—Li Alkyl lithium; R_2CuLi Lithium dialyl copper; R'X should be 1°

Example:

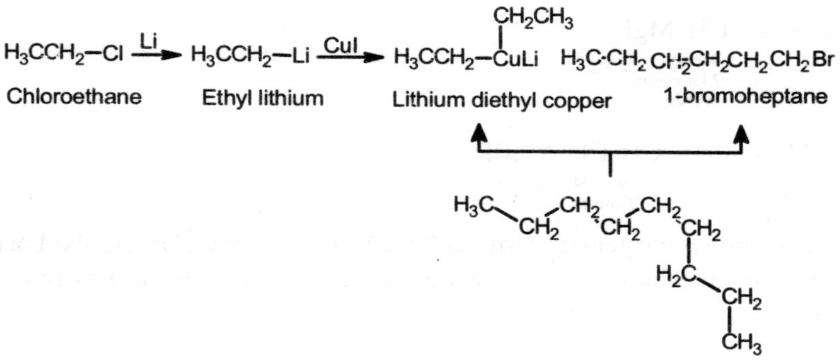

$$H_3CCH_2-Cl \xrightarrow{Li} H_3CCH_2-Li \xrightarrow{CuI} H_3CCH_2-\overset{\substack{CH_2CH_3\\|}}{Cu}Li$$

Chloroethane → Ethyl lithium → Lithium diethyl copper

Lithium diethyl copper + $H_3C\text{-}CH_2CH_2CH_2CH_2CH_2Br$ (1-bromoheptane) → $H_3C-CH_2-CH_2-CH_2-CH_2-CH_2-CH_2-CH_2-CH_3$

n-nonane

$$\underset{\text{tert-butyl chloride}}{H_3C\overset{\substack{CH_3\\|}}{\underset{\substack{|\\Cl}}{C}}\text{-}CH_3} \xrightarrow{Li\ CuI} H\overset{\substack{CH_3\\|}}{\underset{\substack{|\\CH_3}}{C}}-CH_2-\overset{\substack{CH_3\\|\\HC-CH_3\\|}}{Cu}\overset{2-}{}Li$$

$+\ H_3C\text{-}CH_2CH_2CH_2CH_2Br$ (1-bromopentane) →

$$H_3C-\overset{\substack{CH_3\\|}}{\underset{\substack{|\\CH_3}}{C}}CH_2CH_2CH_3$$

2,2-dimethylpentane

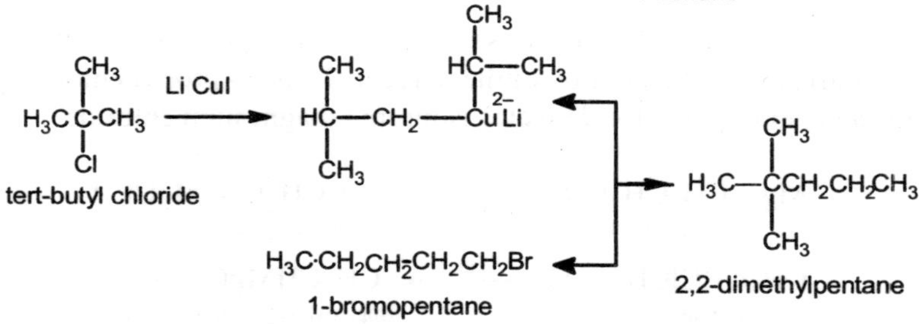

by far the most important of these methods is the hydrogenation of alkenes. When shaken under a slight presence of hydrogen gas in the presence of a small amount of catalyst, alkenes are converted smoothly and quantitatively into alkanes of the same carbon skeleton. The method is limited only by the availability of the proper alkene. This is not a very serious limitation; as we shall see alkanes are really prepared chiefly from alcohols, which in turn can be readily synthesized in a wide variety of sizes and shapes.

Relation of an alkyl halide, either via the Grignard Reagent or directly with metal and acid, involves simply the replacement of a halogen atom by a hydrogen atom; the carbon skeleton remains intact. This method has about the same opportunity as the previous method since like alkenes alkyl halides are generally prepared from alcohols. Where either method could be used, the hydrogenation of alkanes would probably be preferred because of its simplicity and higher yield.

The coupling of alkyl halides with organ metallic compounds is the only one of these methods in which carbon - carbon bonds are formal and a new bigger carbon skeleton is generated.

The Grignard Reagent; An Organometallic Compound

When a solution of an alkyl halide in dry ethyl ether $(C_2H_5)_2O$ is allowed to stand over turnings of metallic magnesium, a vigorous reaction takes place; the solution turns cloudy, begins to boil and the and the magnesium metal gradually disappears. The resulting solution is known as Grignard's Reagent after Victor Grignard (of the University of Lyons) who received Nobel price in chemistry 1912 for this discovery.

It is the most useful and versatile reagents known to the chemist

$$CH_3I + Mg \xrightarrow{\text{Ether}} \underset{\substack{\text{Methyl}\\\text{Manesium}\\\text{Iodide}}}{CH_3MgI}$$

$$\underset{\text{Ethyl Bromide}}{CH_3CH_2Br + Mg} \xrightarrow{\text{ether}} \underset{\substack{\text{Ethyl Magnesium}\\\text{Bromide}}}{CH_3CH_2MgBR}$$

The Grignard reagent has the general formula RMgX and the general name alkyl magnesium halide. The carbon - carbon magnesium bond is covalent electrons from electropositive magnesium; the magnesium halogen bond is essentially ionic

$$R : Mg^+ :\ddot{\underset{..}{X}}:$$

Since magnesium becomes bonded to the same carbon that previously held halogen, the alkyl group remains intact during the precipitation of the reagent. Thus *n* - propyl chloride yields *n* - propyl magnesium chloride and isopropyl chloride yields isopropyl magnesium chloride

$$Mg + \underset{\text{n-propyl alcohol}}{CH_3CH_2CH_2Cl} \xrightarrow{\text{ether}} \underset{\text{n-propyl magnesium chloride}}{CH_3CH_2CH_2MgCl}$$

$$\underset{\text{Isopropyl chloride}}{CH_3CHClCH_3} + Mg \xrightarrow{\text{ether}} \underset{\text{Isopropyl magneisum chloride}}{CH_3CHMgClCH_3}$$

The Grignard reagent is the best known member of a broad class of substance called organometallic compounds in which carbon is bonded to a metal; lithium potassium sodium zinc mercury, lead thallium almost nay known metal. Each kind of organometallic compound has of course, its own set of properties, and its particulars used of compound has, of course, its own set of properties, and its particular uses depend on there. But what ever the metal it is less electronegative than carbon and the carbon - metal bond – like the one in the Grignard Reagent is highly polar. Although the organic group is not a full fledged ***carbanions – an anion is which carbon carries negative charge***, its nevertheless has considerable carbanions character. As well we see organometallic compounds owe their enormous useful ness chiefly to one common quality; they can serve as a source from which carbon is readily transferred with its electrons

$$\overset{\delta}{R}-\overset{\delta}{Ni^{+}}$$

The Grignard Reagent is highly reactive it reacts with numerous inorganic compounds including water; carbon dioxide, and oxygen and with most of organic compounds in many ways of these cases the reaction provides the best way to make a particular class of organic compound.

The reaction with water of form an alkaline is typical of the behaviour of the Grignard reagent – and may of the more reactive organometallic compounds – towards acids. In view of the marked carbanions character of the alkyl group, we may consider the Grignard reagent to be the magnesium salt, RMgX, of the extremely weak acid R – H.

$$RMgX + \underset{\substack{\text{Stronger}\\ \text{Acid}}}{HOH} \rightarrow \underset{\substack{\text{Weak}\\ \text{Acid}}}{R-H} + Mg(OH)X$$

The reaction is simply the displacement of the weaker acid, R – H, from its salt by the stronger acid HOH.

An alkane is such a weak acid that is displaced from the Grignard reagent by compounds that we might ordinarily consider to be very weak acids themselves or possibly not acids at all. Any compound containing hydrogen attached to oxygen or nitrogen is tremendously more acidic than an alkaline and therefore can decompose the Grignard Reagent for an example ammonia or methyl alcohol.

$$RMgX + \underset{\substack{\text{Stronger}\\ \text{Acid}}}{NH_3} \rightarrow \underset{\substack{\text{Wea ker}\\ \text{Acid}}}{R-H} + Mg(NH_2)X$$

$$RMgX + \underset{\substack{\text{Stronger}\\ \text{Acid}}}{CH_3OH} \rightarrow \underset{\substack{\text{Wea ker}\\ \text{Acid}}}{R-H} + Mg(COOH_3)X$$

For the preparation of an alkane one acid is as good as another, so we naturally chose water as the most available and convenient

Coupling of Alkyl Halides with Organometallic Compounds

To make an alkane of higher carbon number than the starting material requires formation of carbon - carbon bonds must, direct by the coupling together two alkyl groups. The most versatile method of doing the late 1960 by J Corey and Herbert House Harvard University and Massachusetts Institute of Technology.

Coupling takes place in the reaction between a lithium dialkyl copper R_2CuLi and an alkyl halide R′X (R′ stands for an alkyl group that may be some as or different from R)

$$R_2CuLi + R'X \longrightarrow R\text{-}R' + RCu + LiX$$

Lithium Dialkyl copper; Alkyl Halide; → Alkane

Alkyl lithium RLi is prepared from an alkyl RX_1 in much the same way as Grignard reagent. To it is added cuprous halide CuX and then finally alkyl halides RX and R′X

$$R\text{—}X \xrightarrow{Li} R\text{—}Li \xrightarrow{CuX} R_2CuLi \xrightarrow{R'X} R\text{—}R'$$

R—Li: Alkyl Lithium; R_2CuLi: Lithium

Halogenation

As we might expect halogenation of the higher alkanes is essentially the same as the halogenation of methane. It can be complicated, however, by the formation of mixtures of isomers.

Under the influence of ultra - violet rays or at 250-400°C chlorine, bromine converts alkanes into chloroalkanes (alkyl chloride) (alkyl chloride) or bromo alkanes an equivalent amount of hydrogen chloride or hydrogen bromide is formed at the same time. When diluted with an inert gas, an in an apparatus designed to carry away the heat produced, fluorine has recently been found to give similar results, as with methane, identification does not that place at all.

Depending upon which hydrogen atom is replaced and any of a number isomeric products can be formed from a single alkane. Ethane can yield only one **halothane;** propane *n*- butane and isobutene can yield two isomers each – *n*– pentane can yield three isomers and iso pentane four isomers. Experiment shows that on halogenation an alkane yields a mixture of all possible isomeric replacements. For example for chlorination

$$H_3CCH_3 \xrightarrow[\text{light } 25°C]{\text{Chlorine}} H_3CCH_2\text{—}Cl$$

chloroethane, b.p 13°C

$$H_3CCH_2CH_3 \xrightarrow[\text{light } 25°C]{\text{Chlorine}} H_3CCH_2CH_2\text{—}Cl + H_3CCH(Cl)\text{-}CH_3$$

propane; 1-chloropropane 45%; 2-chloropropane 55%

$$H_3CCH_2\cdot CH_2CH_3 \xrightarrow[\text{light } 25°C]{\text{Chlorine}} H_3C\cdot CH_2CH_2CH_2Cl + H_3CCH_2CH(Cl)\cdot CH_3$$

butane; 1-chlorobutane 28% b.p 78°C; 2-chlorobutane 72% b.p 68°C

$$\underset{\text{isobutane}}{H_3CH\overset{CH_3}{\overset{|}{C}}CH_3} \xrightarrow[\text{light } 25°C]{\text{Chlorine}} \underset{\substack{\text{1-chloro-2-methylpropane} \\ \text{b.p } 69°C,\ 64\%}}{H_3C\overset{H_3C}{\overset{|}{C}}HCH_2-Cl} + \underset{\substack{\text{2-chloro-2-methylpropane} \\ \text{b.p } 51°C,\ 36\%}}{H_3C\overset{CH_3}{\overset{|}{\underset{Cl}{\underset{|}{C}}}}-CH_3}$$

Bromination gives the corresponding bromides but in different proportions

$$\underset{\text{ethane}}{H_3CCH_3} \xrightarrow[\text{light } 127°C]{\text{Bromine}} \underset{\text{bromoethane}}{H_3CCH_2Br}$$

$$\underset{\text{propane}}{H_3C\text{-}CH_2CH_3} \xrightarrow[\text{light } 127°C]{\text{Bromine}} \underset{\text{1-bromopropane}}{H_3CCH_2\text{-}CH_2Br} + \underset{\text{2-bromopropane}}{H_3C\underset{Br}{\underset{|}{C}}H\text{-}CH_3}$$

$$\underset{\text{isobutane}}{H_3C\overset{CH_3}{\overset{|}{C}}H\text{-}CH_3} \xrightarrow[\text{light } 127°C]{\text{Bromine}} \underset{\text{1-bromo-2-methylpropane}}{H_3C\overset{H_3C}{\overset{|}{C}}HCH_2\text{-}Br} + \underset{\text{2-bromo-2-methylpropane}}{H_3C\overset{CH_3}{\overset{|}{\underset{Br}{\underset{|}{C}}}}CH_3}$$

Although butyl chloride and bromination yield mixture of isomers, the results given above show that the relative amounts of the various isomers differ markedly depending upon the halogens used chlorination gives mixtures in which no isomers greatly predominates, in bromination by contrast, one isomer may predominate to such extent as to be almost the only product, making up 97 - 99% of the total mixture. In bromination, there is a high degree of selectivity as to which hydrogen atoms are to be replaced which reuse *exceptions halogenation of alkanes is not suitable for the laboratory preparation of alkyl halides.* Conveniently and form readily available alternative ways to make alkyl halides conveniently and from readily available precursors.

We begin our study of organic reaction with halogenation of methane and other alkanes, not for its utility in laboratory synthesis; synthesis in only one aspect of organic chemistry. But this reaction offers only an easily understood able approach to principles underlying all reactions we shall study. Alkanes are simple compounds. The mechanism is well - understood and based upon evidence that we can readily grasp. We can deal regularly and quantitatively with the matter of relative rules and orientation. Since values of E_{act} and DH are accurately known. The nature of the transition state is unclouded by uncertainty as to the role played by the solvent. Finally the study of the free radicals is, in itself an important part of organic chemistry.

On an industrial scale, chlorination of alkanes is important. For many purposes – for example use as a solvent, a mixture of isomers is just as suitable as and much cheaper than a pure compound. It may be even worthwhile, when necessary, to separate a mixture of isomers if each isomer can then be marked.

Mechanism

Halogenation of alkanes proceeds by the same mechanism as halogenation of methane.

(1) $X_2 \xrightarrow{250-400°C \text{ or U-Rays}} 2x$ (Chain – initiation step)

(2) $X + 2H \rightarrow HX + R$

(3) $R + X_2 \rightarrow Rx + X$

(2) and (3): chain propegatory step

then (2) (3) (2) (3) etc until finaily a chain is terminated.

A halogen atom obstructs hydrogen from the alkane (RH) to form an alkyl radical (R). The radical in turn abstracts a halogen atom from a halogen molecule to yield the alkyl halide (RX). Which alkyl halide is obtained depends upon which alkyl radical is formed

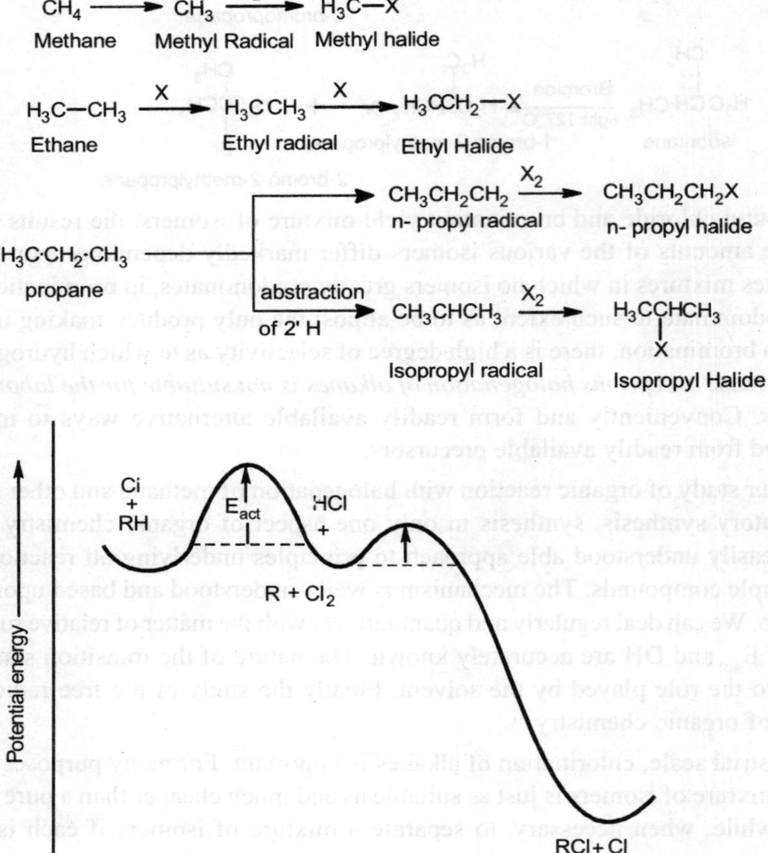

Fig. 5.22. The Potential Energy changes during the progress of reaction: chlorination of an alkane. Formation of the radical is the rate - controlling step

This in turn depends upon the alkane and which hydrogen atom is abstracted from it. For example, n-propyl halide is obtained from a n-propyl radical, formed from propane by abstraction of a passing hydrogen; isopropyl halide is obtained from an isopropyl radical formed by an secondary hydrogen.

How fast an alkyl halide is formed depends upon how fast the alkyl radical is formed. Here also, as was the care with methane of the two chain propagatory steps, step (2) is more difficult than step (3) and hence controls the rate of overall reactor. Formation of the alkyl radical is difficult but once formed the radical is ready converted into the alkyl halide.

Orientation of Halogenation

With us background let us turn to the problem of orientation, that is, let us examine the factor that determines wherein a molecule reaction is not likely to occur. It is a problem that we shall encounter again and again, where we study a compound that offers more than one reactive site to attach by a reagent. It is an important problem, because orientation determines what product we obtain.

As an example let us take chlorination of propane, the relative rates at which n-propyl radicals and isopropyl radicals are formed. If say iso propyl radicals are formed faster than isopropyl chlorides, then isopropyl chloride will form faster, and will make up a larger fraction of the product. As we can see, n-propyl radicals are formed by abstraction of primary hydrogens and isopropyl radicals by the abstraction of secondary hydrogen.

Thus orientation is determined by the relative competing reactions.

Propane $\xrightarrow{Cl^-}$ n-propyl radical $\xrightarrow{Cl_2}$ n-propyl chloride

Propane $\xrightarrow{Cl^-}$ (Abstraction of 2° H) Isopropyl radical $\xrightarrow{Cl_2}$ Isopropyl chloride

In this case we are comparing the rate of abstraction of hydrogen with the rate of abstraction of secondary hydrogen. What are the factors that determine the rates of these two reactions and which of these factors may the two reactions differ?

First of all, there is the collisions frequency; this must be the same for the two reactions, since both involve collisions and the same particles; a propane molecule and a chlorine atom.

Next, there is the probability factor. If a primary hydrogen is to be abstracted the propane molecule must be so oriented at the time of collision that the chlorine atoms strikes a primary hydrogen; if a secondary hydrogen is to be abstracted, the propane must be obtained that the chlorine collides with a secondary hydrogen. Since there are six primary hydrogen and only two secondary hydrogen in each molecule, we might estimate that the probability factor furnishes abstraction of primary hydrogen by the ratio 6: 2 or 3:1

Conceding only frequency and our guess about probability factor can predict that chlorination of propane would yield n-propyl chloride and propyl chloride and isopropyl chloride in the ratio 3:1. As shown in the reaction of hydrogenation, however, the two chlorides are formed in rightly equal amounts that are in the ratio of about 1 : 1 or 3 : 2. The proportion of isopropyl chloride is about three times as great as predicted. Evently about three times as many collisions with secondary hydrogens are successful as collisions with primary hydrogen.

Chlorination of isobutene presents a similar problem. In this case of abstraction of one of the nine primary hydrogens leads to the formation of isobutyl chloride, where as abstraction of a single *tert*-hydrogen leads to the formation of the *tert*-butyl chloride. We would estimate, then that the probability factor favours.

isobutane $\xrightarrow{Cl_2}$ abstraction 1°H → Isobutyl radical $\xrightarrow{Cl_2}$ Isobutyl chloride

isobutane $\xrightarrow{Cl_2}$ abstraction 3°H → tert-butyl radical → tert-butyl chloride

Formation of isobutyl chloride by the ratio of 4:1. The experimental results given under the sub-head halogenation of alkanes shows that the ratio is roughly 2:1 or 9: 4.5. Evidently, about times as many collisions with the tertiary hydrogen are successful as collisions with the primary hydrogens are successful as probably means that E_{act} is less for abstraction of a tertiary hydrogen than abstraction of a primary hydrogen and, in fact even less than for abstraction of a secondary hydrogen.

Study of the chlorination of a great many alkanes has that there are typical results. After allowance there are typical results. After allowance is made for differences in the probability factor, the ratio of

abstraction of hydrogen atoms is always found to follow these sequence 3°>2°>1°. At room temperature, for example, the relative rates per hydrogen atom are 5.0: 3.8:1.0. Using these values we can predict quite well the ratio of isomeric chlorination products from a given alkane. For example.

$$CH_3CH_2CH_2CH_3 \xrightarrow{Cl_2\ \text{light}} \underset{\text{n-butyl chloride}}{CH_3CH_2CH_2CH_2Cl} + \underset{\text{sec-butyl chloride}}{CH_3CH_2CHClCH_3}$$

$$\frac{\text{n-butyl chloride}}{\text{sec-butyl chloride}} = \frac{\text{no of 1°H}}{\text{no of 2°H}} \times \frac{\text{reactivity of 1°H}}{\text{reactivity of 2°H}}$$

$$= \frac{6}{4} \times \frac{1.0}{3.8}$$

$$= \frac{6}{15.2} \text{ equivalent to } \frac{28\%}{72\%}$$

In spite of the differences in reactivity, chlorination rarely yields a great hold of any single isomer. In nearly every alkaline, as in the examples we have studied, the less reactive hydrogen are the more numbers; their lower reactivity is compensated by a higher probability factor, with the result that appreciable amounts of every isomer are obtained.

The same sequence of reactivity 3°>2°>1° is found in bromination; but with normally larger reactivity ratios At 127°C for example, the relative rates per hydrogen atom are 1600:82:1. Here differences in reactivity are so marked as vastly to out weigh probability factors.

Relative Reactivities of Alkanes Towards Halogens

The best way to measure the relativities of different compounds towards the same reagent is by the method of competition, since this permits an exact quantitative comparison under identical reaction conditions. Equimolecular amounts of two compounds to be mixed together and allowed to react with a limited amount of a particular reagent, since there is not enough reagent for both compounds, the two complete with each other. Analysis of the reaction products shows which compound has consumed more of the reagent and hence is more reactive.

For example, if equimolecular amounts of methane and ethane are allowed to react with a small amount of chlorine, about 400 times as much ethyl chloride as methyl chloride is obtained, showing that ethane is 400 times more reactive than methane. When allowance is made for the relative numbers of hydrogens in two kinds of molecules, we see that each hydrogen of ethane as about 200 times as reactive as hydrogen of methane.

$$CH_3Cl \xleftarrow{CH_4} \underset{\text{light } 25°C}{Cl_2} \xrightarrow{C_2H_6} \underset{400}{C_2H_5Cl}$$

Data obtained from similar studies of other compounds are constituents with this simple generalization; the relativity of hydrogen depends upon its class and not upon the alkane to which it is attached. Each primary hydrogen of propane for example, is about as easily obtained as each primary hydrogen in n-butane or isobutene, each secondary hydrogen of propane, about as easily as each secondary hydrogen of n-butane or n-pentane, and so on.

The hydrogen atoms of methane, which fall into a special class, are less reactive than primary hydrogen as shown by the above competition with ethane.

Ease of Abstraction of Hydrogen Atoms Energy Activation

At this stage we can summaries effect of structure on hydrogenation of alkanes in the following way. The controlling steps in halogenation is abstraction of hydrogen by a halogen atom

$$R-H+X \rightarrow H-X+R$$

The relative ease with which the different classes of hydrogen atom are abstracted is

Ease of abstraction of hydrogen atoms $3° > 2° > 1°. > CH_4$

This sequence (a) to the various hydrogens within a single alkane and hence governs the orientation of reaction and (b) to the hydrogens of different alkanes and hence governs relative reactivities.

Earlier we concluded that these differences in ease of abstraction that these differences in rate between closely related reactions are probably due to differences of E_{act}. By study of halogenation of a series of temperatures, the values of E_{act} listed in table below. In arrangement with tentative conditions, the increasing rate of reaction along the series methyl 1°, 2°, 3° is parallel by a decreasing $E_{act.}$ In chlorination the differences in E_{act} like the differences in rate are small; in bromination both differences are large.

Energies of Activation Kcal/mol

$$R-H+X \rightarrow R+H-X$$

R	X = Cl	X = Br
CH_3	4	18
1°	1	13
2°	0.5	10
3°	0.1	7.5

We have that the larger the E_{act} of a reaction, the larger the increase in rate brought about by a given rise in temperature. We have just found the differences in ratio of abstraction among primary, secondary and tertiary hydrogen are due to the differences in E_{act} We predict, therefore, that a rise in temperature should seed up abstraction of primary hydrogen (with the largest E_{act}) least; the three classes of hydrogen should then display more nearly the same reactivity. Thus leaving out effect has indeed been observed as the temperature is raised, the relative ratio per hydrogen atom charge from 5.0: 3.8: 1.0 towards 1: 1:1. At very high temperatures virtually every collision has enough energy for abstraction of even primary hydrogens. It is generally true that when the temperature is raised a given reagent becomes less selective in the position of its attack; conversely, as the temperature is lowered it becomes more selective.

How can we account for the effect of structure on ease of abstraction of hydrogen atoms? Since this is a matter of E_{act} we must look for our answer, as always in the transition state. To do this, however, we must first sift our focus from the hydrogen atom being abstracted to the radical being formed.

Stability of Free Radicals

In the table of homolysis we find the homolytic dissociation energies of the bonds that hold hydrogen atoms to a number of groups. These are the DH values for the following reactions:

$H_3C—H \longrightarrow CH_3 + H$ $\Delta H = 104$ Kcal

$H_3CCH_2—H \longrightarrow H_3C—CH_2—CH_2 + H$ $\Delta H = 98$ Kcal
A 1° radical

$H_3CCH_2CH_2—H \longrightarrow H_3CCH_2CH_2—H$ $\Delta H = 98$ Kcal
A 1° radical

$H_3C—CH(H)CH_3 \longrightarrow H_3C\text{-}CHCH_3 + H$ $\Delta H = 95$ Kcal

$H_3CC(CH_3)(H)CH_3 \longrightarrow H_3C\text{-}C(CH_3)CH_3 + H$ $\Delta H = 92$ Kcal
A 3° radical

By definition, this bond dissociation energy is the amount of energy that must be supplied to convert a mole of alkane into radicals and hydrogen atoms. As we can see, the amount of energy needed to form the various classes of radicals decreases in the order

$CH_3 > 1° > 2° > 3°$

$R - H \rightarrow R + H$ DH = homolytic bond dissociation energy

If less energy is needed to form one radical than another, it can only mean that relative to the alkane from which is formed the one radical contains less energy than the other this is to say, is more stable.

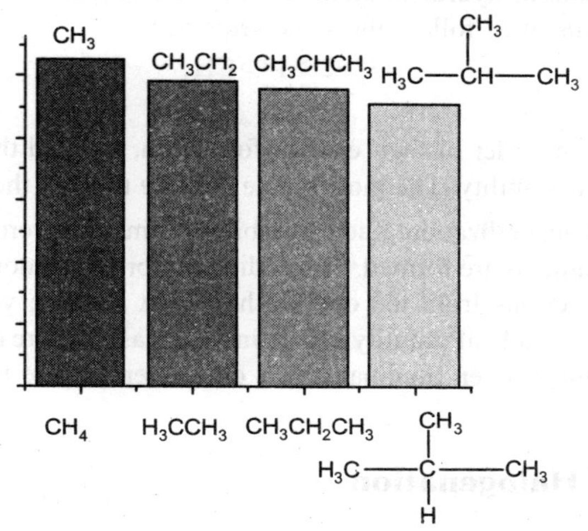

Fig. 5.23. The stabilities of free radicals (the plots are aligned with each other for easy comparison.

We are not attempting to compare the absolute energy of say methyl alcohol and ethyl radicals; we are simply saying the difference in energy between methane and methyl radicals. When we compare stabilities of free radicals, it must be understood that out standard for each radicals is the alkane from which it is formed. As well we shall see this precisely, the kind of stability that we are only interested in.

Stability of free radical $3° > 2° > 1°. > CH_4$

Ease of Formation of Free Radicals

Let us return to the halogenation of alkanes orientation and reactivity, we have seen are governed by the relative ease with which the different classes of hydrogen atoms are abstracted. But by the definition, the hydrogen being abstracted and the radical being formed belong to the same class. Abstraction of primary hydrogen yields a primary radical; abstraction of secondary hydrogen yields a secondary radical and so on.

$$CH_3CH_2CH_2—H + Br \longrightarrow H—Br + CH_3CH_2CH_2$$

$$H_3C\underset{\displaystyle H}{\overset{|}{C}}H{\cdot}CH_3 + Br \longrightarrow H—Br + CH_3CHCH_3$$

$$H_3C—\underset{\displaystyle H}{\overset{\displaystyle CH_3}{C}}—CH_3 + Br \longrightarrow H—Br + H_3C—\underset{\displaystyle H}{\overset{\displaystyle CH_3}{C}}—CH_3$$

If the ease of abstraction of hydrogen atoms follows the sequence $3° > 2° > 1°. > CH_4$, then the ease of formation of free radicals must follow the same sequence

Ease of formation of free radical $3° > 2° > 1°. > CH_4$

In listing free radicals in order of their ease of formation, we find that we have at the same time listed them in order of their stability. The more stable the free radical, the more easily it is formed.

This is an easily useful generalization. Radical stability seems to govern the orientation and reactivity in many reactions where radicals are formed. The addition of bromine atoms to alkanes for examples is a quite different set of reactions from the one we have just studied; yet there too, orientation and reactivity can be governed by radical stability. (Even in these cases where other factors steric hindrance, polar effects are significant or even dominant, it is convenient to use radical stab'lity as a point of departure.

Transition State of Halogenation

Is it reasonable that more stable radical shall be formed more easily? We have already seen that these differences in reactivity towards halogen atoms are due chiefly to differences in E_{act}, the more stable the radical, then, the lower the E_{act} for its formation. This, in turn, means that the more stable the radical, the more stable the transition state leading to its formation – both stabilities being measured, as they must be, against the some standard, the reactants.

> Remember E_{act} is the difference in energy content between reactant and transition state.

Examination of the transition state show that this is exactly what we would expect, as we saw before, the hydrogen - halogen bond is partly formed and the carbon - hydrogen bond is partly broken. To the extent that the bond is broken, the alkyl group process character of the free radical it will become, factor that tend to stabilize the free radical and which in terms tend to stabilize the incipient free radical in the transition state.

$$-\overset{|}{\underset{|}{C}}-H + X \longrightarrow \left[-\overset{|}{\underset{|}{C}}\overset{\sigma}{-\!\!-}H\overset{\sigma}{-\!\!-}X\right] \longrightarrow -\overset{|}{\underset{|}{C}} + H-X$$

Reactants Hydrogen has odd electron — Transition state carbon aquaring free - radical character — Product carbon has odd electron

We have seen that the stabilities of free radicals it will become. *Factors that tend to stabilize the free radical tend to stabilize the incipient free radical in the transition state.*

We have seen that the stabilities of free radicals follow the sequencing 3°>2°>1°>CH_3. A certain factor causes the energy difference between isobutene and *tert*- butyl radical for example, to be smaller than propane and the incipient isopropyl radical in its transition rate (see Fig. 5.24).

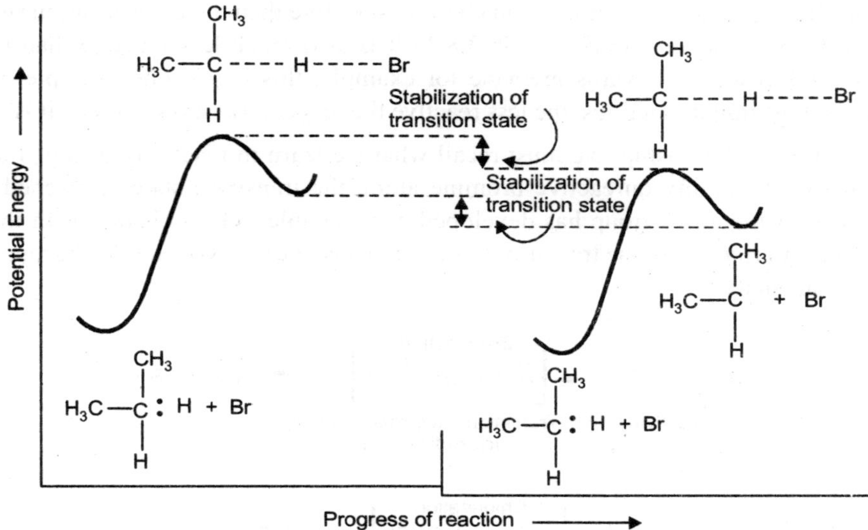

Fig. 5.24. Molecular structure and reactivity: free-radical substitution. The stability of the transition state parallels the stability of the radical; the more stable radical is formed faster (The plots are aligned with each other for easy comparison).

Orientation and Reactivity

Through our study of the organic chemistry, we shall approach the problems of orientation and reactivity in the following way.

Both problems involve comparing the rate of closely related reactions, in the case of orientation, reactions at different sites in the same compound, in the case of reactivity, reactions with different compounds. For such close reactions various in rate are due mostly to differences in E_{act}, *by definition* E_{act} *is the differences in energy content between reaction and transition state.*

We shall examine the most likely structure for the transition state, then, to see what standard feature affect its stability without at the same time affecting by an equal amount the stability of the equal amount of reactants, that is, we shall look for factors that tend to increase or decrease the energy difference between reactants and transition state. Having decided what structural features affect the E_{act}, we shall compare the transition state for the reactions whose rates whose rates we wish to compare, more stable the transition, the faster the reaction.

In many, in not most reactions, where a free radical is formed, as in the present case, the transition state differs from the reactants chiefly in being like the products. It is reasonable, and then that the factor most affecting the E_{act} should be the *radical character* of the transition state. Hence we find that, the more stable the state leading to its formation and the faster the radical is formed.

Reactivity and Selectivity

In this attack on alkanes the bromine atom is much more selective then the chlorine atom (with relative rate factors of 1600:82:1 as compared with 50:3.8:1) It is also much less reactive than the chlorine atom (only 1/375000 reactive) towards methane for example, this is just one example of a general relationship; in a set of similar reactions, the less reactive the reagent, the more relative it is in its attack.

To account for this relationship, we must recall what we learn in reactivity of transition state. In the attack by the comparability unreactive bromine atom, the transition state is reached late in the reaction process, after the alkyl group has developed considerable radical character. In the attack by the highly reactive chlorine atom, the transition state is reached early, when the alkyl group has gained very little radical character.

Bromination

$$R—H + Br \longrightarrow \left[\overset{\sigma}{H_3C}—CH_2\overset{\sigma}{—}Br \right] \longrightarrow R + H—Br$$

low reactivity
high selectivity

Transition state reached late
much radical charater

Chlorination

$$R—H + Cl \longrightarrow \left[H_3C—CH_2—Cl \right] \longrightarrow R + H—Cl$$

High reactivity
Low selectivity

Transition state reached early
little radical charater

Now 'by selectivity' we mean here the differences in rate at which the various classes of free radicals are formed, a more stable free radical is formed faster, we said because the factor that stabilizes it delocalization of the incipient radical in the transition state, the isopropyl radical for example is 3 kcal more stable than the *n*-propyl radical if the radical were completely formed in the transition state. The isopropyl radical, for example is 3 Kcal more stable than *n*-propyl radical, if the radical are completely formed in the transition state the difference in E_{act} would be 3 Kcal. Actually in bromination the difference in E_{act} is 3 Kcal, equal with the limits of experimental error, to the maximum potential stabilization indicating as we expected, a great deal of radical character. In chlorination, by constant the difference in E_{act} is only 0.5 Kcal, indicating only very slight radical character.

A similar a situation exists for reactions of other kinds. What ever the factor responsible for differences in stability among a set of transition states – whether it delocalization of and odd electron or perhaps a charge in crowding of the atoms – the factor with operate more effectively when the transition state is more fully developed that is, when the urgent is less reactive.

Non-Rearrangement of Free Radicals Isotopic Tracer

Our interpretation of orientation was based on an assumption that we have not yet purified that the relative amcunts of isomeric halides we find in the product reflect the reactive rates at which various free radicals were formed form the alkane from isobutene for example, we obtained twice as much isobutyl chloride as *tert-* butyl chloride, and we assume from this that, by abstraction of hydrogen isobutyl radicals are formed twice as first as *tert-* butyl radicals.

Yet how do we know, in this case, that every isobutyl radical that is formed ultimately yields a molecule of isobutyl chloride? Suppose some isobutyl radicals were to change -- by rearrangement of atoms – into *tert-* butyl radicals which then react with chlorine to yield – *tert-* butyl chloride. This supposition is not so far fetched.

$$H_3C-\underset{H}{\overset{CH_3}{C}}-CH_3 \longrightarrow H_3C-\underset{H}{\overset{CH_3}{C}}-CH_3 \xrightarrow[\text{Does not happen}]{\text{rearrangement}} H_3C-\overset{CH_3}{\dot{C}}-CH_3 \longrightarrow H_3C-\underset{Cl}{\overset{CH_3}{C}}-CH_3$$

Isobutane — Isobutane — Isobutyl radical — 2-chloro-2-methylpropane

We, in our present innocence, might think, the doubt it rises is a very real one. We shall shortly see that another kind of reactive intermediate particle, the carbocation, is very prone to rearrange, with less stable ions readily changing into stable ones.

HC Brown and **Glen Russell** (Iowa State University) decided to test the possibility that free radicals, like carbocations, might rearrange, and does the chlorination of isobutene as a good test case, because of the large difference in stability between *tert-* butyl and isobutyl radicals can indeed take place, it should certainly happen here.

What the problem comes down to this; does every abstraction of tertiary hydrogen lead to *tert-* butyl chloride? This we say we could never know, because all hydrogen atoms are exactly alike. But are they? Actually three isotopes of hydrogen exist A portion of ordinary hydrogen ^2H or D *Deuterium,* tertiary hydrogen and ^3H or T, Tritium Protium and deuterium are distributed in nature in the ratio of 500:1*.

Brown and Russell prepared the deuterium labeled isobutene

$$(CH_3)_3C\text{-}D \text{ (isobutane-}d_1) \xrightarrow{Cl_2} \begin{cases} D\text{—}Cl + H_3C\text{—}\dot{C}(CH_3)\text{—}CH_3 \longrightarrow H_3C\text{—}C(CH_3)(D)\text{—}CH_3 \\ H\text{—}Cl + H_3C\text{—}C(CH_3)(D)\text{—}\dot{C}H_2 \xrightarrow{Cl_2} H_3C\text{—}C(CH_3)(D)\text{—}CH_2Cl \end{cases}$$

photochermically chlorinated it and analyzed the product. The DCl: HCl ratio (determined by the Mass spectrophotometer) was found to be equal (within the experimental limit) to the *tert-* butyl chloride isobutyl chloride ratio; clearly every abstraction of a tertiary - butyl chloride and every abstraction of a primary hydrogen (Protium) gave a molecule of isobutyl chloride. Re arrangement of the intermediate free radicals did not occur.

All the existing evidence indicates quite strongly that, although rearrangement of free radicals occasionally happens it is not very common and does not involve simple radicals.

The work of Brown and Russell is just one example of the way in which we can gain insight into a chemical reaction by using isotopically labeled compounds.

---

*tritium the unstable radioactive isotope is present in traces but very pure deuterium is available, at moderate prices called Deuterium oxide (D_2O called heavy water).

CHAPTER

6

Petroleum

Introduction

Petroleum (Greek = *Petrinos* = Stone; Latin = *Oleum* = oil) is an oil obtained from the earth, it is often known as rock oil or mineral oil, in order to differentiate it from the vegetable or animal oil, it differs in composition petroleum is not only one of the important illuminating agent and automobile fuel of the present day, but is also the source of a large number of organic substance of a great commercial value.

Occurrence

Petroleum is found well bellow the surface of the earth in various parts of the world. The important oil fields are Saudi Arabia, USA, Venezuela, Russia (Baku), Iran, Mexico, East Indies, Iraq Rumania, Argentina Columbia, India (Assam, Gujarat) Andaman for refineries

The top ten countries producing and consuming petroleum in the world as given in Fig. 6.1.

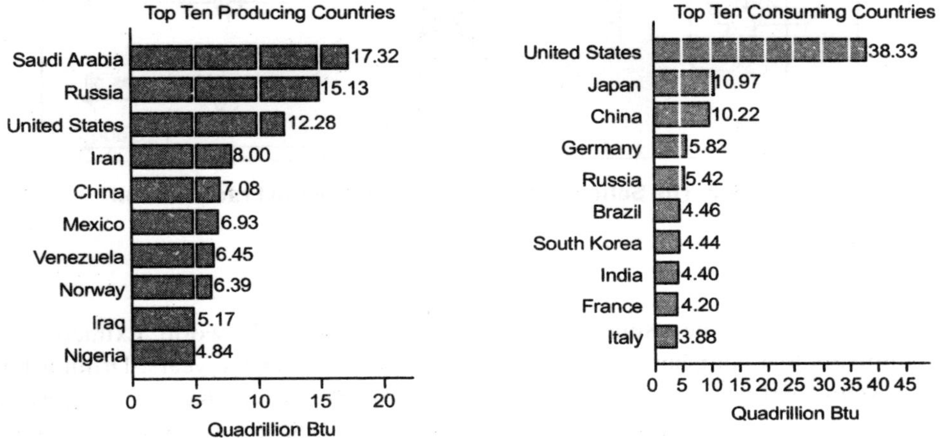

Fig. 6.1.

Two theories have been proposed to explain the occurrence of oil the earth.

(1) ***Organic Theory:*** It has been shown by **Enyler** that the oil is formed by the decomposition of the remains of living creatures (animals and plants) in earth. This view has been confirmed by experimental facts. Petroleum possess slight optical activity and is associated with compounds containing nitrogen and hydrogen

(2) ***Inorganic Theory***: The oil either issues from the earth in the form of a spring or brought to the surface by boring and pumping. The oil is usually accompanied by a natural gas, which consists mostly of the lower paraffin hydrocarbon; the oil also contains in it some sandy soil water and tarryð substance and possesses an unpleasant odour.

The oil cannot be used as such and has to be refined before it is put in the market. The crude oil is carried from the oil field to a distant refining station either through underground pipes by means of big oil tank ships.(the petroleum fractionation is shown in the Fig. 6.2.

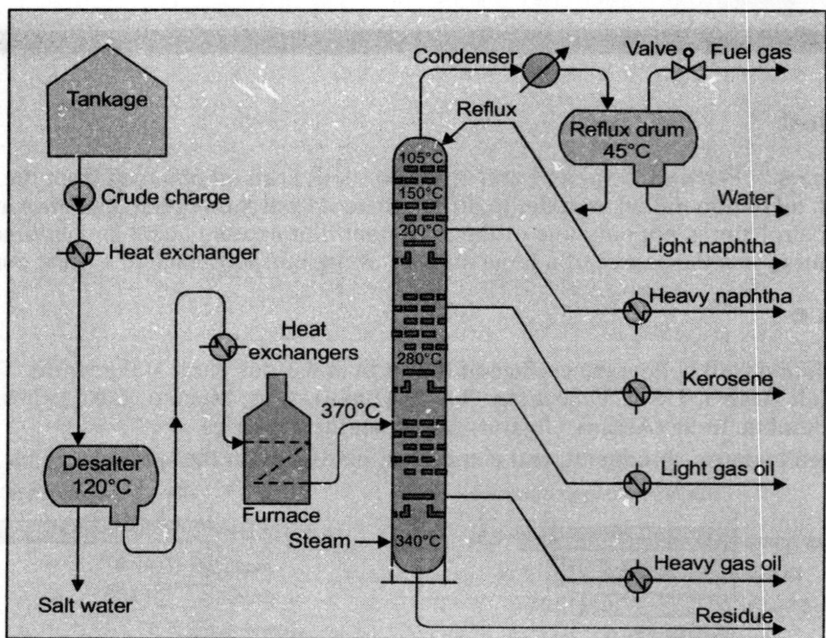

Fig. 6.2. The Schematic Diagram of Petroleum Fractionation

Extraction

Petroleum has been a valuable resource for a long time throughout the world. Extraction of petroleum has been a process that companies have been trying to refine over the years. Drilling for petroleum effects ecosystems and it is for that reason that new processes and enhanced recovery methods are being created. This way it eliminates the evidence of past drilling equipment and the amount of land effected in the processes. In some regions, old rigs is one the few things that remains a problem from in this new age of technology.

Basic Composition of Petroleum

- Carbon — 85-90%
- Hydrogen — 9-15%
- Nitrogen — 0.0.1%
- Sulfur-ppm — 4%
- Oxygen — ppm

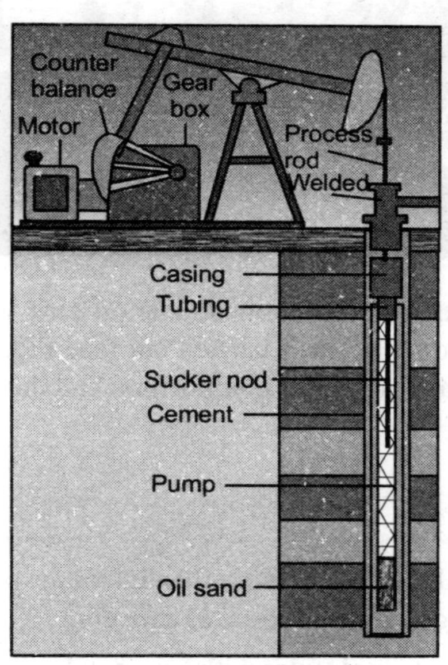

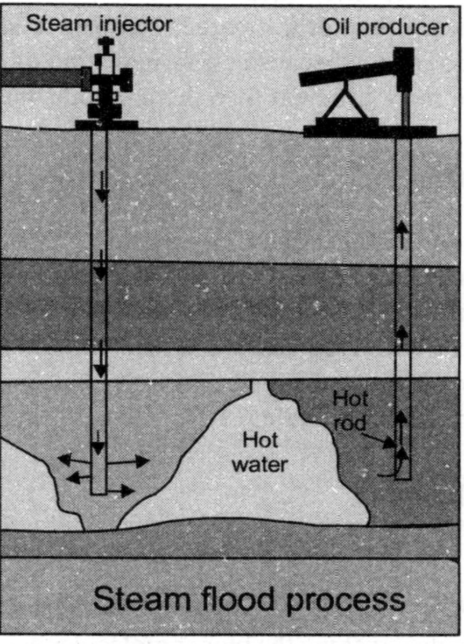

Fig. 6.3. The Petroleum Extraction Steam Flood Process.

With all these new methods of oil extraction, the price had decreased significantly over the past decade alone. Since the first oil drill in Titusville, PA to the present, there are many things that have contributed to the drop in petroleum prices. In 1984, a barrel of petroleum cost roughly \$30, while in 2005 the price is roughly \$45. Even though there is a rise in the price, the worth of the dollar has increased far beyond that, making the price 20 years ago seem less but actually be more. The amount of petroleum extracted now in a single day is amazing. What we were able to extract in one year a century ago, we now can do in one single day! That is why concerns have risen as to when we will run out of petroleum. It has been estimated that the world will run out of oil in the next 63 to 95 years to come.

Drilling rigs have been around for years. People are constantly creating new petroleum extraction methods since the discovery of petroleum. However, it has been found that the easiest way to extract petroleum is by drilling and that is why people are constantly forming new kinds of drilling techniques. From rigs powered by a personos foot, to drills powered to mulls, to Colonel Drakeos steam engine powered drill, to rotary drills, to cable-tool rigs, to modern day hydraulically powered motors used to power drill rigs, new, better and more efficient drills are being created more often then we can even imagine. With every new drill created, companies saver time and money because every new kind of drill is made more efficient then the previous one.

Refining

The crude oil is refined by repeated process of fractional distillation, the underivable impurities are removed and various products of commercial importance are obtained. The oil first warmed to expel the dissolved gases and then continuous fractional distillation is carried out (see Fig. 6.2). The various products obtained thus from the petroleum. Their boiling points composition and the important uses are given in Table 6.1.

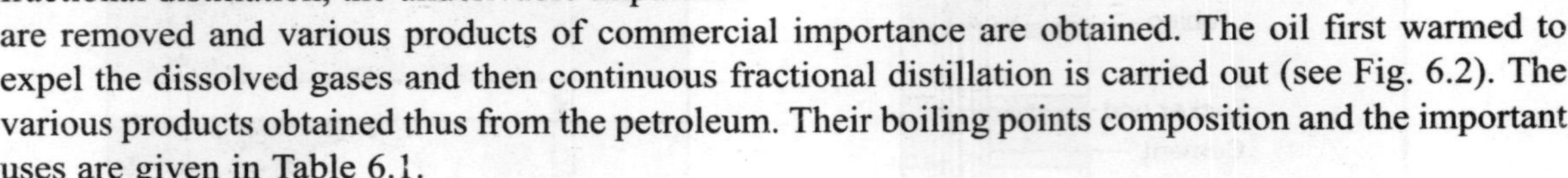

Cracking

The process in which the liquid hydrocarbons with high boiling points are heated *under pressure to decompose them into lower hydrocarbons with low boiling points is known as cracking.*

During cracking the bigger molecules are broken down into molecules with lesser molecular weights.

$$C_{12}H_{16} \rightarrow C_7H_{16} + C_5H_{10}$$

(b.p 214°C) (b.p 98°C) (b.p 38°C)

$$C_{15}H_{32} \rightarrow C_6H_4 + C_5H_{12} + C_3H_6 + C$$

b.p 127°C b.p 69°C b.p 36°C (b.p 48°C)

Table 6.1 : Petroleum Fractionations

No	Name	B.P.	Approximate Composition	Main Uses
I.	**GASES** On Condensation (a) Cymogene (b) Rhigolene	0 - 18° up to 0°C 0-19°C	CH_4, $C_2H_6C_2H_8$ C_4H_{10} and C_5H_{12}	In ice making and as fuel and as fuelan also used as solvent
II.	**CRUDE NAPTHA** On re fractionation gives (a) Petroleum ether (b) Petroleum Gasoline (c) Ligroin (d) Benzene	 18-70°C 70-90°C 90-120°C 120-150°C	$C_6H_{12}C_9H_{20}$ C_5H_{14} C_2H_{25} C_8H_{18} and C_8H_{18} C_7H_{16} C_8H_{18} and C_9H_{20}	As solvent for factory oils, varnishes. Ass fuel for motors, dry cleaning making spirit and metal polish.
III.	**KEROSENE OIL** OR **ILLUMINATING OIL** This function is wasted with conc. Sulphuric acid, Sodium Hydroxide and finally redistilled when pure kerosene oil is obtained	200-300°C	C_1-C_{16} paraffins	For illumination and a fuel as insecticide and for making oil gas in the lab
IV.	**HEAVY OIL** On strong cooling it gives (a) Paraffin (b) Liquid product which on further distillation gives (i) Fuel oil (ii) Lubricating oil	280-360°C Paraffin with more than 20 carbon also and other high molecular hydrocarbon	C_{13}-C_{29} paraffins and other hydrocarbons	As fuel in diesel engines As fuel in industrial and heavy machineries. And for cracking into petrol

In modern distillation plants an arrangement id provided for heating the higher boiling fractions to convert them into lower boiling petrol which is greater demand.

The process of cracking may be carried out in the vapour or liquid phase, with or without a catalyst. The important catalyst use for this purpose is aluminium silicate, alumina, aluminium chloride metallic copper or iron.

The unsaturated hydrocarbons, produced during cracking are easily oxidised in air and this process an unpleasant odour. They may be reduced with hydrogen in the presence of finely divided nickel. The natural petrol.

The process of cracking is of commercial importance as petrol, which is in great demand as modern fuel, may be obtained from the cheap heavy oils for which these petrol - gas by cracking.

Synthetic Petroleum

It is estimated that the sources of the world will not be sufficient to meet for long the very great demand for petrol and gasoline.

Methods have therefore been tried to prepare the petrol synthetically. Petroleum like hydrocarbon has been made artificially by the hydrogenation of coal which goes long way in solving the petrol problem.

It is refined by distillation and serves a very good substitute for petrol. This method is called **Bergius Process**. The yield can be increased by the use of a catalyst (such as aluminium trichloride or ferric oxide) and replacing hydrogen gas by water gas (a mixture of CO and H_2)*

Flash Point

If the kerosene oil is to be used for illuminating purposes, it should not contain low boiling constituents (naphtha etc.), which may cause an explosion during its burning in lamps. In order to check the use of influence by purified for lighting purposes. The government has a fixed fuel standard – Flash point which must be determined before the oil is put in the market.

The flash point of oil is defined as "the lowest temperature" at which it gives of sufficient amount of vapour for form an explosive mixture with air and gives a flash of light when a flame is brought near it.

The flash point is fixed in different countries by the Government and various according to the climatic conditions prevailing their, flashing point permitted in India is 44°C in France 35°C and in England 22.8°C.

The sample point of oil is determined by a special apparatus called Abel's Apparatus (Fig. 6.4).

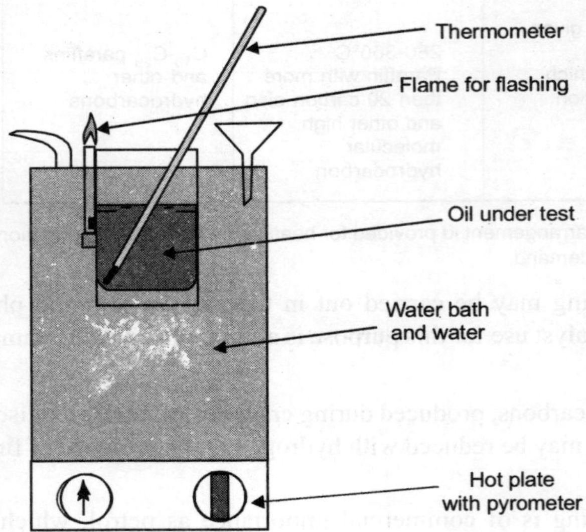

Fig. 6.4. The Abel's Apparatus

*Water gas is a mixture of carbon monoxide and hydrogen and can be prepared by passing over white hot coke

$$H_2O + C \rightarrow \underbrace{CO + H_2}_{\text{Water gas}}$$

The sample of the oil is taken in an small copper beaker B which is heated electrically in a specially constructed water bath A thermometer (T) is introduced into the oil. A small flame is brought near the beaker (flame can be controlled).

As soon as the vapour over the surface of the oil ignites with a momentary flash of the light, the temperature is observed, this is called the flashing point.

A Modern Refinary

The choice of a site for erecting petroleum depends on a number of factors, the chief of which is demands of the nearby regions for petroleum products. It is naturally desirable that there be sources of petroleum near the refinery. But a technical and economical analysis has shown that it is a cheaper to transport petroleum than commercial petroleum products from the refineries near the petroleum fields to the place of their consumptions. That is why present new refineries are going up in regions having absorbed no petroleum deposits, but with a very great demand for petroleum products.

The demand of an economical region for petroleum products of a definite the choice of the direction of petroleum processing at a refinery.

The quality of the feed stock is no longer so significant as previously because process have been developed that enable most petroleum products including high – grade ones to be produced from any petroleum.

Depending on the assortment of products obtained and the nature of the flow diagram, we designated refineries producing fuel, fuel and oil and petrochemical production process. Refineries are also characterized according to the depth of processing the petroleum, ie the depth of processing the petroleum that is the level of cutting light – end products, we shall discuss the treat different petroleum in flow diagrams (Fig. 6.5).

Processing of Petroleum for Fuel Production with Low Level of Cutting Light End Production

Refineries following this diagram (see Fig. 6.5) were erected in 1970 in regions having no other sources of organic fuel (coal, natural gas) and where fuel oil is used for supplying power plants.

The petroleum is first **desalted and dehydrated** and then it is distilled in the atmospheric distillation plants with the production of the gasoline, kerosene and diesel distillates. The gasoline distillate is divided into fractions. One of the fractions is subjected to the catalytic reforming with the production of a concentrate from which by products are subsequently separated. The other fractions, the heavier ones, are also directed to catalytic reforming to produce a high octane component of commercial gasoline.

The middle distillates are improved and purified of sulphur in hydro fining plants. Hydro fining yields high quality jet fuels and sweet diesel fuels. A part of the treated diesel fuel is de waxed with production of liquid paraffin wax and winter diesel fuel.

The gas streams from the primary petroleum distillation plants and catalytic reforming plants are processed in gas plants commercial liquefied gas is obtained, namely propane, butane and isobutene.

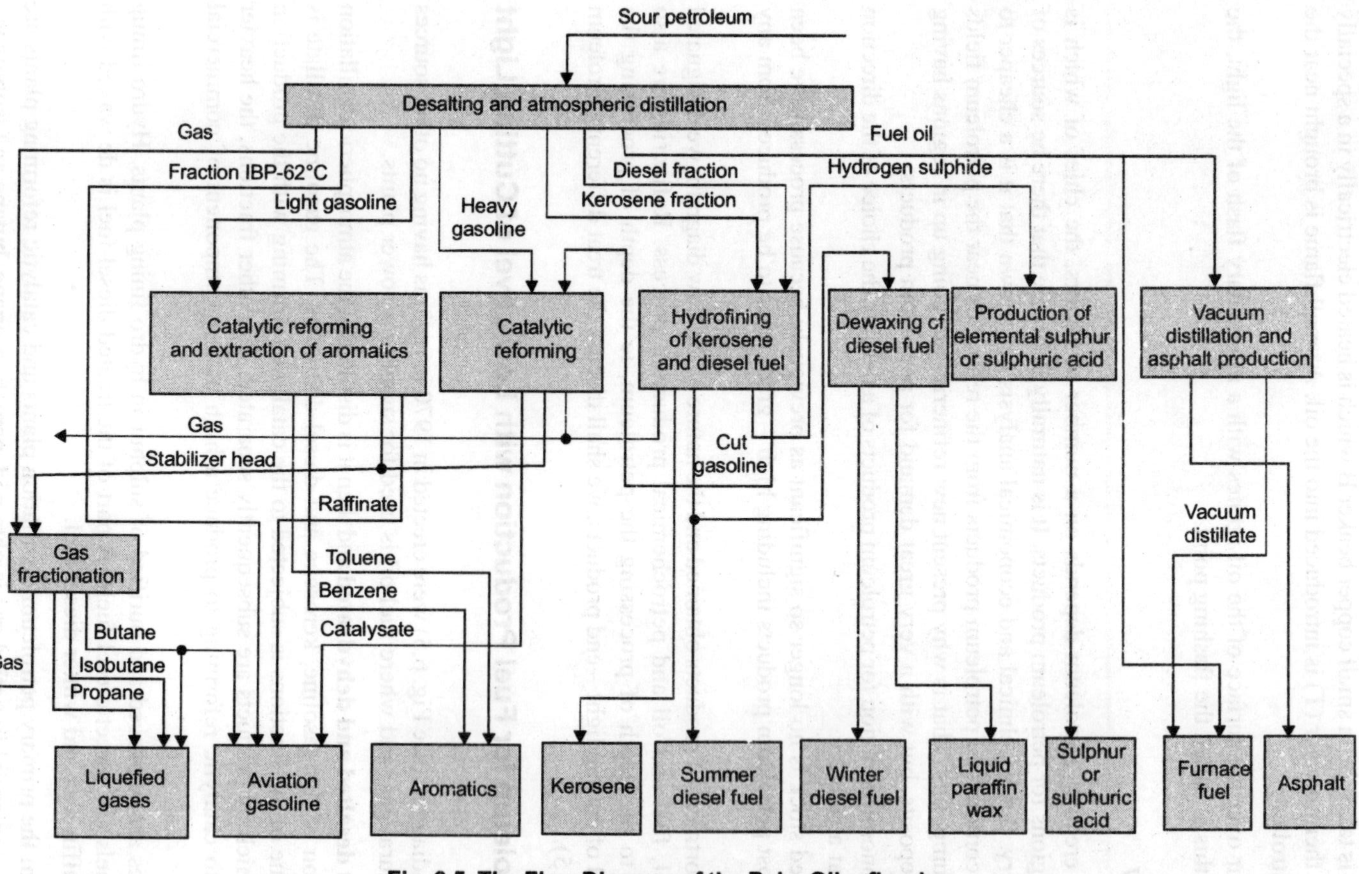

Fig. 6.5. The Flow Diagram of the Baku Oil refineries

At refineries with a small depth of refining, at other refineries the production of various grades of asphalt is organized. The asphalt obtained by oxidation of the tar separated from fuel out in a special unit included in the asphalt plant. The main amount of the fuel oil is delivered to consumers as furnace oil.

The hydrogen sulphide evolved in hydro fining is utilized in plants producing sulphuric acid or sulphur. In connection with the reduction of petroleum resources, at refineries with a low fuel level of cutting light – end products, it is planed to deepen processing, set up plants for cracking, coaking and visbreaking. The yields of light – end products will grow from 40 – 45 to 55 – 60% and more.

Processing of Petroleum for Fuel Production with a High Level of Cutting Light end Products

At refineries with such a flow diagram (Fig. 6.6) primary distillation of the petroleum is performed at atmospheric and vacuum plants. In addition to atmospheric distillates – the 350 - 500°C fraction.

The gasoline kerosene and diesel atmospheric distillates are processed in the same way as at a refinery with to a catalytic cracking plant. The cracking products are gas, gasoline, light and heavy gas oils. The gas is directed to a gas plant, the gasoline is used as a component of commercial gasoline while the light gas out is treated in a mixture with the straight run diesel fraction in a hydro fining plant and is then used a diesel fuel component. The heavy gas oil is the feed stock for producing technical carbon.

Hydro cracking is also used for processing the vacuum distillation in addition to catalytic then the ratio of the products can be changed in hydro cracking with an increasing in the yield of gasoline, kerosene and diesel fuel.

Tar is subjected to one of the thermal process; it is good practice to improve the gasoline of thermal processes by deep hydrogenation and catalytic refining. Light gas oil is employed as a component of gas turbine, motors or furnace fuel, while after hydro fining it can be included in diesel fuel.

The gas streams of thermal process and catalytic cracking contain unsaturated hydrocarbon. The propane – propylene and butane – butylenes fractions separated from this steam are used to produce additional amounts of quality gasoline

Processing of Petroleum for Fuel and Oil Production

With this diagram (Fig. 6.7) several distillates and tar are obtained in the atmospheric and vacuum distillates and tar are obtained in the atmospheric and vacuum distillation plants in addition to the light - end (gasoline, kerosene and diesel) distillate. Next each of the distillate fractions is subjected to (a) selective purifications from resinous and asphaltenic components (b) dewaxing and (c) additional treatment by the hydrogenation on adsorption method

The tar is first treated with propane to extract the asphalt substance. The resulting deasphalttizate fractions (relative treatment, dewaxing, additional treatment).

After additional treatment, the distillate and residue components are directed for compounding. The required grades of lubricating and other oils are produced by changing the ratio of the components and introducing various additives other diagrams also exist for producing oils. In treatment with double

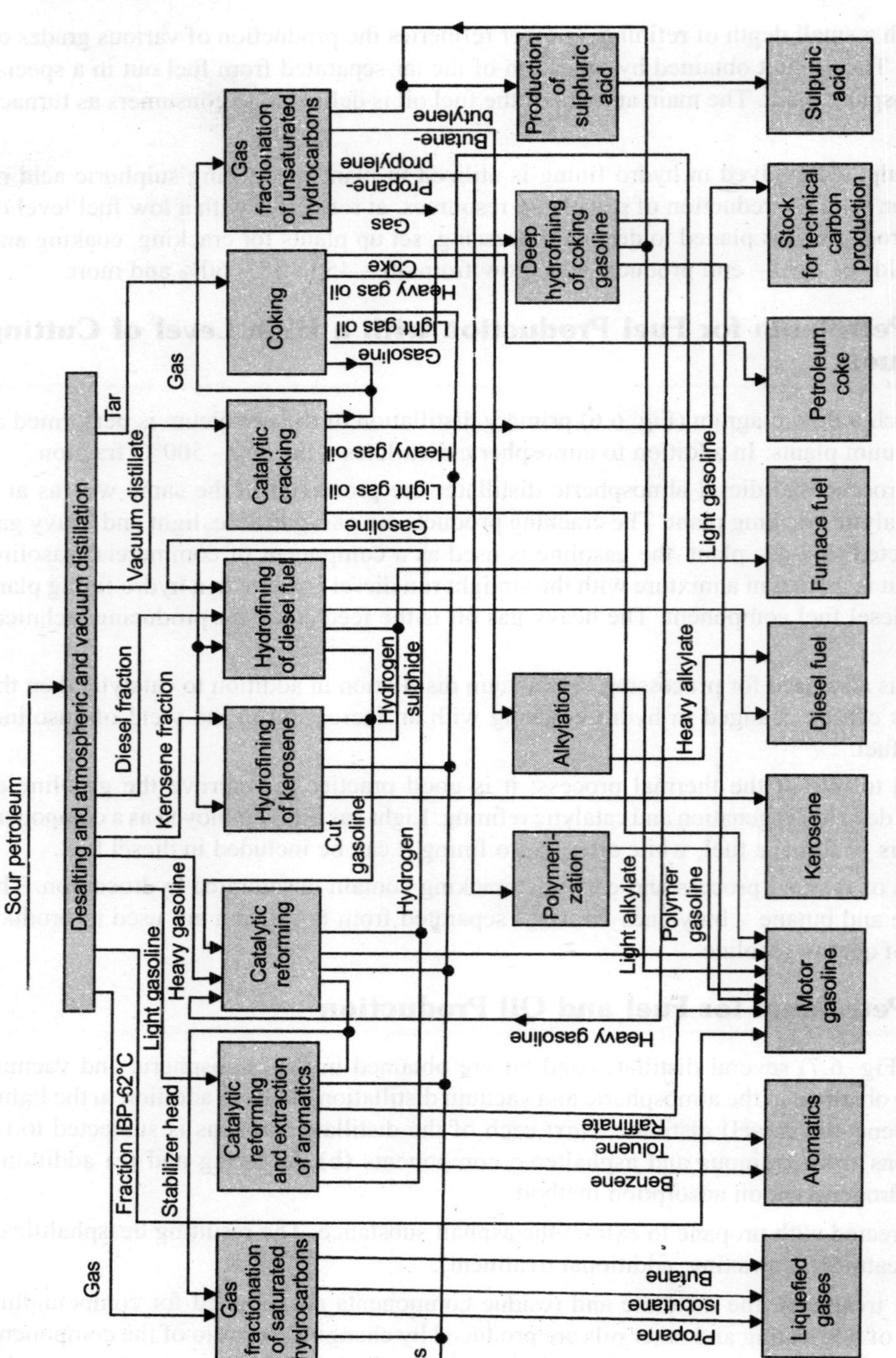

Fig. 6.6. The Processing of the Sour Petroleum for Fuel Production with High Level of Light end Product Cutting

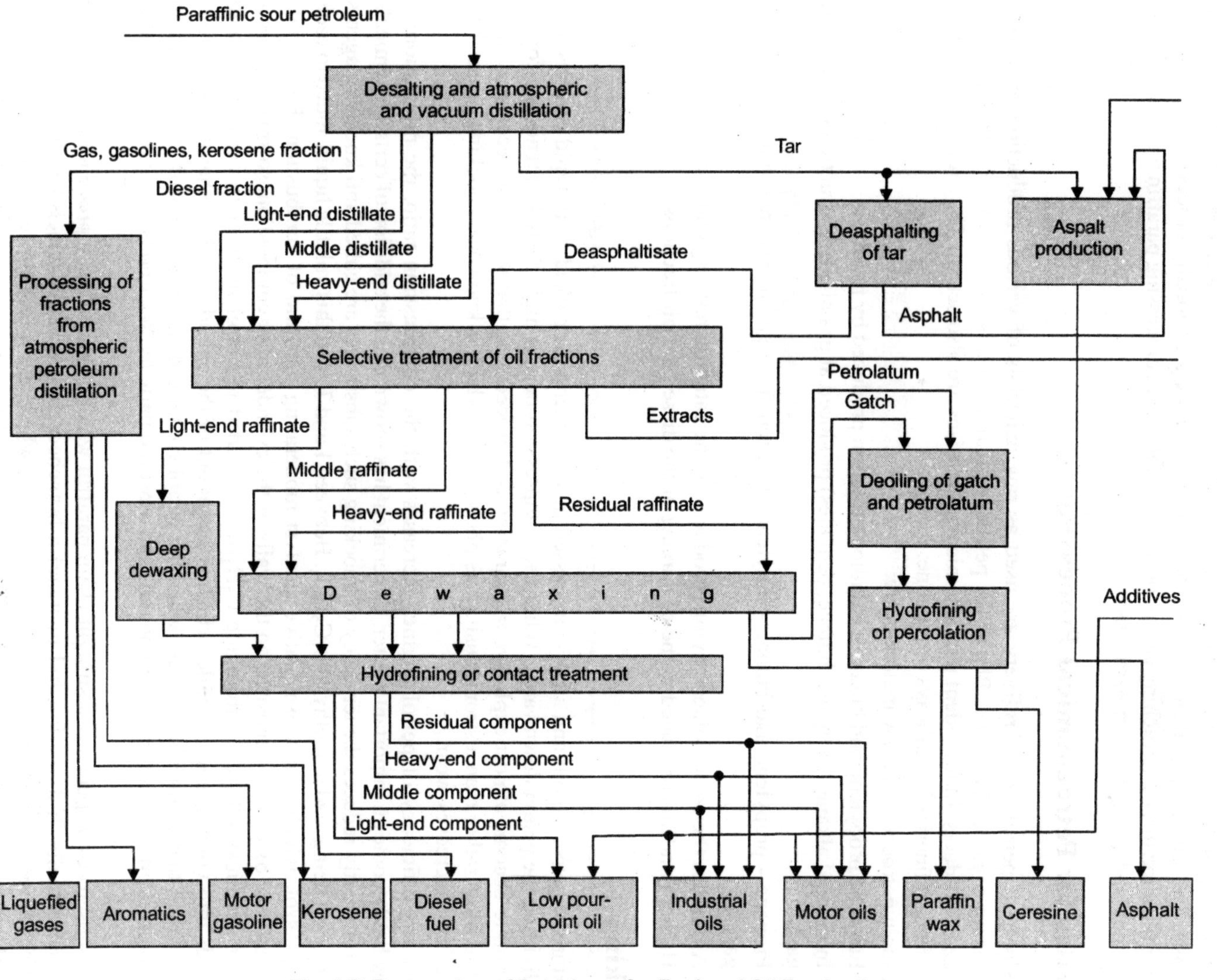

Fig. 6.7. Processing of Petroleum for Fuel and Oil Production

solvents (the Duo-Sol process) the deasphalting and selective treatment of oils are obtained together with the oils. The gatch and petroleum separated in dewaxing are additionally processed in plants for decracking filtration through beaching earths, or hydro fining. After processing paraffin wax is produced form the gatch and ceresine from petroleum.

Refineries for Petrochemical Processing

The products of petroleum of refinery are used as raw materials of various production processes manufacturing organic substance. Plants for petrochemical synthesis, which considerably expands their range of products. Independent petrochemical enterprises are also set up. They receive their raw materials from refineries by railways or pipeline.

The main petrochemical raw materials are:

(1) Ethylene propylene and butane - butadiene fraction produced by pyrolysis
(2) Individual alkanes (propane, butane, pentanes) produced at gas plants by fraction of saturated gas)
(3) Propane - propylene and butane - butylenes fraction from gas plants fractionating unstable gases.
(4) Aromatics (benzene, toluene nylon) produced in the catalytic refining and pyrolysis
(5) Higher paraffins produced in the de waxing of the diesel and oil fractions.

Knocking

If a mixture of petrol vapour and air is compressed in a cylinder and sparked, very rapid combustions occur, dividing the piston downwards. When the compression ratio is high, there is a tendency for the combustion to become almost explosive in character as it proceeds so that the piston receives a sharp and violent blow rather than a strong push. This phenomenon which is known as knocking causes less efficiency and increases wear.

Since the efficiency of petrol engines increases with the compression ratio, the prevention of knocking has become an important matter. Experiment has shown that the addition of certain chemicals to the petrol greatly reduces the tendency of knocking such substances are called anti knocking agents the commonest being lead tetraethyl, $Pb(C_2H_5)_4$. Between 1 and 3 cc of this volatile liquid added to each gallon of petrol, which acts as a negative catalyst moderating the ratio of combustion. Early trials showed that lead oxide was deposited on the walls of the cylinder after combustion, so small amounts of another chemical, ethylene dibromide $C_2H_4Br_2$ are added to petrol in order to convert the lead compounds into volatile lead bromide, which is swept out of the cylinder in the exhaust gases. It has also been found that engines knock less when the petrol contains high proportions of branched chain hydrocarbons. Certain unsaturated compounds and most aromatic hydrocarbons also have a beneficial effect.

Two substances *n*- pentane, a straight chain paraffin which is particularly prone to knocking and iso- octane (actually 2, 3, 4 trimethyl pentane) a branched chair paraffin which knocks only under extreme conditions, have been chosen as standards by which the knocking tendency of any given petrol is the percentage of iso- octane, to give it same knocking tendency as the petrol under test, so that the higher the octane number the better is petrol.

Determination of Octane Number

Laboratory Applications - LabCON-V

LabCON-V, in conjunction with a CFR knock test engine, automatically determines the research or motor octane number of finished gasoline and gasoline components. LabCON-V provides the precision and accuracy demanded by today's testing requirements. LabCON-V provides the means to generate a more accurate octane number than previously available.

Fig. 6.8. The LabCON – V Instrument

A specially designed micro valve carburetor automatically adjusts the fuel to air ratio and a proprietary algorithm determines the critical maximum knock intensity. A sample shelf and pumps allow LabCON-V to sequentially rate up to four samples without operator assistance. The LabCON-V system can also perform toluene standardization fuel ratings. It compares the actual ratings with Toluene tables and flags all ratings that fall outside the acceptable limits. Operation is simplified by displaying temperature tuning limits for Research and Motor methods. Additionally, LabCON-V automatically compensates the compression ratio using an internal barometric pressure sensor. LabCON-V increases production efficiency by allowing the personnel to operate more than one engine, or to focus on other laboratory tasks. The system has minimal calibration requirements, therefore can be operated with minimal maintenance.

Features

- Determines Research or Motor Method Octane Numbers
- Utilizes Bracketing or CR Method Rating Approach
- Provides Prototype Tank Calibration Procedure
- Sequentially Rates up to 4 Samples Unattended
- Laboratory Information Management System (LIMS) Compatible
- Automatically Controls Compression Ratio and Fuel to Air Ratio
- Automatically Compensates Compression Ratio Using Internal Barometric Pressure Sensor
- Provides Menu Driven Operator Interface with Graphical Knock Intensity Trace
- Includes Sample Shelf with 4 Sample Probes
- Includes Engine Stability Check Routine
- Provides Context Sensitive Online Help
- Maintains Sample History by Sample Identifier
- Printer sold separately.

LabCON-V Software Version 4.10

This version was released in January 2005. The barometric CR compensation algorithm was revised to match the ASTM 2699 and 2700 barometric compensation tables exactly. It also corrects the manual barometer input to allow values below 22. The advanced settings will now be saved when the dialogue window is closed.

LabCON-V Software Version 4.09

This version was released in May 2004. It provides changes in the mouse routine to allow for the new touch screen controller in Advantech units s/n 516 and higher.

LabCON-V Software Version 4.08

This version was released in March 2004 and changes the Proto Calibration routine to comply with ASTM D 2885-03. This revision counts proto-line as one complete cycle so the Total Cycles parameter should be modified to include 10 delta octane data points.

LabCON-V Software Version 4.07

This version was released in February 2004. It removes reference to the obsolete calibration slug in the calibration routine. It also changes the temperature tuning limit algorithm to match the revised ASTM 2699 and 2700 standards.

LabCON-V Software Version 4.06

This version was released in July 2003. The revision fixes a problem with the date where a LabCON unit left on over a two day or greater period, with no activity, will lose track of the day. The new software correctly increments the date when two consecutive "midnight" occur. This problem is specific to the Advantech style computer (S/N 500+).

LabCON-V Software Version 4.05

This version was released in March 2003 and adds an option to enter PRF and sample micro valve averages manually from the Initialize Micro valve screen. This allows for manual entry of micro valve averages before and after testing of fuels with unusual micro valve positions (premium gasoline, gasoline components etc.)

LabCON-V Software Version 4.04

This version was released in December 2002 and includes a manual screen saver as well as clarification to engine temperature warnings. The CR method's tolerance for KI values not requiring manual set point adjustment has been opened to 30-70 KI.

LabCON-V Software Version 4.03

This version was released in September 2000 and includes a bug fix for error message memory handling as well as clarification to help messages.

Prediction of FCCU Gasoline Octane and Light Cycle Oil Cetane Index

Background

Octane is a measurement of the knocking characteristics of a gasoline as determined in a laboratory engine by a standard ASTM test method. Research Octane number and Motor Octane number are the two tests employed and both can be run on the same test engine. The differences are that the two tests are run under different engine operating variables, for example, the engine speed used for the research method is 600 rpm compared to 900 rpm, for the motor method.

The Octane Number of a particular gasoline sample is determined by blending n-heptane, which has 0 octane, and iso-octane, which has 100 octane, in the correct proportion to produce the same knock intensity, as the sample being run in the test engine. The percent of iso-octane in the blend is then assigned as the Octane Number of the test sample.

The cetane number is also an ASTM test, used to measure the ignition qualities of diesel fuels. The higher the number the better the quality. The cetane index is a calculated value using the ASTM mid boiling point and API gravity, and correlates well with cetane number. The cetane index is used if the volume of sample is too small or a test engine is not available.

A diesel fuel with a high cetane index has a low spontaneous ignition point, that is, combusts at lower temperature from the heat of compression of the diesel engine. A high cetane index of FCCU light cycle oil (LCO) is desirable since the higher the index, the more LCO can be blended into the more valuable road diesel fuel.

Many factors affect the gasoline octane and LCO cetane index. Feed quality, catalyst type and unit operating conditions all play major roles in affecting the quality of FCCU products. The purpose of this report is to show ways to predict the changes in FCC gasoline octane and FCC LCO cetane index from changes in feed quality, unit operating conditions and product qualities.

Factors Affecting Gasoline Octane

One of the major variables affecting gasoline octane, is reactor temperature. As a rule an increase of 10 deg. F in reactor temperature generally increases the Research Octane of FCC gasoline 0.6 RONC. While Motor Octane is not as dramatically affected by reactor temperature directionally some improvement in Motor Octane should occur when increasing reactor temperature. Motor Octane is affected more by hydrocarbon types. At higher reactor temperature, paraffins crack leaving a higher olefin and aromatic content gasoline. The increased olefins are the major improvers to Research Octane and although olefins generally lower the Motor Octane, there is some slight improvement from the increased aromatic content.

An indication of the change in hydrocarbon type in the gasoline can be seen by the change in the gasoline bromine number and aniline point. The bromine number is an indication of olefin content. A higher number indicates a higher olefin content. The aniline point indicates aromatic content with a lower aniline point indicating a higher aromatic content.

Feed Quality Can Have a Major Impact on FCC Gasoline

The other major factor affecting FCC gasoline octane is feed quality. Generally, the less paraffinic feed, the higher the gasoline octane. Feed paraffinicity or lack of it can be indicated by a number of tests. Aniline point, 'K' factor and gravity are all rough indicators of feed type.

Aniline point is the temperature at which a 50/50 mix of sample and aniline are mutually soluble. Since aromatics are readily soluble in aniline, a lower temperature is needed to achieve solubility. Thus, aniline point is a rough measure of aromaticity, a lower number indicating a higher aromatic content.

The 'K' factor also measures the hydrocarbon chemical nature. Feeds with 'K' factors above 11.8 indicate a higher paraffin content. Feeds with 'K' factors below 11.8 are generally more aromatic. 'K' factor is a calculated number based on boiling point and gravity. Altamont crude which yields a gas oil with a high 'K' factor produces a gasoline with a relatively low octane number. On the other hand, a crude such as Alaskan North Slope, which produces a cat feed with a relatively low 'K' factor, yields a gasoline fairly high in octane.

Another indicator of feed quality, although a very general indicator but one easily obtained, is API gravity. API gravity is a measurement unique to the oil industry. The chemical nature of the feed determines the API gravity. Aromatics are much denser than paraffins with the same boiling point and therefore, will have a lower API gravity.

A higher aromatic content feed (or lower paraffin content feed) as discussed above should generally produce a higher octane gasoline. Figures 6.9 and 6.10 are commercial data plots showing the effect of changes in feed gravity on the FCC gasoline Research and Motor Octane, respectively. The lower API gravity feed (or higher aromatic content) will increase FCCU gasoline octane 0.3 RONC and 0.15 MONC for every 1.0 deg API change. Altamont feed has a gravity in the 35 deg API to 40 deg API range, whereas Alaskan North Slope cat feed is usually 20 deg API or lower.

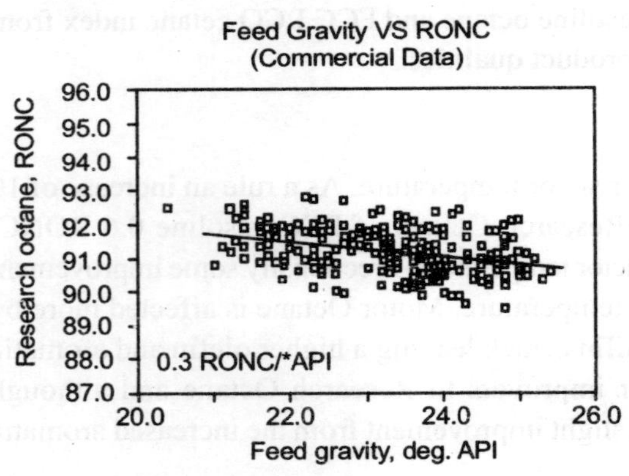

Fig. 6.9. The Feed Gravity Plot

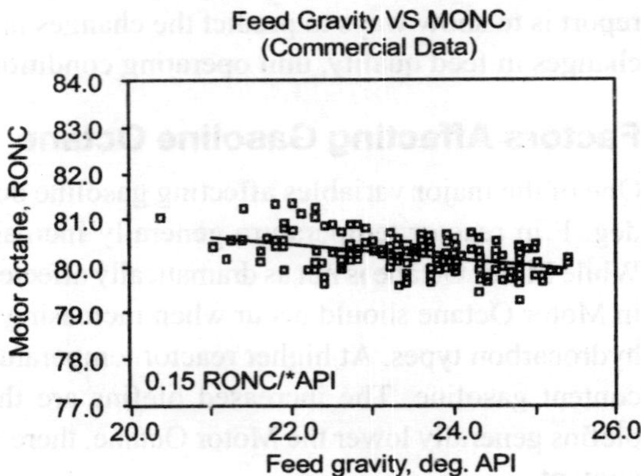

Fig. 6.10. The Plot

Other factors which affect gasoline octane within the control of the refiner and independent of reactor temperature and feed quality are Reid Vapor Pressure (RVP) and gasoline cut points as well as FCC Catalyst choice. The RVP of gasoline indicates butane content and generally increases gasoline octane 0.2 RONC for every 1.0 psi increase in RVP due to the amount of butanes dissolved in the gasoline. Changes in the cut point of gasoline also affect the octane of FCC gasoline. Various tests have shown that the front end (0-30% point) of FCC gasoline is higher in octane than the middle cut (40-60%) and tail end (90%+) in an FCC gasoline whose boiling range is C-5 to 430 deg F end point. A lowering of the FCC gasoline end point and increase in the RVP, as usually occurs during the winter heating oil season, will result in a higher FCC gasoline octane.

Catalyst Choice Can Improve Octane

Catalyst choice is also within the control of the refiner, however, catalyst inventories cannot be changed as quickly as RVP or cut points. The amount of rare earth or hydrogen exchanged zeolite as well as the degree of ultrastability of the catalyst have different effects on the ultimate octane of the FCCU gasoline. A non-rare earth ultrastable catalyst will produce a higher octane gasoline than a non-ultrastable rare earth exchanged catalyst.

Compared to a rare earth exchanged gasoline type catalyst, a partial rare earth exchanged, ultrastable catalyst can improve gasoline octane by 1.0 to 2.0 research numbers. A completely non-rare earth, ultrastable catalyst can improve the octane 2.0 to 3.0 research numbers clear or higher.

Of course, certain known yield changes occur with octane catalysts such as a slight decrease in gasoline yield and an increase in C_3 and C_4 olefins. The refiners own product requirements determine whether a gasoline or octane catalyst or a catalyst in between is used.

The table below gives a listing of the octane improvement a refiner may expect to see using various Engelhard Octane Catalysts when compared at constant conditions to **ULTRASIV 260** which is a more traditional rare earth exchanged gasoline catalyst.

Catalyst Series	Zeolite Type	Rare Earth%	Relative* RON
ULTRASIV 260	REY	<2.2	Base
OCTISIV PLUS 100	USY/REY	<1.5	+1.5
OCTISIV PLUS 500	USY	<0.1	+2.0
OCTISIV PLUS 600	USY/REY	<1.0	+1.5
DYNASIVA PLUS 800	USY/REY	<1.3	+1.0
OCTIDYNE 1100	USY	<0.1	+2.0
ULTRADYNE PLUS 1500	USY/REY	<1.3	+1.5

*RON improvement relative to **UL TRASIV 260** performance.

Factors affecting LCO Cetane Index Directly Relate to Paraffin Content

Cetane index is different from cetane number only in that the former is a calculation. The cetane index correlates well with the cetane number in diesel fuels which have not had "improvers" added. Paraffins exhibit the highest cetane index, aromatics the lowest. Therefore, high cetane index LCO has a high 'K' factor as well.

The cetane index of LCO is calculated using the 50% boiling point and API gravity. The higher the 50% point and/or API gravity, the higher the cetane index. Figure 6.11 is a graphical representation of the cetane index calculation. The graph shows that the cetane index changes 1.2 numbers for every 1.0 deg API change in gravity (at a constant 50% point) or 1.0 numbers for every 10°F change in 50% point (at a constant API gravity). Figure 6.12 shows the effect of feed gravity on LCO gravity. An increase of 1.0 deg API in feed gravity increased the LCO gravity approximately 0.75 deg API.

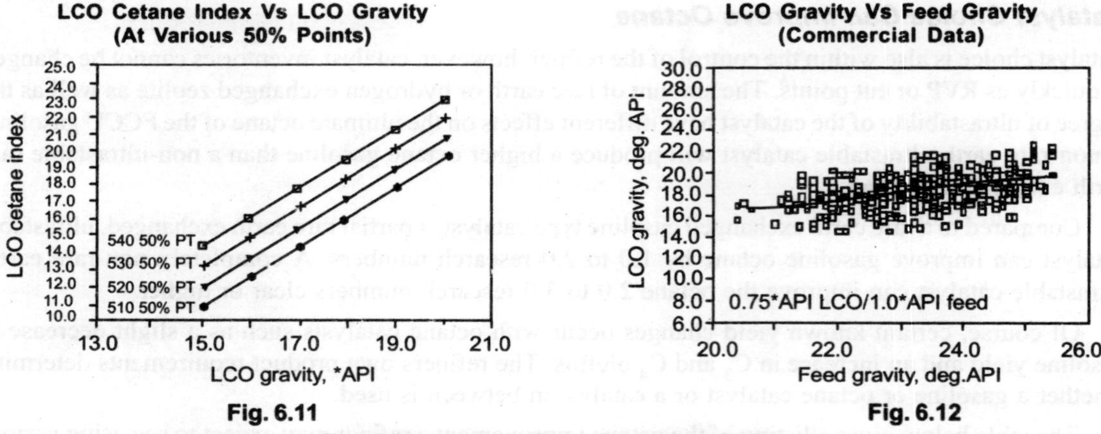

Fig. 6.11 and 6.12. The LCO Cetane Index

Combining the data in Figures 6.9 and 6.10, means that a 1.0 deg API increase in the feed gravity should increase the LCO cetane index 0.9 numbers as shown in Fig. 6.13.

High paraffin content crudes, such as Altamont, yield gas oil feeds which produce an FCC LCO with very high cetane indices. The LCO produced from this feed normally have cetane indices in the mid-40s. Conversely, crudes with low paraffin content, such as Alaskan North Slope, yield gas oils which produce an FCC LCO with low cetane indices. An LCO produced from Alaskan North Slope feed will have a cetane index in the low 20s.

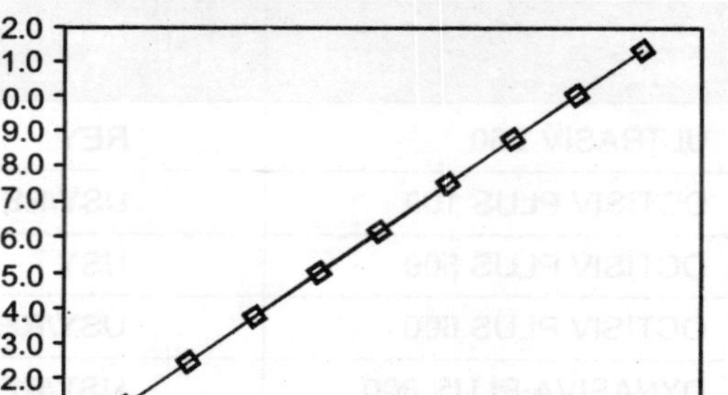

Fig. 6.13. The Feed Gravity Index.

Octane Catalysts Will also Improve Cetane Index

Feeds which produce a high octane gasoline generally produce a low cetane index LCO. The reason is that the hydrocarbon types which affect gasoline octane and LCO cetane index are inversely related. Steps taken to improve gasoline octane, on one particular feed, will generally improve LCO cetane index as well. For instance, increasing reactor temperature improves octane by cracking the paraffins in the gasoline boiling range leaving a higher percentage of olefins and aromatics which are higher in octane. Long chain paraffins in the slurry boiling range will crack as well to shorter chain paraffins in the LCO boiling range. This improves the LCO cetane index. Reducing the rare earth content of a cracking catalyst will improve the LCO cetane index as well as the octane of the gasoline fraction. A reduced rare earth level means a reduction in hydrogen transfer thus leaving the hydrogen in the LCO fraction. This improves the paraffinicity and hence the cetane index of the LCO. At the same time hydrogenation of olefins in the gasoline fraction is reduced and octane in increased. These reactions are illustrated by the reaction mechanism shown below:

$$3C_nH_{2n} + C_mH_{2m} \rightarrow 3C_nH_{2n+2} + C_mH_{2m-6}$$

Gasoline Olefins + LCO Naphthenes → Gasoline Parafins + LCO Aromatics

Comparing the performance of two widely used FCC catalysts, this effect of catalyst rare earth concentration on octane and cetane index is clearly demonstrated in the commercial data comparison that follows:

	Rare Earth, %	RON	Cetane Index
UL TRASIV 260	1.8	92.1	28.9
Catalyst AC	3.0	90.4	26.0
Δ	-1.2	+2.1	+2.9

The above data were taken from a commercial unit operating at constant reactor temperature unit conversion and FCC feed stock.

In conclusion it should be noted that the relationship of feed gravity to gasoline octane and LCO cetane index naturally can vary from the data presented here. However, as a general rule, a 1.0 deg API increase in feed gravity should decrease the gasoline Research Octane 0.3 RONC, decrease Motor Octane 0.15 RONC and increase the LCO cetane index 0.9 numbers.

Uses of Petroleum

The modern petroleum industry began in 1859, when the American oil pioneer E. L. Drake drilled a producing well on Oil Creek in Pennsylvania at a place that later became Titusville. Many wells were drilled in the region. Kerosene was the chief finished product, and kerosene lamps soon replaced whale oil lamps and candles in general use. Little use other than as lamp fuel was made of petroleum until the development of the gasoline engine and its application to automobiles, trucks, tractors, and airplanes. Today the world is heavily dependent on petroleum for motive power, lubrication, fuel, dyes, drugs, and many synthetics.

During the late 19th cent., many of the modern oil companies were created: John D. Rockefeller invested in a Cleveland oil refinery during the Civil War and in 1870 created Standard Oil, which refined about 95% of the United States' oil in 1880. In 1911, Standard Oil was declared an illegal monopoly and split into 34 companies, including Esso (renamed Exxon in 1972), Mobil, Chevron, Atlantic Richfield (later ARCO), and Amoco. Texaco (founded in 1902), Shell (1907), and British Petroleum (1909) were also established in this period. As the auto industry vastly increased the demand for gasoline refined from oil, oil companies expanded their search for new reserves. In the 1930s oil companies began exploiting a huge E Texas oil field that would eventually produce 4 billion barrels of oil. Chevron, Texaco, Exxon, and Mobil expanded their reserves by purchasing the rights to the extensive Saudi Arabian oil fields for only $50,000 from the king. In 1946 oil replaced coal as the world's most popular energy source.

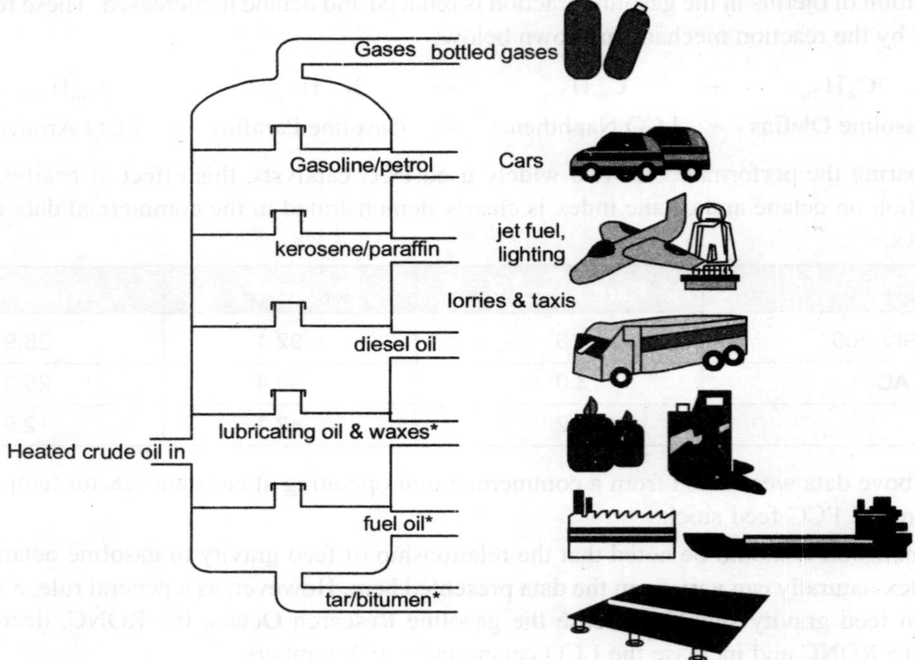

Fig. 6.14. The Different Uses and Applications of Petroleum

During World War II (1939-1945), the American oil industry proved its ability to increase production and develop specialized products quickly. Huge quantities of oil were produced and converted into fuels and lubricants. Such new refining processes as catalytic cracking and alkylation vastly increased the output of high-octane aviation gasoline. The United States supplied over 80 per cent of the aviation gasoline used by the Allies during the war. American refineries also manufactured butadiene, used in making synthetic rubber; toluene, an ingredient in TNT; medicinal oils to treat the wounded and many other military necessities.

CHAPTER

7

Unsaturated Hydrocarbons

There are two main series of unsaturated hydrocarbons – the olefines and the acetylene. The former contain a double bond and the latter is triple bond.

OLEFINES, ALKANES OR ETHYLENIC HYDROCARBONS

They are unsaturated hydrocarbons of the general formula C_nH_{2n}. The following are the general formulas of this series:

Ethylene	C_2H_4	Pentylene	C_5H_{10}
Propylene	C_3H_6	Hexylene	C_6H_{12}
Butylene	C_4H_8		

All the hydrocarbons of this series contain two of their carbon atoms linked together with a double bond.

$$-\overset{|}{C}=\overset{|}{C}-$$

This type of double bond is also known as an ethylenic bond or knowledge and is characteristic of olefines.

The typical and parent hydrocarbon of the series is ethylene.

$$H-\underset{}{\overset{H}{\overset{|}{C}}}=\overset{H}{\overset{|}{C}}-H$$

Ethylene

$$H_2C=\underset{H}{\underset{|}{\overset{}{C}}}-\overset{H}{\overset{|}{\underset{H}{\underset{|}{C}}}}-H$$

Propylene

$$H-\overset{H}{\overset{|}{\underset{H}{\underset{|}{C}}}}-\underset{H}{\underset{|}{C}}=\underset{H}{\underset{|}{C}}-\overset{H}{\overset{|}{\underset{H}{\underset{|}{C}}}}-H$$

Butylene

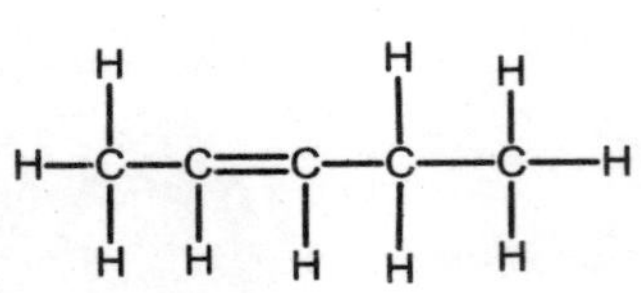

Pentylene

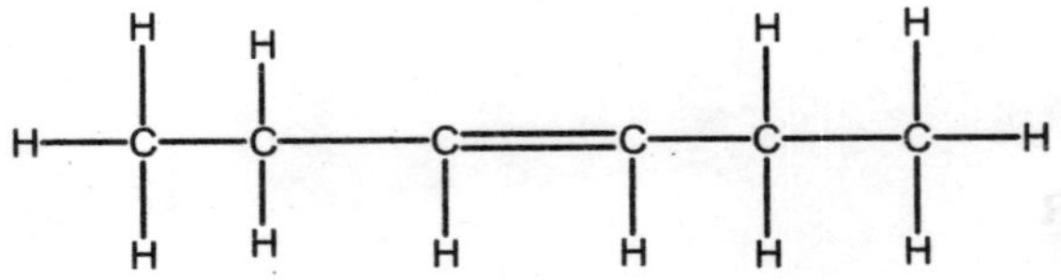

Hexylene

Occurrence

Olefines are found in petroleum, natural gas, coal gas and wood gas. The presence of olefines in these substances is probably due to the cracking processes.

Methods of Preparation - A Gerneral Discussion

Olefines can be prepared by the following general methods:

(1) Dehydration of Alcohols by means of Sulphuric or Phosphoric acid, oxalic acid, and analysis zinc chloride or boric anhydride:

$$C_nH_{2n+1}OH \xrightarrow{-H_2O} C_nH_{2n}$$

Alcohols Olefines

$$CH_3CH_2OH \xrightarrow{-H_2O} CH_2 = CH_2$$

Ethyle Alcohol Ethylene

$$CH_3CH_2CH_2OH \xrightarrow{-H_2O} CH_3CH = CH_2$$

Propyl Alcohol Propylene

Pure alumina (Al_2O_3) and thoria can also be used as dehydrating agents. Olefines may be prepared by passing the vapour of an alcohol over alumina or thoria heated to 300°C

(2) Removal of halogen acid from a halogen chloride particularly the iodides by the action of alcoholic caustic alkalies.

$$C_nH_{2n+1}I + KOH\ (alcoholic) \rightarrow C_nH_{2n} + KI + H_2O$$

$$C_2H_5I + KOH\ (alcoholic) \rightarrow C_2H_4 + KI + H_2O$$

(3) Electrolysis of sodium or potassium salts of dibasic acids of succinic acid series: This method is similar to Kolbe's Reaction for the preparation of paraffins.

The products of electrolysis are:

At the anode: Ethylene gas and carbon dioxide, and

At cathode: Potassium Hydroxide and Hydrogen

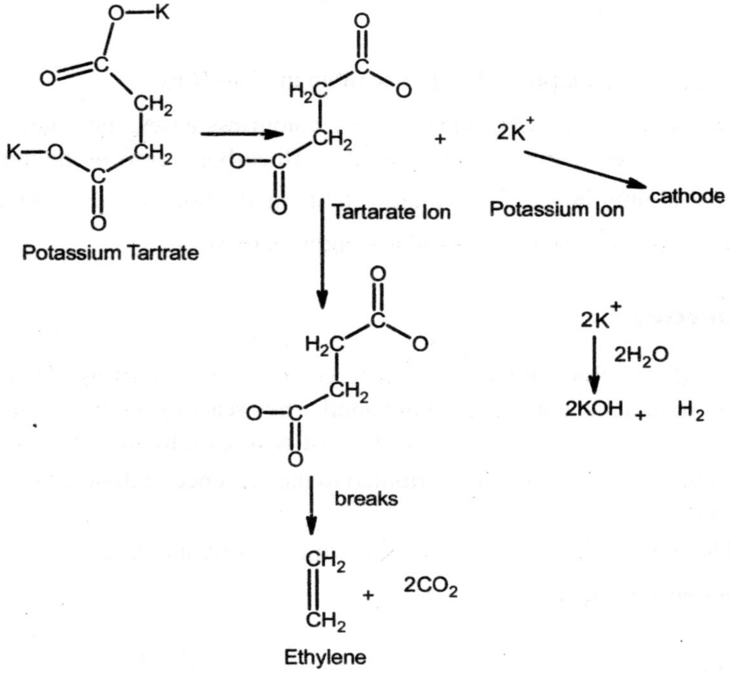

Potassium Tartrate → Tartarate Ion + $2K^+$ → cathode (Potassium Ion)

Tartarate Ion —breaks→ $CH_2{=}CH_2$ + $2CO_2$ (Ethylene)

$2K^+ \xrightarrow{2H_2O} 2KOH + H_2$

(4) ***From the dihalogenic compounds by the action of Zinc***: This is an easy method of preparing pure of olefines. The reaction is similar to Wirtz Reaction.

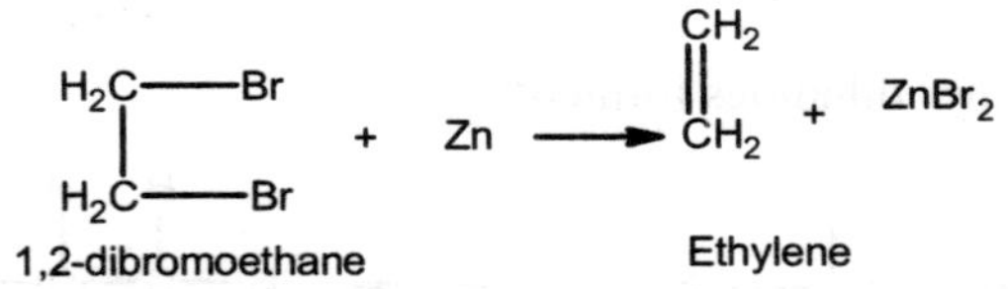

$$H_2C(Br){-}CH_2(Br) + Zn \longrightarrow CH_2{=}CH_2 + ZnBr_2$$

1,2-dibromoethane → Ethylene

(5) ***Reduction of Acetylene Hydrocarbons by hydrogen in the presence of catalyst*** at high temperature (Sabatier and Sanderen's reaction)

$$\underset{\text{Acetylene}}{C_nH_{2n-2}} + 2H \xrightarrow{Ni(200^\circ C)} \underset{\text{Olefines}}{C_nH_{2n}}$$

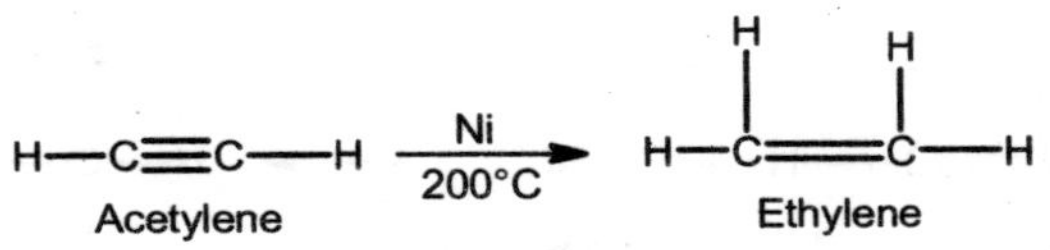

$$\underset{\text{Acetylene}}{H{-}C{\equiv}C{-}H} \xrightarrow[200^\circ C]{Ni} \underset{\text{Ethylene}}{H_2C{=}CH_2}$$

Platinum and palladium can also be used as catalyst in the above reaction.

(6) ***Olefines are also produced by the destructive distillation of organic substance*** such as wood lignite and coal.

Physical Properties

In their physical properties the olefines closely resemble the paraffins:

1. The lower members C_2H_4 C_3H_6 and C_4H_8 are colourless gases, the intermediate member are liquids and the higher members with more than 18 carbon atoms in the molecule are solids.
2. They are very slightly in water, but more soluble in alcohol and other organic solvents.
3. Their boiling points rise as the molecular weight increase.

Chemical Properties

In chemical reaction, the olefines are entirely different from the paraffins. They are very reactive because they are unsaturated, i.e. contain a double bond. They react by addition instead of substitution, i.e. the reagent adds on the olefines molecules instead of removing hydrogen from them.

The chemical properties of olefines are attributed to the presence of double bond which is a sign of weakness and reactivity.

ETHYLENE: The typical olefine shows the following important reaction:

1. ***With hydrogen:*** Paraffins are formed

$$\underset{\text{Ethylene}}{H_2C{=}CH_2} + 2H \xrightarrow[200°C]{Ni} \underset{\text{Ethane}}{H_3C{-}CH_3}$$

2. ***With Halogen:*** Olefines dihalides formed*

$$\underset{\text{Ethylene}}{H_2C{=}CH_2} + Br_2 \longrightarrow \underset{\text{Ethylene Bromide}}{BrH_2C{-}CH_2Br}$$

The decolourization of bromine water in the above reaction is often used as a least for olefines or unsaturation in organic compound.

3. ***With halogen Acid***: Alkyl halides are produced

$$\underset{\text{Ethylene}}{H_2C{=}CH_2} + \underset{\text{Hydrogen Iodide}}{HI} \longrightarrow \underset{\text{Ethyl Iodide}}{H_3C{-}CH_2I}$$

*Chlorine combines mostly readily of the halogens and iodine least readily. On the other HI adds on more readily than HBr and thus more readily than HCl.

4. ***With Hypochlorous Acid***: Chlorohydrins are produced

$$H-\underset{\text{Ethylene}}{\overset{H}{\overset{|}{C}}=\overset{H}{\overset{|}{C}}}-H + \underset{\text{Hypochlorous acid}}{OH-Cl} \longrightarrow \underset{\text{Ethylene Chlorohydrine}}{H-\underset{OH}{\overset{H}{C}}-\underset{Cl}{\overset{H}{C}}-H}$$

Ethylene chlorohydrins is largely used in the glycols,

$$\begin{array}{l} H_2C-OH \\ \quad| \\ H_2C-OH \end{array}$$

which it gives on hydrolysis.

5. ***With sulphuric acid:*** Olefines are absorbed by gold fuming sulphuric acid forming alkyl hydrogen sulphates

$$\underset{\text{Ethylene}}{H-\overset{H}{C}=\overset{H}{C}-H} + \underset{\text{Sulfuric acid}}{H-O-SO_2-O-H} \longrightarrow \underset{\text{Ethyl Hydrogen Sulphate}}{H-\underset{H}{\overset{H}{C}}-\underset{SO_4}{\overset{H}{C}}-H}$$

Ethyl hydrogen sulphate on heating gives ethylene gas and on hydrolysis gives ethyl alcohol. Hence, the reaction can be used for separating the olefines from the paraffins and also for converting the olefines into alcohols.

$$C_2H_5HSO_4 \xrightarrow{\text{Heat}} C_2H_4$$

$$C_2H_5HSO_4 \xrightarrow{\text{Water}} C_2H_5OH + H_2SO_4$$

7. ***With Con Nitric Acid***: Nitro derivatives are formed

$$\underset{\text{Ethylene}}{H-\overset{H}{C}=\overset{H}{C}-H} + \underset{\text{Nitric acid}}{H-O-\overset{+}{N}(=O)-O^-} \longrightarrow \underset{\text{2-nitroethanol}}{H-\underset{HO}{\overset{H}{C}}-\underset{\overset{+}{N}(=O)O^-}{\overset{H}{C}}-H}$$

8. ***With Alcoholic Potassium Permanganate*** Bayer's Reagent Oxidation

$$\underset{\text{Ethylene}}{H-\overset{H}{C}=\overset{H}{C}-H} + H_2O + O \longrightarrow \underset{\text{Ethylene glycol}}{H-\underset{OH}{\overset{H}{C}}-\underset{OH}{\overset{H}{C}}-H}$$

The decolourization of potassium permanganate solution in the above is also used or a test for olefines or unsaturation in organic components.

9. ***With Ozone***: Ozonides are formed

$$H_2C{=}CH_2 + O_3 \rightarrow \text{H–C(H)–O–C(H)–H (with O–O bridge)}$$

Ethylene + Ozone → Ethylene Ozonide

10. ***Combustibility***: Olefines burn with a smoky and luminous flame, forming carbon and water luminous flame, forming carbon dioxide and water.

$$C_2H_4 + 3O_2 \rightarrow 2CO_2 + 2H_2O$$

11. ***Polymerization***: Olefines undergo polymerization, in the presence of sulphuric acid; ethylene polymerizes to form a new molecule of multiple molecular weight.

Test for Ethylene Bond

Bayer's permanganate test ozonolysis and addition of bromine are of general uses of testing the presence of double bonds in unsaturated open-chain compounds. These tests are also shown by acetylenic compounds.

Individual Memembers

ETHYLENE

Formula

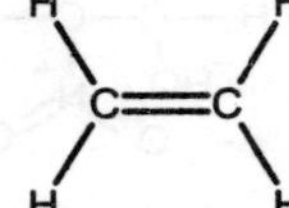

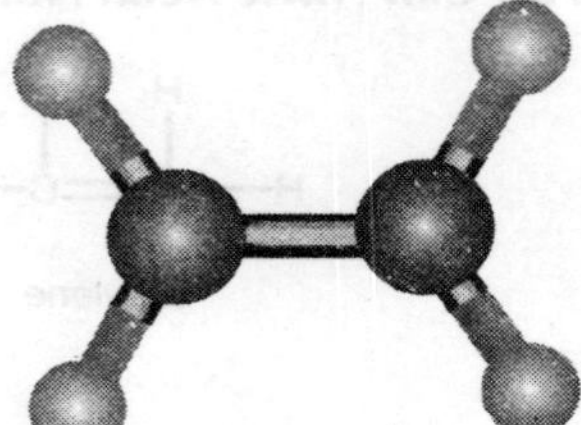

Occurrence

Ethylene occurs in natural gas, coal gas, and woods gas. It is also formed in large quantities, during the cracking of petroleum.

Preparation

Ethylene can be obtained by any of the general methods used for the precipitation of the olefines.

Laboratory Method

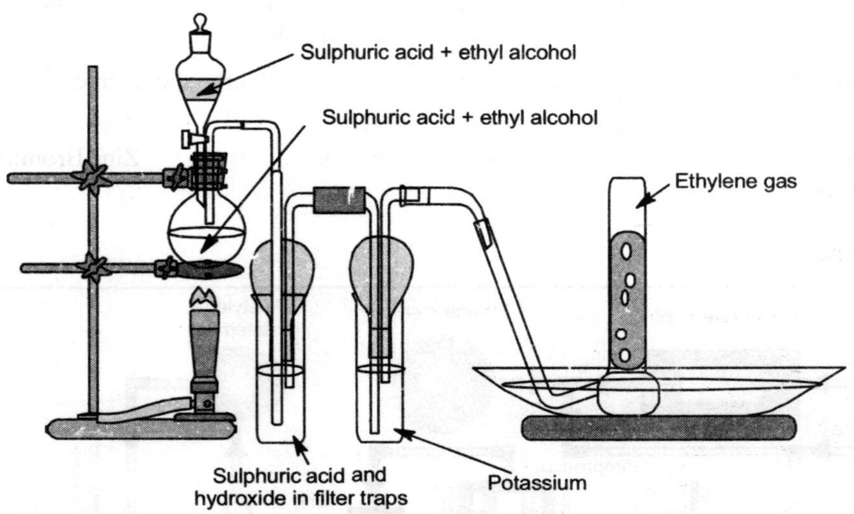

Fig. 7.1. The Laboratory Preparation of Ethylene Gas

Theory: Dehydration of ethyl alcohol. The lab ethylene is dehydrated by conc. Sulphuric acid and ethyl alcohol

Apparatus	Chemicals	Catalyst	Temperature
Flask, burner, stands, delivery tube, gas jar, trough, wash bottles dropping funnel, thermometer.	Sulphuric acid, ethyl alcohol, aluminium sulphate potassium hydroxide	Aluminium sulphate	160°C

Reaction Equation: The reaction is supported to take place in two stages:

C_2H_5OH Ethyl alcohol	+	H_2SO_4 Sulphuric Acid	$\rightleftharpoons$	$C_2H_5HSO_4$ Ethylene Sulphate	+	H_2O Water
$C_2H_5HSO_4$ Ethylene Sulphate	$\rightleftharpoons$	C_2H_4 Ethylene	+	H_2SO_4 Sulphuric acid		

Purification: The gas evolved at 160°C contains sulphur dioxide, carbon dioxide as impurities, so the gas is purified by passing it through sulphuric acid and then potassium hydroxide

Precaution : *The temperature should not be increased above 160°C when sulphuric acid is used and if syrupy phosphoric acid is used, then the temperature must be increased to 200°C.*

Pure Ethylene can also be prepared by two following methods: -

(i) By heating gently ethyl iodide with alcoholic potassium hydroxide*

---

*Alcoholic KOH a molecule of halogen acid from the alkyl halide. Where as aqueous KOH replaces the halogen by—OH group, forming alcohol.

C_2H_5I	+	KOH ®	C_2H_4	+	KI	+	H_2O
Ethyl	+	Potassium	Ethylene		Potassium		Water
Iodide		Hydroxide			Iodide		

(ii) By warming an alcoholic solution of Ethylene Di bromide with granulated zinc

$C_2H_4Br_2$	+	Zn	→	C_2H_4	+	$ZnBr_2$
Ethylene		Zinc	→	Ethylene	+	Zinc Bromide
Di bromide						

Manufacture

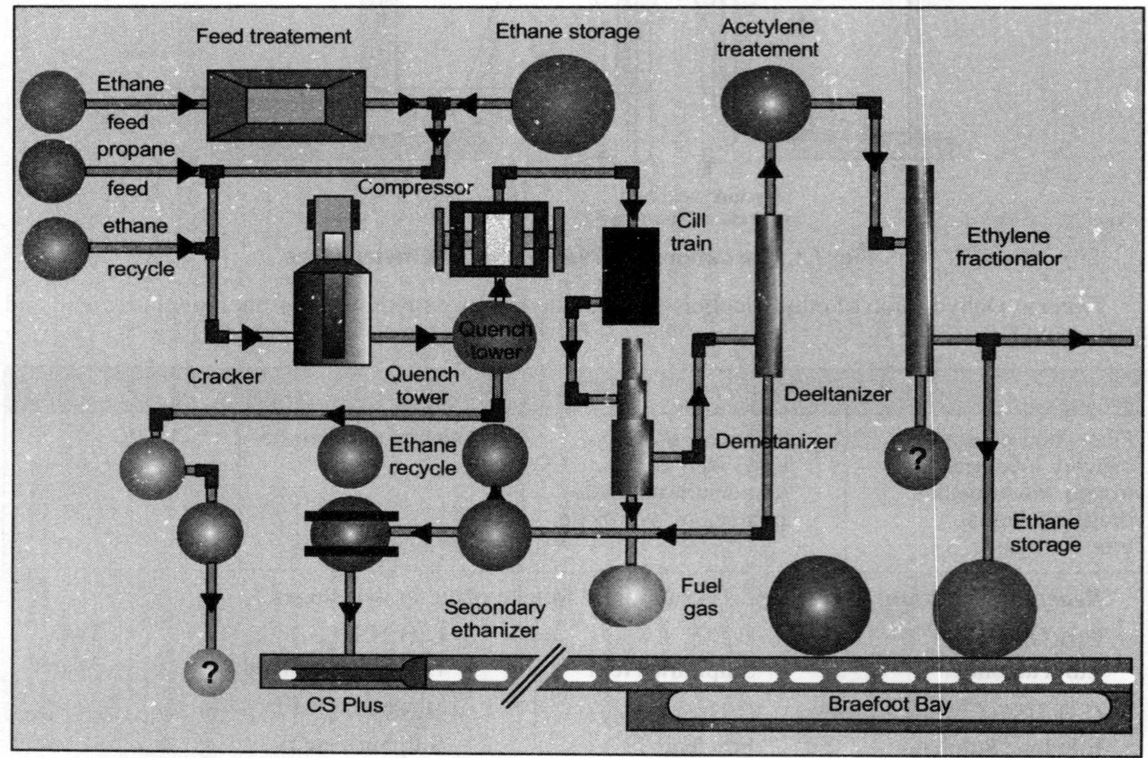

Fig. 7.2. The Industrial Preparation of Ethylene

Ethylene is manufactured by the following methods:

1. By passing alcohol vapour over heated catalyst such as aluminium oxide, thorium oxide or aluminium sulphate or clay balls placed in copper tube

$$C_2H_5OH \xrightarrow{Al_2SO_4} C_2H_4 + H_2O$$

Ethyl Alcohol		Ethylne	+	Water

2. By cracking of lower hydrocarbon present in natural gas

$$\underset{\text{Natural Gas}}{C_2H_5} \xrightarrow{\text{Cracking}} \underset{\text{Ethylene}}{C_2H_4} + \underset{\text{Hydrogen}}{H_2}$$

3. By passing Alcohol vapours through heated syrupy phosphoric acid, that is dehydration of alcohol

$$\underset{\text{Ethyl Alcohol}}{C_2H_5OH} \xrightarrow{-H_2O} \underset{\text{Ethylene}}{C_2H_4}$$

4. By the catalytic controlled hydrogenation of acetylene

$$C_2H_2 + H_2 \xrightarrow{Ni/200^\circ C} C_2H_4$$

Physical Properties

- colourless gas at room temperature and pressure
- Melting point -169°C, Boiling point -104°C
- slightly sweet smell
- flammable
- non-polar molecule: soluble in non-polar solvents and insoluble in polar solvents like water
- reactive: the active site is the double bond
- Readily undergoes addition reactions, for example
- reacts with bromine water (red-brown) to produce colourless 1,2-dibromoethane

Uses of Ethene (Ethylene)

Production of Polythene (polyethylene)	$nCH_2{=}CH_2$	$\xrightarrow[\text{high pressure}]{\text{catalyst}}$	$(CH_2\text{-}CH_2)n$
Production of Industrial Alcohol (ethanol)	$CH_2{=}CH_2$	$\xrightarrow[(2)\ H_2O]{(1)\ H_2SO_4}$	$CH_3\text{-}CH_2OH$
Production of 1, 2-ethanediol (ethylene glycol)	$CH_2{=}CH_2$	$\xrightarrow[(2)\ H_2O]{(1)\ O_2/\text{catalyst}}$	$CH_2OH\text{-}CH_2OH$
Production of Styrene. Styrene can be polymerised to form polystyrene	$CH_2{=}CH_2$	$\xrightarrow{\text{benzene}}$	$CH{=}CH_2$ attached to benzene ring
Production of chloroethane. Chloroethane is used to manufacture tetraethyl lead tetraethyl lead is the lead additive in leaded petrol	$CH_2{=}CH_2 + HCl$	$\longrightarrow$	$CH_3\text{-}CH_2Cl$

Ethene is also used as a plant hormone to control the ripening and colour development of fruit.

Polymerization of Ethylene

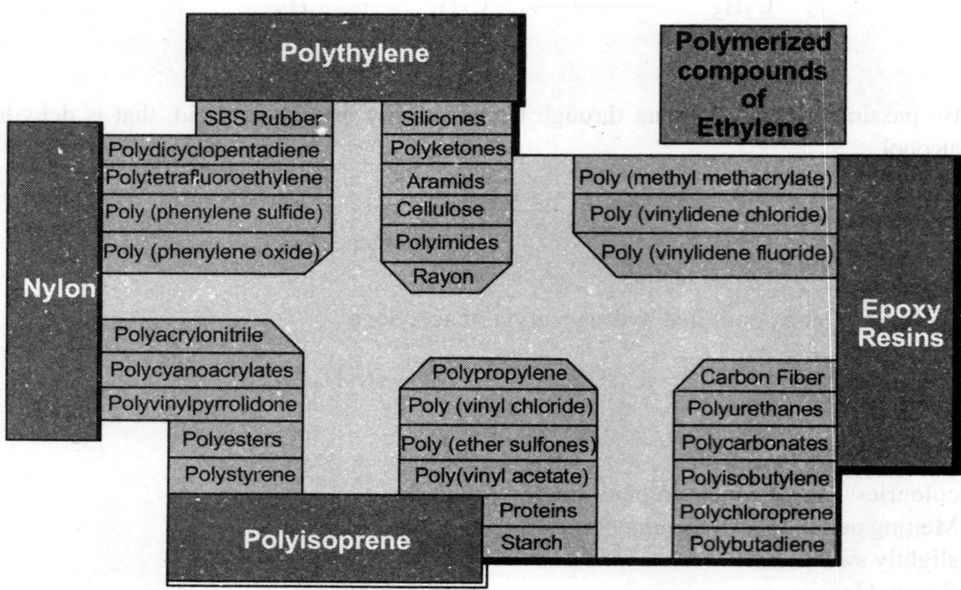

SBS – Rubber

Polystyrene is a tough hard plastic, and this gives SBS its durability. Polybutadiene is a rubbery material, and this gives SBS its rubber-like properties. In addition, the polystyrene chains tend to clump together. When one styrene group of one SBS molecule joins one clump, and the other polystyrene chain of the same SBS molecule joins another clump, the different clumps become tied together with rubbery polybutadiene chains. This gives the material the ability to retain its shape after being stretched. The SBS is made with some really clever chemistry using what is called living anionic polymerization.

We start off using a technique called living anionic polymerization. A living polymerization is a polymerization which takes place without any termination reactions. This means that once all the monomer in your beaker is used up, and has been turned into polymer, the polymer chains are still active. This means that if you put more monomer in the beaker, it would add to the polymer and make the polymers bigger. This lets us do some tricks.

The first thing we have to do is make a chain of living polystyrene. This is done by polymerizing the monomer styrene with an anionic initiator like butyl lithium.

$$CH_3{-}CH_2{-}CH_2{-}CH_2{-}Li + CH_2{=}CH(C_6H_5) \longrightarrow CH_3{-}CH_2{-}CH_2{-}CH_2{-}[CH_2{-}CH(C_6H_5)]_n{-}CH_2{-}\overset{H}{\underset{C_6H_5}{C}}{:}^{-}\ Li^+$$

Remember now this polystyrene chain is living, so if we add a second monomer to it, it will add to the polymer. So we'll add some of the monomer butadiene.

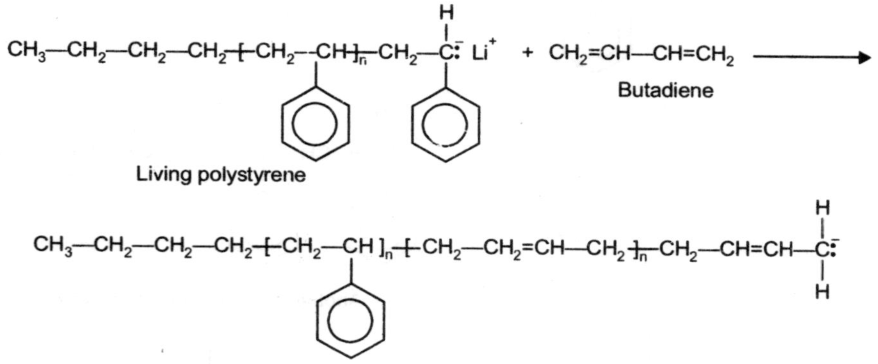

$$CH_3-CH_2-CH_2-CH_2-[CH_2-CH(C_6H_5)]_n-CH_2-CH(C_6H_5):^- \; Li^+ + CH_2=CH-CH=CH_2 \longrightarrow$$

Living polystyrene

Butadiene

$$CH_3-CH_2-CH_2-CH_2-[CH_2-CH(C_6H_5)]_n-[CH_2-CH_2=CH-CH_2]_n-CH_2-CH=CH-CH_2:^-$$

Living styrene-butadiene block copolymer

That gives us a living styrene-butadiene block copolymer.

The next step is obvious: just add more styrene monomer, and get a styrene-butadiene-styrene triblock copolymer. Easy as pie. Funny thing, though. Although butadiene monomer will add to the anion at the end of a polystyrene chain, styrene monomer won't add to the anion at the end of a living polybutadiene chain. This is most inconvenient. To get around this, we do a little trick: we're going to react it with a compound called dichlorodimethylsilane.

Dichlorodimethylsilane

$$CH_3-CH_2-CH_2-CH_2-[CH_2-CH(C_6H_5)]_n-[CH_2-CH=CH-CH_2]_n-CH_2-CH=CH-CH_2:^- + Cl-Si(CH_3)_2-Cl$$

$$\longrightarrow CH_3-CH_2-CH_2-CH_2-[CH_2-CH(C_6H_5)]_n-[CH_2-CH=CH-CH_2]_n-Si(CH_3)_2-Cl$$

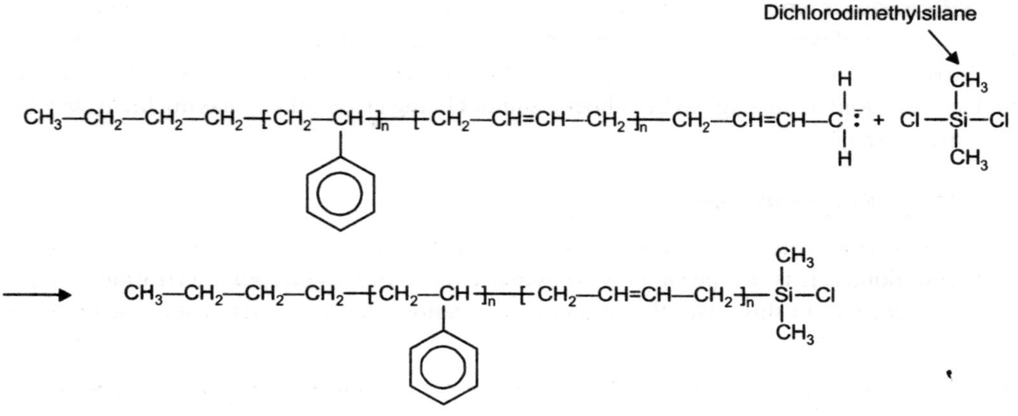

The end copolymer is capped with a dichlorosilane.

The anionic living chain kicks off a chlorine atom from the silane, and we get a chlorosilane end-capped polymer. So what good is that? Now our polymer is no longer living! It's good because we can do something with this chlorosilane end-capped polymer. You see, if we take living polystyrene homopolymer, it will react with the chlorosilane end-capped polymer, just like the styrene-butadiene copolymer reacted with the dichlorodimethylsilane.

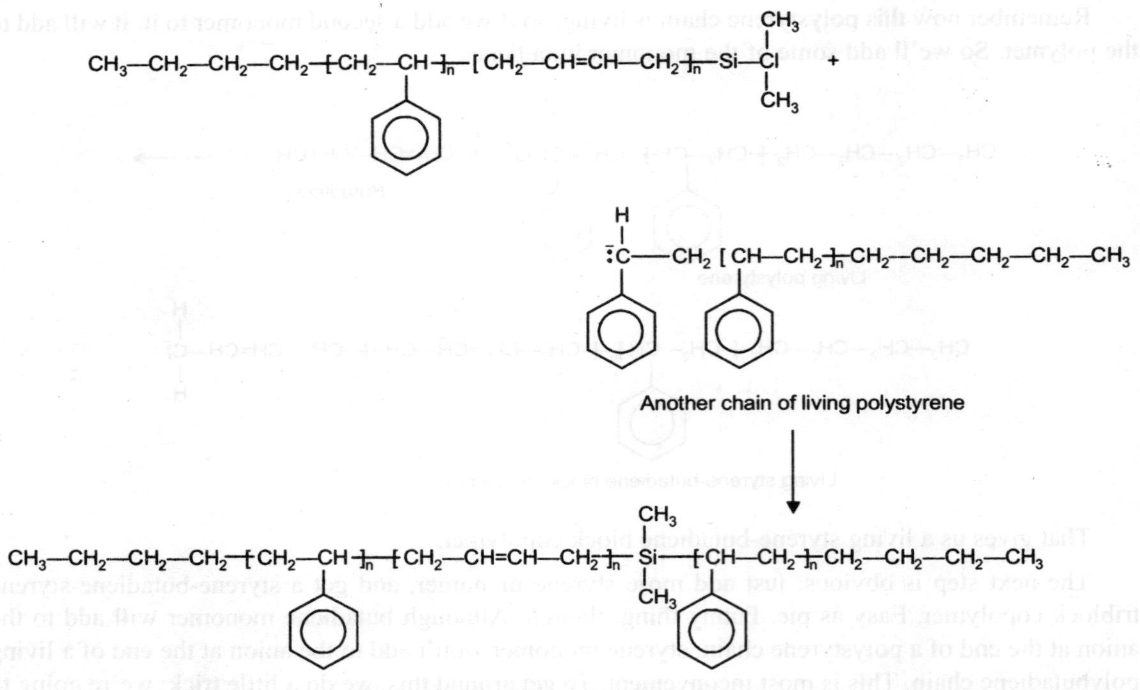

CH_3—CH_2—CH_2—CH_2—[CH_2—CH(C$_6H_5$)]$_n$—[CH_2—CH=CH—CH_2]$_n$—Si(CH_3)$_2$—Cl +

H—C:(C$_6H_5$)—CH_2 [CH(C$_6H_5$)—CH_2]$_n$—CH_2—CH_2—CH_2—CH_3

Another chain of living polystyrene

↓

CH_3—CH_2—CH_2—CH_2—[CH_2—CH(C$_6H_5$)]$_n$—[CH_2—CH=CH—CH_2]$_n$—Si(CH_3)$_2$—[CH(C$_6H_5$)—CH_2]$_n$—CH_2—CH_2—CH_2—CH_3

Styrene-butadiene-styrene triblock copolymer

The SBS is also a type of unusual material called a *thermoplastic elastomer*. These are materials which behave like elastomeric rubbers at room temperature but when heated can be processed like plastics. Most types of rubber are difficult to process because they are crosslinked. But SBS and other thermoplastic elastomers manage to be rubbery without being crosslinked, making them easy to process into nifty useful shapes.

Anionic Vinyl Polymerization

Anionic vinyl polymerization is a method of making polymers from small molecules containing carbon-carbon double bonds. It is a type of vinyl polymerization. In anionic polymerization, the process is begun by an *initiator*. In this case, the initiator is an anion; that is, an ion with a negative electrical charge.

There are a lot of different initiators used in anionic vinyl polymerization, but the most often used is an unassuming little molecule called butyl lithium.

$$CH_3 - CH_2 - CH_2 - CH_2 - Li$$

butyl lithium

Now a little bit of the butyl lithium will always fall apart. Not a lot, but some. It falls apart to form a positive lithium cation and a negative butyl anion. We call an anion like this where the negative charge is on a carbon atom a *carbanion*.

$$CH_3-CH_2-CH_2-CH_2-Li \longrightarrow CH_3-CH_2-CH_2-\overset{H}{\underset{H}{|}}\!\!\!C{:}^- + Li^+$$

A pair of electrons from the butyl anion will be donated to one of the double bond carbon atoms of the monomer. Now this carbon atom already has eight electrons in its outer shell which it shares with the atoms to which it is bonded, so one pair of these electrons, specifically a pair in the carbon-carbon double bond, will leave the carbon atom, and settle on the other carbon atom of the carbon-carbon double bond. This forms a new carbanion, with the negative charge resting on that carbon. The process in which the butyl lithium falls apart, and the butyl anion reacts with a monomer molecule is called *initiation*.

$$CH_3-CH_2-CH_2-CH_2{:}^-\ Li^+ \quad H_2C=CH_2 \longrightarrow CH_3-CH_2-CH_2-CH_2-CH_2-CH_2{:}^-\ Li^+$$

The carbanion now reacts with another monomer molecule in just the same manner as the initiator reacted with the first monomer molecule; another carbanion is generated. This keeps happening, and each time another monomer is added to the growing chain, a new anion is generated allowing another monomer to be added. In this way the polymer chain grows. This adding of monomer after monomer is called *propagation*.

$$CH_3-CH_2-CH_2-CH_2-CH_2-CH_2{:}^-\ Li^+ + H_2C=CH_2 \longrightarrow CH_3-CH_2-CH_2-CH_2-CH_2-CH_2-CH_2-CH_2{:}^-\ Li^+$$

$$\longrightarrow -\!\!\left[CH_2-CH_2\right]_n\!\!-$$

The Chain That Wouldn't Die

Now while you may get this nagging feeling that this can't go on forever, that something has to put a stop to this convenient process. In many cases, the only thing that stops monomers from adding to the growing chain, is that eventually there are no more monomer molecules in the beaker left to add! And even then, if someone came along some time later and dumped *more* monomer into the beaker, they would add to the chain and the chain would grow some more! Some chains of polystyrene have been known to stay active like this for years. In order to stop them, something like water, which reacts with the carbanions, has to be added to the polymer. Systems like this are called *living* anionic polymerizations. This allows us to do some interesting tricks...

Modular Chemistry

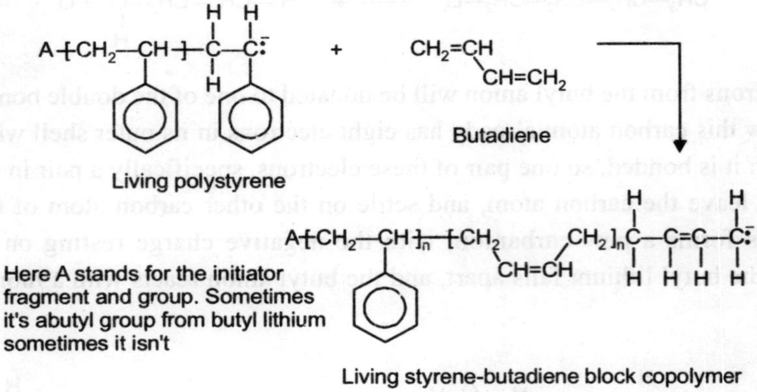

Living polystyrene

Butadiene

Here A stands for the initiator fragment and group. Sometimes it's abutyl group from butyl lithium sometimes it isn't

Living styrene-butadiene block copolymer

Some time ago someone was pondering this beaker full of polymer which would stay active for years; and how if one added additional monomer, this new monomer would add to the living polymer chains, and came up with an idea. Instead of adding the same monomer to the solution filled with living polymer, why not add a different monomer? The result was a polymer whose chains consisted of a long stretch of one type of polymer, and a second long stretch of another polymer. Polymers like this are called block copolymers. For example, a solution of living polystyrene chains will react with butadiene to give a styrene-butadiene block copolymer.

Armids

Aramids are a family of nylons, including Nomex® and Kevlar® . Kevlar® is used to make things like bulletproof vests and puncture resistant bicycle tires. I suppose one could even make bulletproof bicycle tires from Kevlar® if one felt the need.

Blends of Nomex® and Kevlar® are used to make fireproof clothing. Nomex® is what keeps the monster truck and tractor drivers from burning to death should their fire-breathing rigs breathe a little too much fire. Thanks to Nomex® , an important part of American culture can be practiced safely. (Polymers play another part in the monster truck show in the form of elastomers from which those giant tires are made.) Nomex® -Kevlar® blends also protect fire fighters.

Kevlar® is a polyamide, in which all the amide groups are separated by *para*-phenylene groups, that is, the amide groups attach to the phenyl rings opposite to each other, at carbons 1 and 4.

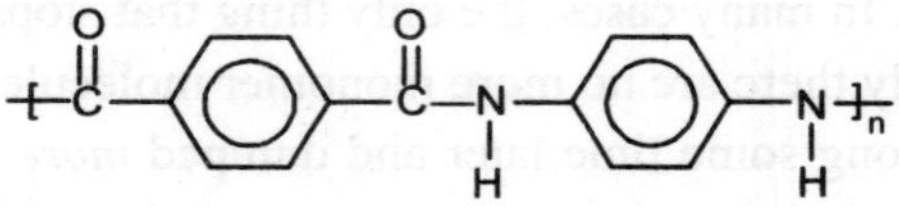

In Kevlar the aromatic groups are all linked into the backbone chain through the 1 and 4 positions. This is called *para*-linkage.

Nomex® , on the other hand, has *meta*-phenylene groups, that is, the amide groups are attached to the phenyl ring at the 1 an 3 positions.

In Nomex the aromatic groups are all linked into the backbone chain through the 1 and 3 positions. This is called *meta*-linkage.

Kevlar® is a very crystalline polymer. It took a long time to figure out how to make anything useful out of Kevlar® because it wouldn't dissolve in anything. So processing it as a solution was out. It wouldn't melt below a right toasty 500°C, so melting it down was out, too. Then a scientist named Stephanie Kwolek came up with a brilliant plan. Click here to find out what it was.

Aramids are used in the form of fibers. They form into even better fibers than non-aromatic polyamides, like nylon 6,6.

Why? Why?

It has to do with a little quirky thing that amides do. They have the ability to adopt two different shapes, or *conformations*. You can see this in the picture of a low molecular weight amide. The two pictures are the same compound, in two different conformations. The one on the left is called the *trans* conformation, and the one on the right is the *cis*- conformation.

a *trans*-amide a *cis*-amide

In Latin, *trans* means "on the other side". So when the hydrocarbon groups of the amide are on opposite sides of the *amide bond*, the bond between the carbonyl oxygen and the amide nitrogen, it's called a *trans*-amide. Likewise, *cis* in Latin means "on the same side", and when both hydrocarbon groups are on the same side of the amide bond, we call it a *cis*-amide.

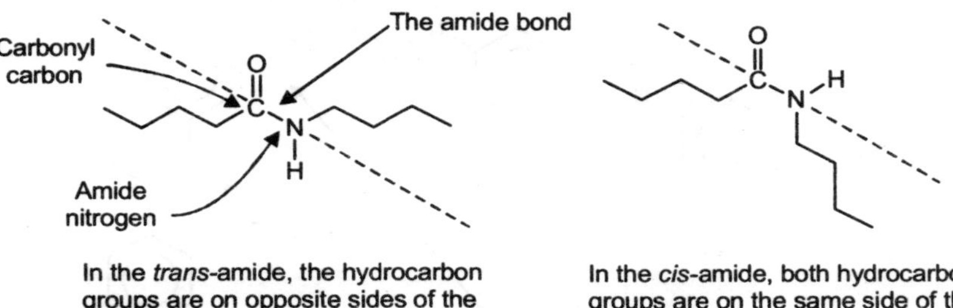

In the *trans*-amide, the hydrocarbon groups are on opposite sides of the amide bond

In the *cis*-amide, both hydrocarbon groups are on the same side of the amide bond

The same amide molecule can twist back and forth between the *cis*- and *trans*-conformations, given a little bit of energy.

The same *cis*- and *trans*-conformations exist in polyamides, too. When all the amide groups in a polyamide, like nylon 6,6 for example, are in the *trans*-conformation, the polymer is fully stretched out in a straight line. this is exactly what we want for fibers, because long straight, fully extended chains pack more perfectly into the crystalline form that makes up the fiber. But sadly, there's always at least some amide linkages in the *cis*-conformation. So nylon 6,6 chains never become fully extended.

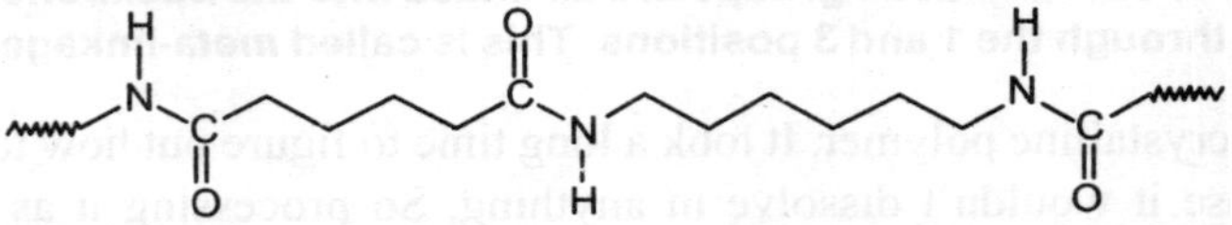

Nylon 6, 6 can stretch into fibres easily when all the amide groups are in the *trans*-conformation . . .

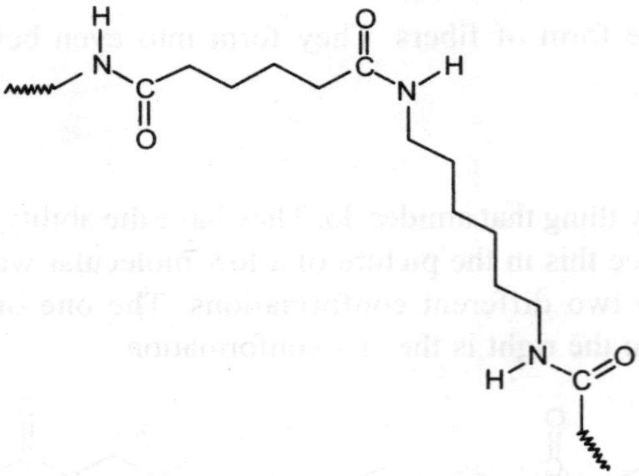

. . . but one *cis* linkage causes a kink which messes everything up!

But Kevlar® is different. When it tries to twist into the *cis*-conformation, the hydrogens on the big aromatic groups get in the way! The *cis*-conformation puts the hydrogens just a little closer to each other than they want to be. So Kevlar® stays nearly fully in the *trans*-conformation. So Kevlar® can fully extend to form beautiful fibres.

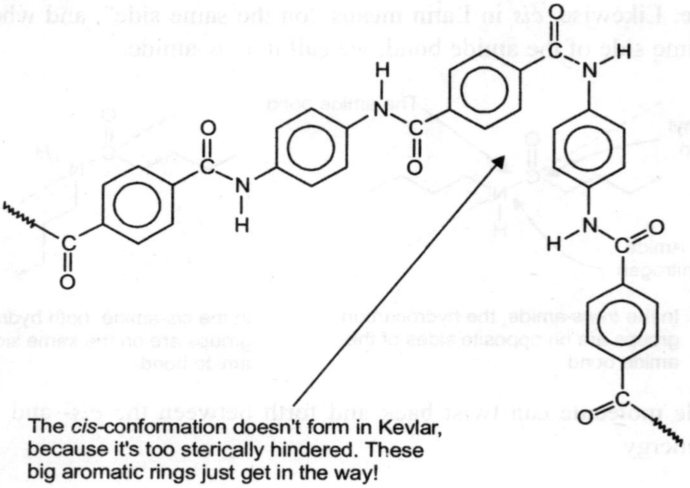

The *cis*-conformation doesn't form in Kevlar, because it's too sterically hindered. These big aromatic rings just get in the way!

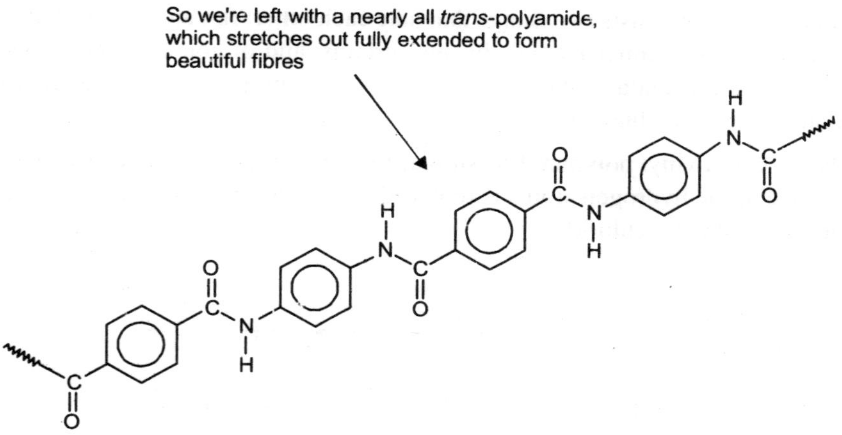

So we're left with a nearly all *trans*-polyamide, which stretches out fully extended to form beautiful fibres

Now it may help to look at a close-up picture of this. Look at the picture below and you can see that when Kevlar® tries to form the *cis*-conformation, there's not enough room for the phenyl hydrogens. So only the *trans*-conformation is usually found.

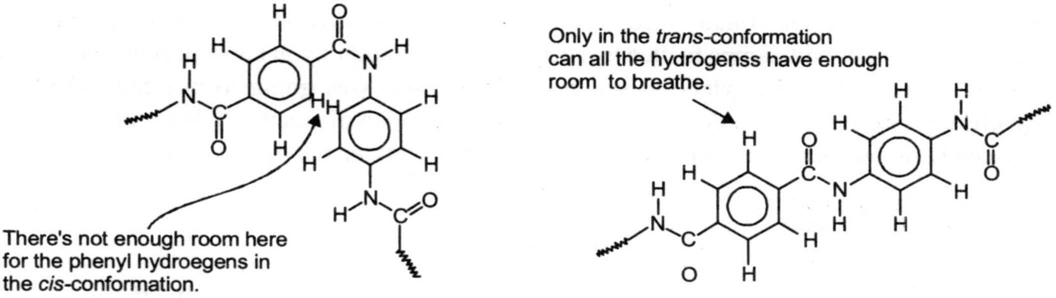

There's not enough room here for the phenyl hydroegens in the *cis*-conformation.

Only in the *trans*-conformation can all the hydrogenss have enough room to breathe.

But there's another polymer that stretches out even better called ultra-high molecular weight polyethylene. It even replaced Kevlar® for making bullet-proof vests! But back to Kevlar®.

Also the phenyl rings of adjacent chains stack on top of each other very easily and neatly, which makes the polymer even more crystalline, and the fibers even stronger.

Poly(vynyl chloride)

Poly(vinyl chloride) is the plastic known at the hardware store as PVC. This is the PVC from which pipes are made, and PVC pipe is everywhere. The plumbing in your house is probably PVC pipe, unless it's an older house. PVC pipe is what rural high schools with small budgets use to make goal posts for their football fields. But there's more to PVC than just pipe. The "vinyl" siding used on houses is made of poly(vinyl chloride). Inside the house, PVC is used to make linoleum for the floor. In the seventies, PVC was often used to make vinyl car tops.

PVC is useful because it resists two things that hate each other: fire and water. Because of its water resistance it is used to make raincoats and shower curtains, and of course, water pipes. It has flame resistance, too, because it contains chlorine. When you try to burn PVC, chlorine atoms are released, and chlorine atoms inhibit combustion.

Structurally, PVC is a vinyl polymer. It is similar to polyethylene, but on every other carbon in the backbone chain, one of the hydrogen atoms is replaced with a chlorine atom. It is produced by the free radical polymerization of vinyl chloride.

$$CH_2{=}CHCl \xrightarrow{\text{Free radical vinyl polymerization}} \left[-CH_2-CHCl- \right]_n$$

Vinyl chloride — Poly(vinyl chloride)

And here, is that monomer, vinyl chloride:

$$\left[-CH_2-CH(Cl)- \right]_n$$

The PVC was one of those odd discoveries that actually had to be made twice. It seems around a hundred years ago, a few German entrepreneurs decided they were going to make loads of cash lighting people's homes with lamps fuelled by acetylene gas. Wouldn't you know it, right about the time they had produced tons of acetylene to sell to everyone who was going to buy their lamps, new efficient electric generators were developed which made the price of electric lighting drop so low that the acetylene lamp business was finished. That left a lot of acetylene laying around.

Leftover acetylene from the acetylene lamp fiasco

$$H-C{\equiv}C-H + HCl \longrightarrow CH_2{=}CHCl \longrightarrow \left[-CH_2-CHCl- \right]_n$$

Vinyl chloride — Poly (vinyl chloride)

So in 1912 one German chemist, Fritz Klatte decided to try to do something with it, and reacted some acetylene with hydrochloric acid (HCl). Now this reaction will produce vinyl chloride, but at that time no one knew what to do with it, so he put it on the shelf, where it polymerized over time. Not knowing what to do with the PVC he had just invented, he told his bosses at his company, Greisheim Electron, who had the material patented in Germany. They never figured out a use for PVC, and in 1925 their patent expired.

Wouldn't you know it, in 1926 the very next year, and American chemist, Waldo Semon was working at B.F. Goodrich when he independently invented PVC. But unlike the earlier chemists, it dawned on him that this new material would make a perfect shower curtain. He and his bosses at B.F. Goodrich patented PVC in the United States (Klatte's bosses apparently never filed for a patent outside

Germany). Tons of new uses for this wonderful waterproof material followed, and PVC was a smash hit the second time around. You see, these polymers are so new that not much has been made out of them yet. But polyketones, which are plastics, by the way, have a lot going for them, and we figure it won't be long before a lot of stuff is made from them.

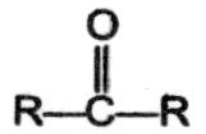

So what's so great about polyketones? Let's find out by taking a look at them. A polyketone is of course a polymer with a *ketone* group in the backbone chain. The polyketones we're going to talk about on this page are based on the picture you see at the top of the page, where *R′* is an ethylene linkage, $-CH_2CH_2-$. Shell has just put this family of polymers on the market and sells them under the name Carilon. This is not to be confused with a *carillon*, which is a musical instrument with differently pitched bells, which are often controlled with a piano-like keyboard. Let's see what one of these polymers looks like, next to a chain of good ol' polyethylene for comparison:

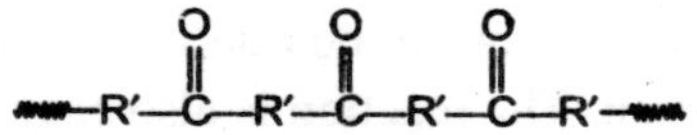

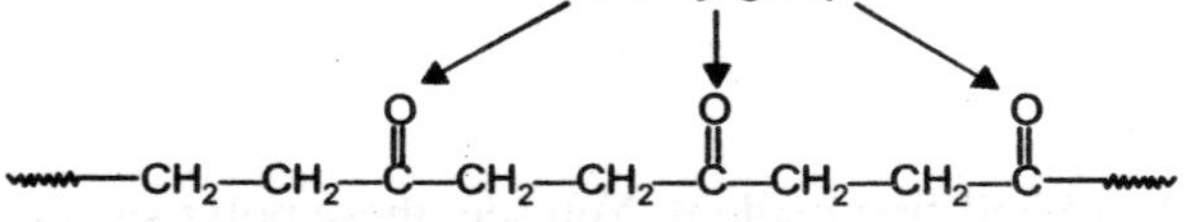

$$R-\overset{\displaystyle O}{\overset{\|}{C}}-R$$

a ketone

$$\sim R'-\overset{\displaystyle O}{\overset{\|}{C}}-R'-\overset{\displaystyle O}{\overset{\|}{C}}-R'-\overset{\displaystyle O}{\overset{\|}{C}}-R'\sim$$

a polyketone

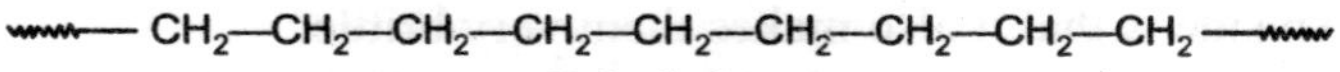

Carbonyl groups

$$\sim CH_2-CH_2-\overset{\displaystyle O}{\overset{\|}{C}}-CH_2-CH_2-\overset{\displaystyle O}{\overset{\|}{C}}-CH_2-CH_2-\overset{\displaystyle O}{\overset{\|}{C}}\sim$$

A polyketone

$$\sim CH_2-CH_2-CH_2-CH_2-CH_2-CH_2-CH_2-CH_2-CH_2\sim$$

Polyethylene

It's not much different. The only difference is that the polyketone has those carbonyl groups in it. But those carbonyls do a lot. You see, carbonyl groups are very polar. This is because oxygen is electronegative and draws electrons away from the carbon atom. So oxygen has a slight negative charge and the carbon has a slight positive charge. These polar carbonyl groups are attracted to each other, and very strongly at that. This attraction is so strong that while polyethlyene melts at a mere 140°C, the polyketone doesn't melt until 255 °C.

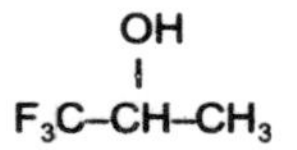

$$\overset{\displaystyle O^{\delta^-}}{\overset{\|}{C_{\delta^+}}}$$

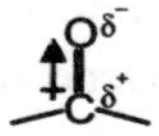

The oxygen sucks electrons away from the carbon, so the carbonyl group is polar.

These polyketones have another interesting property. They won't dissolve in much of anything. To dissolve it you have to use something really weird like hexafluoroisopropanol. This comes in handy should you want to make a car part out of this stuff. A car part is no good if it dissolves when you spill gasoline on it. But that's no problem with polyketones

$$F_3C-\overset{\displaystyle OH}{\overset{|}{C}}H-CH_3$$

hexafluorisopropanol

This is all very nice, but really, now, what's the big deal? There are lots of high performance plastics out there. There are poly(ether sulfones), poly(phenylene sulfide), polyimides... We've seen high performance before. What makes polyketones so special?

This is what makes polyketones so special. To make polyketones, you take ethylene gas and carbon monoxide, and react them with a palladium(II) catalyst:

$$H_2C{=}CH_2 + C{\equiv}O \xrightarrow{\text{Pd(II) catalyst}} {+}CH_2{-}CH_2{-}\overset{\overset{O}{\|}}{C}{+}_n$$

Ethylene   Carbon monoxide   A polyketone

Ethylene is dirt cheap; it's the monomer for polyethylene. Carbon monoxide is dirt cheap, too. You make carbon monoxide every time you burn wood, or a candle, or just about anything containing carbon. The reaction is also easy to carry out. Most high performance polymers are difficult to make, because the chemistry has to be *just* right for the reactions to work. Cheap monomers, and easy chemistry make polyketones rather cheap, a whole lot cheaper than other high performance plastics. In time the price may come down to less than a dollar per pound! This is why we think a lot of stuff is going to be made out of polyketones in the future.

Here are some models of the monomers if you want to play with them. Ethylene is on the left and carbon monoxide is on the right.

One Last Detail

There was a problem with this polyketone when Shell first made it. You see, these polymers are very highly crystalline, that is, the polymer chains are packed together in a very orderly fashion. This makes the polyketones very strong, but the down side is that it also makes them very brittle. the chemists went to work and came up with a solution. They decided to throw a little bit of propylene in with the ethylene and the carbon monoxide when they make the polymer. Propylene reacts just like ethylene does, so we end up with a polymer that looks like this:

$CH_2{=}CH{-}CH_3$

Propylene

Ethylene linkage

$$\sim CH_2{-}CH_2{-}\overset{O}{\overset{\|}{C}}{-}CH_2{-}CH_2{-}\overset{O}{\overset{\|}{C}}{-}CH_2{-}CH_2{-}\overset{O}{\overset{\|}{C}}{-}CH_2{-}\underset{CH_3}{\underset{|}{CH}}{-}\overset{O}{\overset{\|}{C}}{-}CH_2{-}CH_2{-}\overset{O}{\overset{\|}{C}}{-}\sim$$

Every so often these methyl groups show up along the backbone chain.

Every now and then you get an extra methyl group attached to one of the ethylene units. These methyl groups get in the way when the polymer chains try to pack into crystals. They can still pack, but not as well as before. The bad part is that the melting temperature drops to about 220 °C, because the crystals aren't as strong. But the good part is that the polymer is now a lot tougher and less brittle. This polyketone made with ethylene and carbon monoxide.

Carbon Fiber

Carbon fibre is a polymer which is a form of graphite. Graphite is a form of pure carbon. In graphite the carbon atoms are arranged into big sheets of hexagonal aromatic rings. The sheets look like chicken wire.

Carbon fibre is a form of graphite in which these sheets are long and thin. You might think of them as ribbons of graphite. Bunches of these ribbons like to pack together to form fibers, hence the name carbon *fibre*.

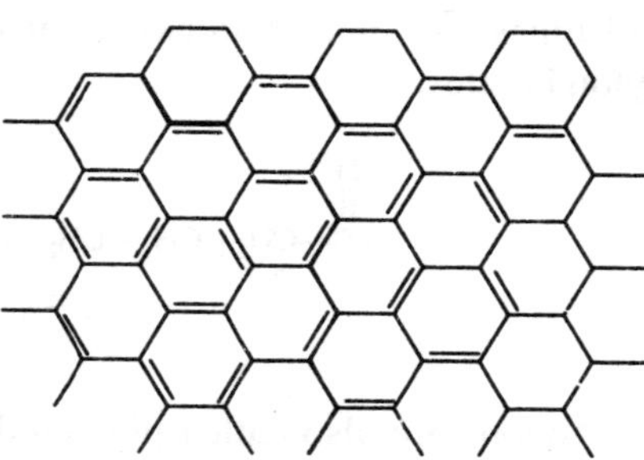

A section of a sheet of graphite

These fibers aren't used by themselves. Instead, they're used to reinforce materials like epoxy resins and other thermosetting materials. We call these reinforced materials *composites* because they have more than one component.

Carbon fibre reinforced composites are very strong for their weight. They're often stronger than steel, but a whole lot lighter. Because of this, they can be used to replace metals in many uses, from parts for airplanes and the space shuttle to tennis rackets and golf clubs.

$$-\!\!\left[CH_2-\underset{\substack{|\\ C\equiv N}}{CH}\right]_n\!\!-$$

Polyacrylonitrile

Carbon fibre is made from another polymer, called polyacrylonitrile, by a complicated heating process.

Polychloroprene

Polychloroprene is usually sold under the trade name Neoprene. It's especially resistant to oil. It was the first synthetic elastomer, or rubber to be a hit commercially. It was invented by the Arnold Collins, while working under the same fellow who invented nylon, Wallace Carothers.

Polychloroprene is made from the monomer chloroprene, believe it or not, like this:

$$H_2C=CH-C(Cl)=CH_2 \longrightarrow -\!\!\left[CH_2-C(Cl)=CH-CH_2\right]_n\!\!-$$

Chloroprene Polychloroprene

Chloroprene is a diene monomer, that is, it has two carbon-carbon double bounds

Chloroprene has two double bonds, so we call it a *diene*. Polychloroprene has properties similar to those of other diene polymers like polyisoprene and polybutadiene. If you want to know more about diene polymers.

Nylon

Nylons are one of the most common polymers used as a fibre. Nylon is found in clothing all the time, but also in other places, in the form of a thermoplastic. Nylon's first real success came with its use in women's stockings, in about 1940. They were a big hit, but they became hard to get, because the next year the United States entered World War II, and nylon was needed to make war materials, like parachutes and ropes. But before stockings or parachutes, the very first nylon product was a toothbrush with nylon bristles.

$$-\!\!\left[\overset{O}{\overset{\|}{C}}-CH_2-CH_2-CH_2-CH_2-\overset{O}{\overset{\|}{C}}-NH-CH_2-CH_2-CH_2-CH_2-CH_2-CH_2-NH\right]_n$$

Amide group

Nylons are also called polyamides, because of the characteristic amide groups in the backbone chain. Proteins, such as the silk nylon was made to replace, are also polyamides. These amide groups are very polar, and can hydrogen bond with each other. Because of this, and because the nylon backbone is so regular and symmetrical, nylons are often crystalline, and make very good fibres.

$$-\!\!\left[\overset{O}{\overset{\|}{C}}-CH_2-CH_2-CH_2-CH_2-\overset{O}{\overset{\|}{C}}-NH-CH_2-CH_2-CH_2-CH_2-CH_2-CH_2-NH\right]_n$$

Six carbon atoms — Six carbon atoms

The nylon in the pictures on this page is called nylon 6,6, because each repeat unit of the polymer chain has two stretches of carbon atoms, each being six carbon atoms long. Other nylons can have different numbers of carbon atoms in these stretches.

Nylons can be made from diacid chlorides and diamines. Nylon 6,6 is made from the monomers adipoyl chloride and hexamethylene diamine.

Cl–C(=O)–(CH₂)₄–C(=O)–Cl + H₂N–(CH₂)₆–NH₂

Adipoyl chloride — Hexamethylene diamine

$\longrightarrow$ –[C(=O)–(CH₂)₄–C(=O)–NH–(CH₂)₆–NH]–

Nylon 6,6

This is one way of making nylon 6,6 in the laboratory. But in a nylon plant, it's usually made by reacting adipic acid with hexamethylene diamine:

Adipic acid + Hexamethylene diamine → Nylon 6,6

Another kind of nylon is nylon 6. It's a lot like nylon 6,6 except that it only has one kind of carbon chain, which is six atoms long.

ε-caprolactam → $-[CH_2-CH_2-CH_2-CH_2-CH_2-C(=O)-N(H)]_n-$

Nylon 6

It's made by a ring opening polymerization form the monomer caprolactam. Nylon 6 doesn't behave much differently from nylon 6,6. The only reason both are made is because DuPont patented nylon 6,6, so other companies had to invent nylon 6 in order to get in on the nylon business

Poly(methyl methacrylate)

Poly(methyl methacrylate), which scientists call PMMA, is a clear plastic, used as a shatterproof replacement for glass. The barrier at the ice rink which keeps hockey pucks from flying in the faces of fans is made of PMMA. The chemical company Rohm and Haas makes windows out of it and calls it Plexiglas. Ineos Acrylics also makes it and calls it Lucite. Lucite is used to make the surfaces of hot tubs, sinks, and the ever popular one piece bathtub and shower units, among other things.

When it comes to making windows, PMMA has another advantage over glass. PMMA is more transparent than glass. When glass windows are made too thick, they become difficult to see through. But PMMA windows can be made as much as 13 inches (33 cm) thick, and they're still perfectly transparent. This makes PMMA a wonderful material for making large aquariums, whose windows must be thick in order to contain the high pressure millions of gallons of water. In fact, the largest single window in the world, an observation window at California's Monterrey Bay Aquarium, is made of one big piece of PMMA which is 54 feet long, 18 feet high, and 13 inches thick (16.6 m long, 5.5 m high, and 33 cm thick).

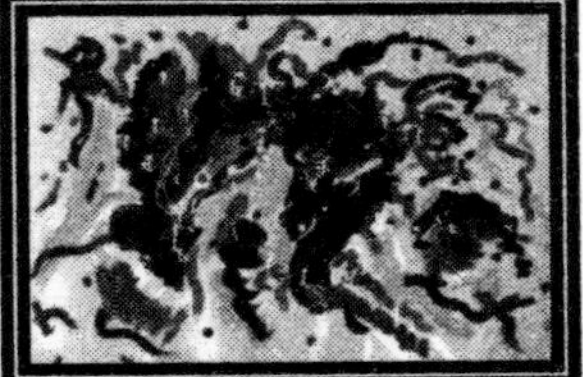

PMMA is also found in paint. The painting on your right, *Acrylic Elf* was painted by Pete Halverson with acrylic paints. Acrylic "latex" paints

often contain PMMA suspended in water. The PMMA doesn't dissolve in water, so dispersing PMMA in water requires we use another polymer to make water and PMMA compatible with each other.

But PMMA is more than just plastic and paint. Often lubricating oils and hydraulic fluids tend to get really viscous and even gummy when they get really cold. This is a real pain when you're trying to operate heavy equipment in really cold weather. But when a little bit PMMA is dissolved in these oils and fluids, they don't get viscous in the cold, and machines can be operated down to -100 °C (-150 °F), that is, presuming the rest of the machine can take that kind of cold!

The PMMA is a vinyl polymer, made by free radical vinyl polymerization from the monomer methyl methacrylate.

Free radical vinyl polymerization

Methyl methacrylate → Poly (methyl methacrylate)

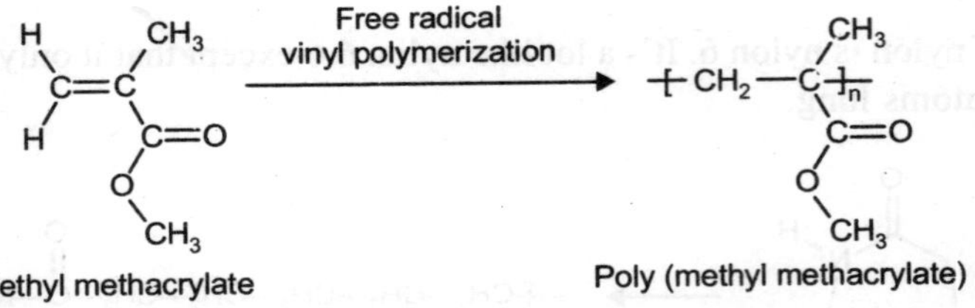

As you can see from the above picture, the structure of methyl methacrylate kind of looks like Massachusetts. But Massachusetts doesn't polymerize, because there's only one. Here's a better, 3-D look at the monomer methyl methacrylate:

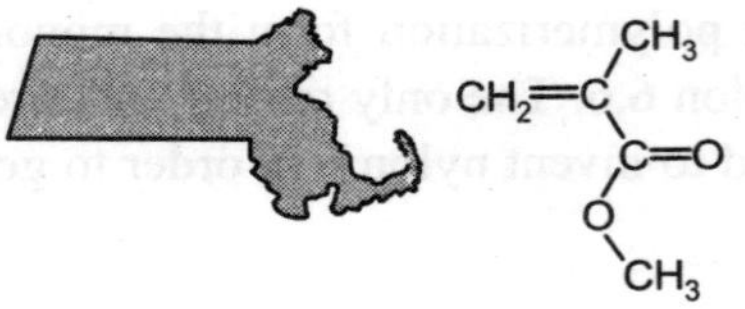

Poly(ether sulfones)

Poly(ether sulfones), PES for short, are high performance polymers. The most popular is one made by Union Carbide called Udel™. Poly(ether sulfones) act a lot like polycarbonates, but they're a lot more heat resistant.

Poly(ether sulfones) also can stand up well to water and steam, so they're used to make things like cookware and medical products that need to be sterilized between uses.

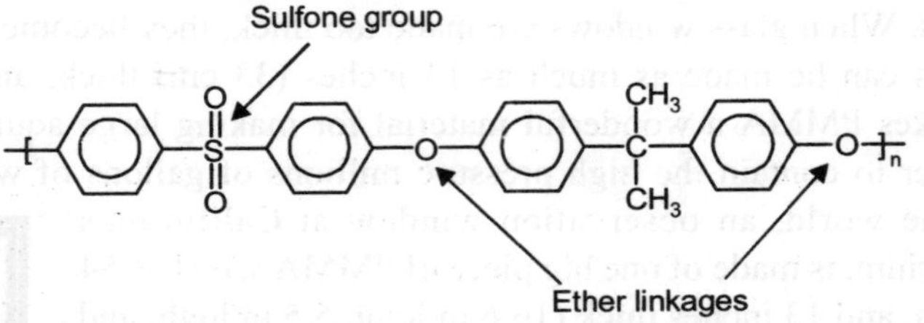

Sulfone group

Ether linkages

Poly(ether sulfones get their name because they have *ether* groups and *sulfone* groups in their backbone chains. The PES polymers have high glass transition temperatures, or T_g's, because the sulfone groups are so stiff. In fact, poly(phenyl sulfone) is so stiff that it doesn't have a glass transition temperature! It stays hard as a rock right up until it decomposes at around 500 °C.

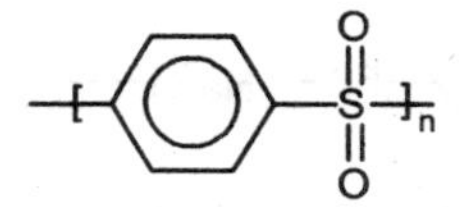

Poly(phenylene sulfone)

This is bad. This means that it can't be processed. In order to make it processable, we have to make that chain a little more flexible so the polymer will soften at a reasonable temperature. We do this by putting flexible groups in the backbone chain, namely ether linkages.

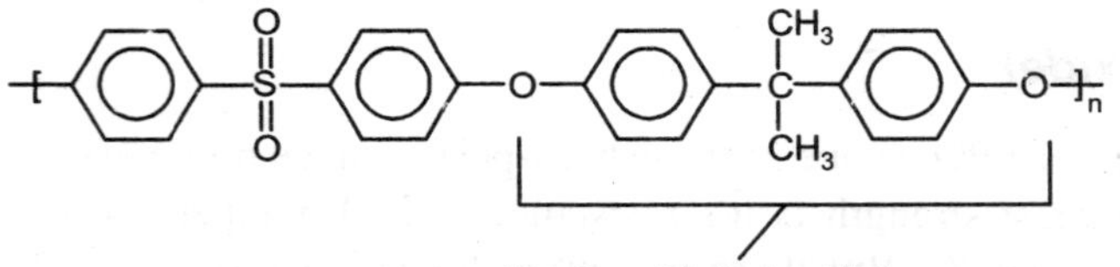

Bisphenol A derived linkage

Udel™ uses a bisphenol A derived linkage to make the chain flexible. You'll notice that this polymer has two ether linkages in the repeat unit. For this reason Udel™ is more properly called a poly(ether sulfone).

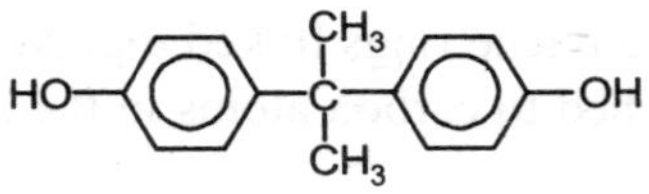

Bisphenol A

We call this a bisphenol A linkage because it's based on the monomer bisphenol A, but more on that later.

This brings the *Tg* down to 190 °C. By playing around and using different kinds of flexible groups we can alter the *Tg* further. If we use a less flexible linkage, the *Tg* jumps to 230 °C.

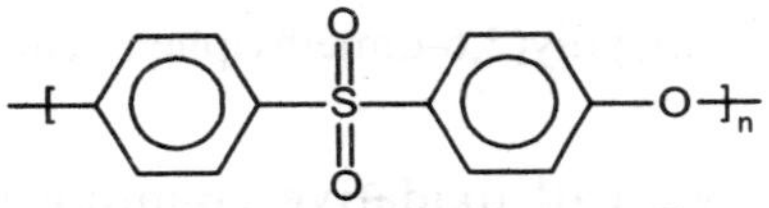

ICI's Polyether Sulfone 200 P™

This PES is made by Imperial Chemical Industries, and they call it by the romantic name of Polyether Sulfone 200 P™. You'll notice that unlike Udel™, this polymer has only one ether linkage between sulfone groups. This makes the chain stiffer and the *Tg* higher.

But you ask, "How do we make a wonderful poly(ether sulfone)?" Like this: One reacts the sodium salt of bisphenol A (remember bisphenol A?) and reacts it with di-*para*-fluorophenylsulfone in a solvent like dimethylsulfoxide (DMSO) at between 130 and 160 °C.

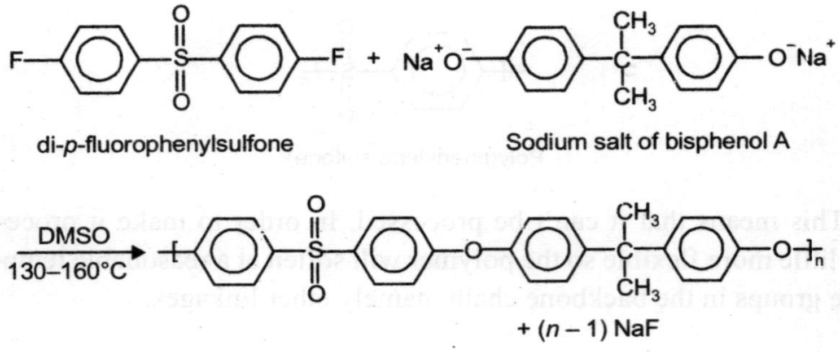

di-*p*-fluorophenylsulfone + Sodium salt of bisphenol A

DMSO, 130–160°C → + (*n* – 1) NaF

Poly(phenylene oxide)

Poly(phenylene oxide), or PPO, is one of those high-performance polymers we like to call *engineering thermoplastics*. It's biggest strength is it's resistance to high temperatures. It has a very high glass transition temperature, 210 °C. But there is a price for being resistant to heat. Most polymers are processed at high temperature in a liquid-like state. But if your polymer won't become liquid-like at reasonable temperatures, you can't process it! For this reason, PPO is often made into blends with high-impact polystyrene (HIPS for short). Blending PPO with HIPS makes the PPO easier to process, plus it gives PPO some resilience. PPO needs this toughening because by itself PPO can be brittle in some situations. General Electric makes PPO/HIPS blends and sells them under the name Noryl™.

Structurally, PPO is made of phenylene rings linked together by ether linkages in the 1,4 or *para*-positions, with a methyl group attached to carbon atoms in the 2 and 6 positions.

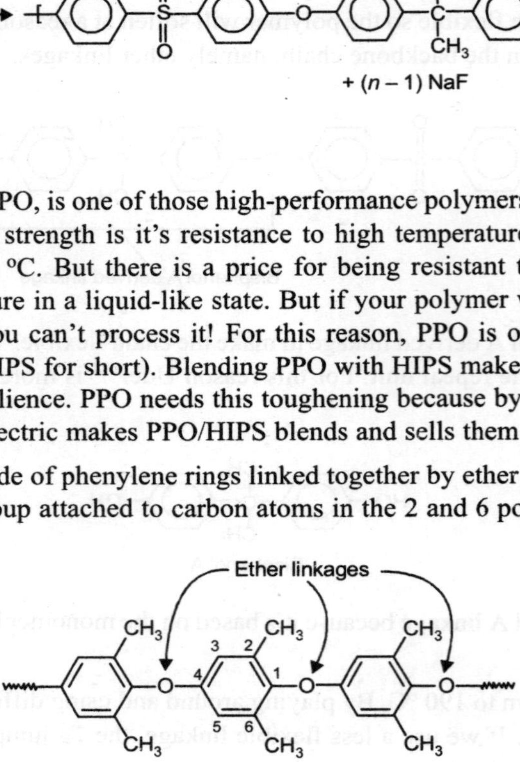

Ether linkages

This polymer should really be called poly(2,6-dimethylphenylene oxide), but we call it poly(phenylene oxide) because we're lazy.

The PPO is made by what we call oxidative coupling polymerization of the monomer 2,6-dimethylphenol. Water is a by-product, and thus this is a condensation polymerization

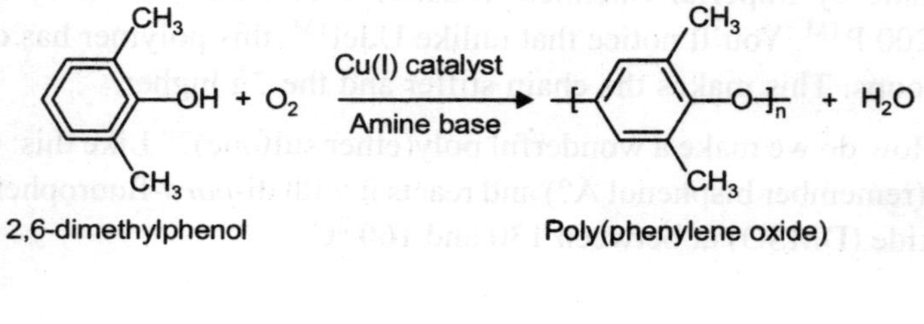

Cu(I) catalyst, Amine base

2,6-dimethylphenol → Poly(phenylene oxide) + H_2O

Poly(phenylene sulfide)

Poly(phenylene sulfide), or PPS, is one of those really high-performance plastics that is very strong and can resist very high temperatures. How high? PPS doesn't melt until around 300 °C. It's also flame resistant. People in the plastics business call high performance plastics like PPS *engineering thermoplastics* when they want to feel like big shots. The PPS is expensive, so it's used only when good heat resistance is needed. Electrical sockets, and other electrical components are made of PPS. So are certain parts of cars, microwave ovens, and hairdryers.

Here's what you do: You have to react *para*-dichlorobenzene and sodium sulfide in a polar solvent like *N*-methyl pyrrolidone.

$$Cl-C_6H_4-Cl \quad + \quad Na_2S \longrightarrow -[C_6H_4-S]_n-$$

p-dichlorobenzene Sodium sulfide Polyphenylene sulfide

This gives you a low molecular weight PPS, which is a good thing if you want to use it as a coating. But if you want to use it as a material, you need to heat it in oxygen, and this will give you higher molecular weight.

Poly(vinyl acetate)

Poly(vinyl acetate), or PVA for short, is one of those low-profile behind-the-scenes polymers. It isn't blatantly obvious where it's found, as is the case with polyethylene or polystyrene. PVA likes to hide. But it's everywhere, if you're willing to look for it. It's a polymer that rewards one who is willing to look beyond the surface. One place PVA can be found hiding is between two pieces of wood that are glued together. PVA is used to make wood glues, as well as other adhesives. Paper and textiles often have coatings made of PVA and other ingredients to make them shiny.

The PVA is a vinyl polymer, as if you couldn't guess from the name. It's made by free radical vinyl polymerization of the monomer vinyl acetate.

$$CH_2{=}CH(-O-C(=O)-CH_3) \xrightarrow{\text{Free radical vinyl polymerization}} -[CH_2-CH(-O-C(=O)-CH_3)]_n-$$

Vinyl acetate Poly (vinyl acetate)

One interesting place where PVA likes to hide is in paint cans. So what is PVA doing hiding in your paint? PVA is the latex in acrylic latex paint. So what does that mean? Let me explain. Before PVA can hide in a can of paint, we have to react it with NaOH and methanol like this:

$$-[CH_2-CH(O-C(=O)-CH_3)]_n- \xrightarrow[\text{methanol}]{NaOH} -[CH_2-CH(O-H)]_n-$$

Poly(vinyl acetate) Poly(vinyl alcohol)

Poly(vinylidene chloride) is polymer that's only used for one thing, but it's an important thing. PVDC, as we call it for short, is the plastic wrap that food comes in at the grocery store, and that you put over your dish of casserole that you're taking to the pot luck supper. Dow Chemical makes this stuff, and calls it *Saran*.

Poly(vinylidene chloride) is a vinyl polymer. As you might be guessing, it is made from the monomer vinylidene chloride, using free radical vinyl polymerization like this:

$$CH_2=CCl_2 \longrightarrow -[CH_2-CCl_2]_n-$$

Vinylidene chloride Poly(vinylidene chloride)

Polyacrylonitrile (ABS plastic)

Polyacrylonitrile is used for very few products an average consumer would be familiar with, except to make another polymer, carbon fibre. Homopolymers of polyacrylonitrile have been uses as fibres in hot gas filtration systems, outdoor awnings, sails for yachts, and even fibre reinforced concrete. But mostly copolymers containing polyacrylonitrile are used as fibres to make knitted clothing, like socks and sweaters, as well as outdoor products like tents and such. If the label of some piece of clothing says "acrylic", then it's made out of some copolymer of polyacrylonitrile. Usually they're copolymers of acrylonitrile and methyl acrylate, or acrylonitrile and methyl methacrylate:

$$-[CH_2-CH(C\equiv N)]_n-[CH_2-CH(C(=O)-O-CH_3)]_m-$$

Poly(acrylonitrile-*co*-methyl acrylate)

$$-[CH_2-CH(C\equiv N)]_n-[CH_2-C(CH_3)(C(=O)-O-CH_3)]_m-$$

Onitile-*co*-methyl methacrylate)

Also, sometimes we make copolymers of acrylonitrile and vinyl chloride. These copolymers are flame-retardant, and the fibres made from them are called *modacrylic* fibres.

$$-[CH_2-CH(C\equiv N)]_n-[CH_2-CH(Cl)]_m-$$

Poly(acrylonitrile-*co*-vinyl chloride)

But the slew of copolymers of acrylonitrile doesn't stop there. Poly(styrene-*co*-acrylonitrile) (SAN) and poly(acrylonitrile-*co*-butadiene-*co*-styrene) (ABS), are used as plastics.

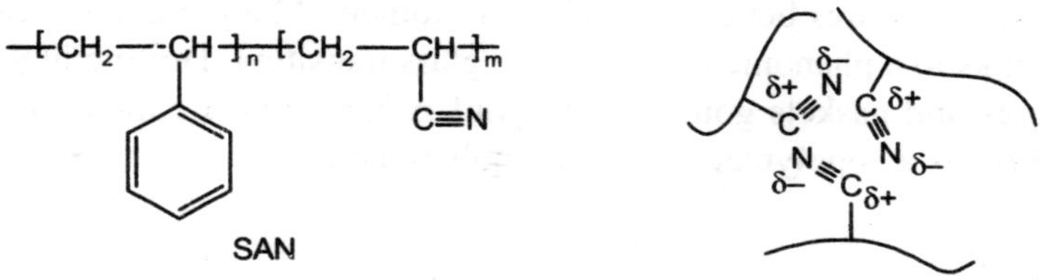

$$-\!\!\left[CH_2-CH\right]_n\!\!-\!\!\left[CH_2-CH(C{\equiv}N)\right]_m\!\!-$$

SAN

The SAN is a simple random copolymer of styrene and acrylonitrile. But ABS is more complicated. It's made by polymerizing styrene and acrylonitrile in the presence of polybutadiene. Polybutadiene has carbon-carbon double bonds in it, which can polymerize, too. So we end up with a polybutadiene chain with SAN chains grafted onto it, like you see below :

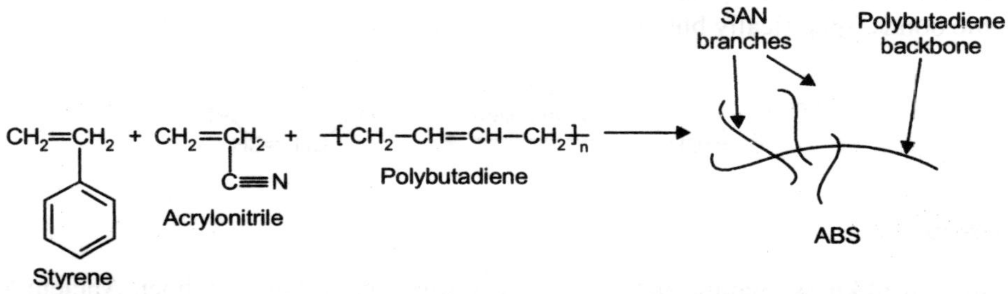

SAN branches

Polybutadiene backbone

$$CH_2{=}CH_2 + CH_2{=}CH_2 + \left[CH_2-CH{=}CH-CH_2\right]_n \longrightarrow$$

Styrene

Acrylonitrile

Polybutadiene

ABS

The ABS is very strong and lightweight. It is strong enough to be used to make automobile body parts, but it is so light that Wassana can lift this front bumper fascia over her head with only hand! Using plastics like ABS makes automobiles lighter, so they use less fuel, and therefore they pollute less.

The ABS is a stronger plastic than polystyrene because of the nitrile groups of its acrylonitrile units. The nitrile groups are very polar, so they are attracted to each other. This allows opposite charges on the nitrile groups to stabilize each other like you see in the picture on the left. This strong attraction holds ABS chains together tightly, making the material stronger. Also the rubbery polybutadiene makes ABS tougher than polystyrene.

Polyacrylonitrile is a vinyl polymer, and a derivative of the acrylate family of polymers. It is made from the monomer acrylonitrile by free radical vinyl polymerization.

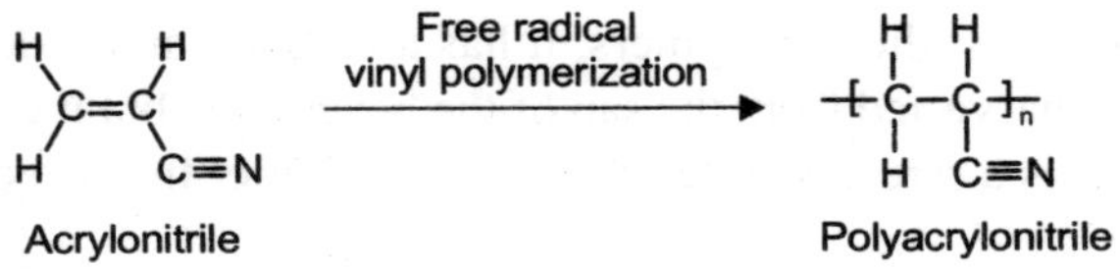

$$H_2C{=}CH(C{\equiv}N) \xrightarrow{\text{Free radical vinyl polymerization}} \left[CH_2-CH(C{\equiv}N)\right]_n$$

Acrylonitrile

Polyacrylonitrile

Polybutadiene

Polybutadiene was one of the first types of synthetic elastomer, or rubber, to be invented. It didn't take a great a degree of imagination to come up with, as its very similar to natural rubber, polyisoprene. It is good for uses which require exposure to low temperatures. Tires treads are often made of polybutadiene copolymers. Belts, hoses, gaskets and other automobile parts are made from polyubutadiene, because it stands up to cold temperatures better than other elastomers. Many polymers can become brittle at low temperatures thanks to a phenomenon called the glass transition. Driving in the winter can be bad enough with out hoses and gaskets going out on you! A hard rubber called poly(styrene-butadiene-styrene), or SBS rubber is a copolymer containing polybutadiene.

Butadiene has two carbon-carbon double bonds

Polybutadiene is a diene polymer, that is a polymer made from a monomer containing two carbon-carbon double bonds, specifically butadiene. It is made by Ziegler-Natta polymerization.

$CH_2{=}CH{-}CH{=}CH_2 \xrightarrow{\text{Zieglar-Natta polymerization}} {-}[CH_2{-}CH{=}CH{-}CH_2]_n{-}$

Polyisoprene

One of the most well known natural polymers is polyisoprene, or natural rubber. Ancient Mayans and Aztecs harvested it from the hevea tree and used it to make waterproof boots and the balls which they used to play a game similar to basketball. It is what we call an elastomer, that is, it recovers its shape after being stretched or deformed. Normally, the natural rubber is treated to give it crosslinks, which makes it an even better elastomer.

Isoprene has two carbon-carbon double bonds

Polyisoprene is diene polymer, which is a polymer made from a monomer containing two carbon-carbon double bonds. Like most diene polymers, it has a carbon-carbon double bond in its backbone chain. Polyisoprene can be harvested from the sap of the hevea tree, but it can also be made by Ziegler-Natta polymerization. This is a rare example of a natural polymer that we can make *almost* as well as nature does.

$CH_2{=}C(CH_3){-}CH{=}CH_2 \xrightarrow{\text{Zieglar-Natta polymerization}} {-}[CH_2{-}C(CH_3){=}CH{-}CH_2]_n{-}$

Polypropylene

Polypropylene is one of those rather versatile polymers out there. It serves double duty, both as a plastic and as a fibre. As a plastic it is used to make things like dishwasher-safe food containers. It can do this because it doesn't melt below 160 °C, or 320 °F. Polyethylene, a more common plastic, will anneal at around 100 °C, which means that polyethylene dishes will warp in the dishwasher. As a fibre, polypropylene is used to make indoor-outdoor carpeting, the kind that you always find around swimming pools and miniature golf courses. It works well for outdoor carpet because it is easy to make colored polypropylene, and because polypropylene doesn't absorb water, like nylon does.

Structurally, it is a vinyl polymer, and is similar to polyethylene, only that on every other carbon atom in the backbone chain has a methyl group attached to it. Polypropylene can be made from the monomer propylene by Ziegler-Natta polymerization and by metallocene catalysis polymerization.

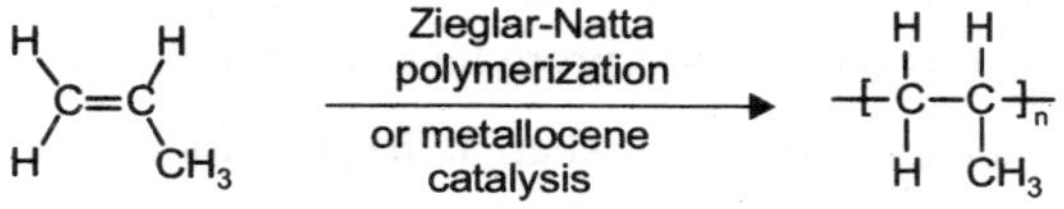

$$\underset{\text{Propylene}}{H_2C=CH(CH_3)} \xrightarrow[\text{or metallocene catalysis}]{\text{Zieglar-Natta polymerization}} \underset{\text{Polypropylene}}{\left[-CH_2-CH(CH_3)-\right]_n}$$

This is what the monomer propylene really looks like:

Research is being conducted on using metallocene catalysis polymerization to synthesize polypropylene. Metallocene catalysis polymerization can do some pretty amazing things for polypropylene. Polypropylene can be made with different tacticities. Most polypropylene we use is *isotactic*. This means that all the methyl groups are on the same side of the chain, like this:

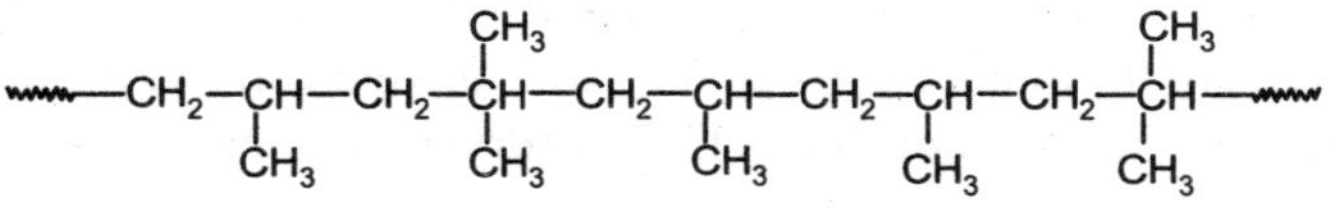

$$\sim CH_2-CH(CH_3)-CH_2-CH(CH_3)-CH_2-CH(CH_3)-CH_2-CH(CH_3)-CH_2-CH(CH_3)\sim$$

isotactic polypropylene

But sometimes we use *atactic* polypropylene. *Atactic* means that the methyl groups are placed randomly on both sides of the chain like this:

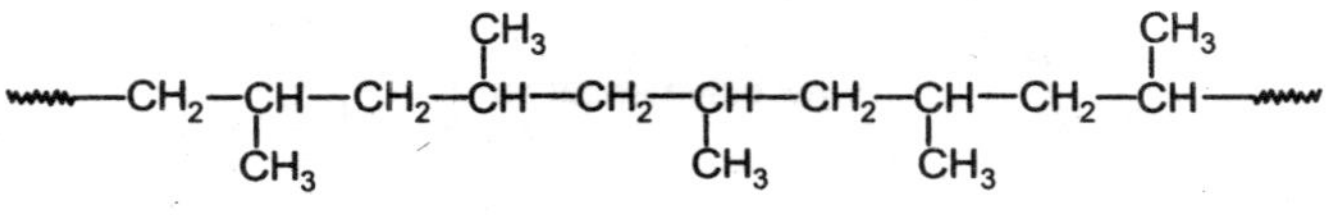

$$\sim CH_2-CH(CH_3)-CH_2-CH(CH_3)-CH_2-CH(CH_3)-CH_2-CH(CH_3)-CH_2-CH(CH_3)\sim$$

atactic polypropylene

However, using special metallocene catalysts it is believed that we can make polymers which contain blocks of isotactic polypropylene and blocks of atactic polypropylene in the same polymer chain, as is shown in the picture below:

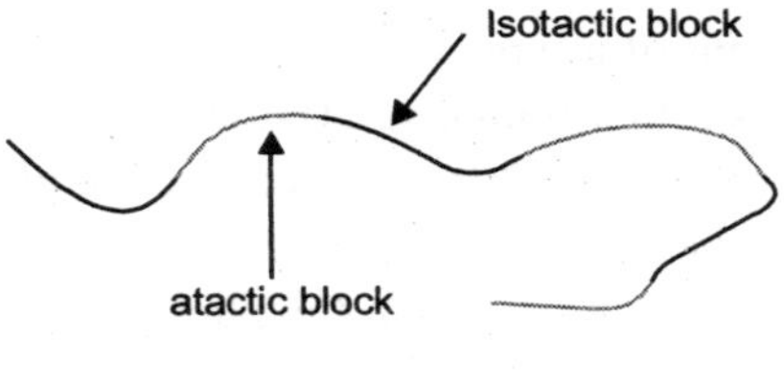

Isotactic block

atactic block

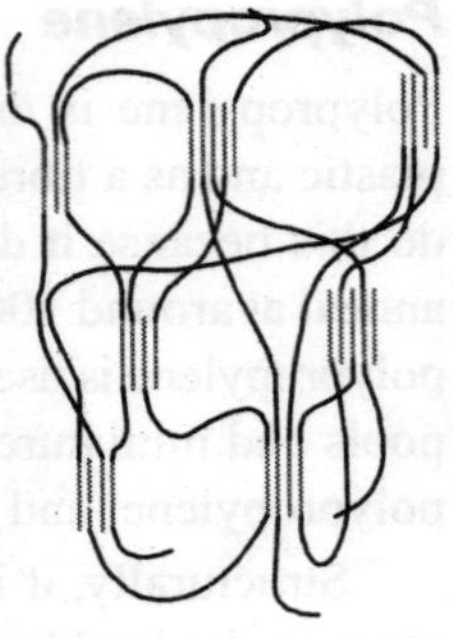

This polymer is rubbery, and makes a good elastomer. This is because the isotactic blocks will form crystals by themselves. But because the isotactic blocks are joined to the atactic blocks, each little hard clump of crystalline isotactic polypropylene will be tied together by soft rubbery tethers of atactic polypropylene, as you can see in the picture on the right.

To be honest, atactic polypropylene would be rubbery without help from the isotactic blocks, but it wouldn't be very strong. The hard isotactic blocks hold the rubbery isotactic material together, to give the material more strength. Most kinds of rubber have to be crosslinked to give them strength, but not polypropylene elastomers.

Elastomeric polypropylene, as this copolymer is called, is a kind of thermoplastic elastomer. However, until the research is completed, this type of polypropylene will not be commercially available.

The polypropylene which you can buy off the shelf at the store today has about 50 - 60% crystallinity, but this is too much for it to behave as an elastomer.

Polystyrene

Polystyrene is an inexpensive and hard plastic, and probably only polyethylene is more common in your everyday life. The outside housing of the computer you are using now is probably made of polystyrene. Model cars and airplanes are made from polystyrene, and it also is made in the form of foam packaging and insulation (Styrofoam^{TM} is one brand of polystyrene foam). Clear plastic drinking cups are made of polystyrene. So are a lot of the molded parts on the inside of your car, like the radio knobs. Polystyrene is also used in toys, and the housings of things like hairdryers, computers, and kitchen appliances.

Polystyrene is a vinyl polymer. Structurally, it is a long hydrocarbon chain, with a phenyl group attached to every other carbon atom. Polystyrene is produced by free radical vinyl polymerization, from the monomer styrene.

Styrene —(Free radical vinyl polymerization)→ Polystyrene

This is a better picture of what the monomer styrene looks like:

Polystyrene is also a component of a type of hard rubber called poly(styrene-butadiene-styrene), or SBS rubber. The SBS rubber is a thermoplastic elastomer. There's a new kind of polystyrene out there, called 'syndiotactic polystyrene'. It's different because the phenyl groups on the polymer chain are attached to alternating sides of the polymer backbone chain. "Normal" or atactic polystyrene has no order with regard to the side of the chain on which the phenyl groups are attached.

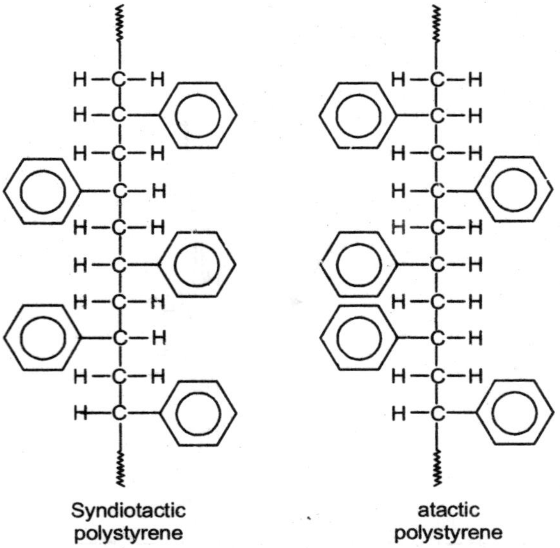

Syndiotactic polystyrene

atactic polystyrene

Syndiotactic polystyrene has a regular structure, so it canpack into crystal structures. The irregular atactic polystyrene can't.

You can see the new syndiotactic polystyrene alongside the old atactic polystyrene. The new syndiotactic polystyrene is crystalline, and melts at 270 °C. But it's a lot more expensive! Syndiotactic polystyrene is made by metallocene catalysis polymerization. But there are still some fun things you can do with old fashioned atactic polystyrene. What would happen if we were to take some styrene monomer, and polymerize it free radically, but let's say we put some polybutadiene rubber in the mix. Take a look at polybutadiene, and you'll see that it has double bonds in it that can polymerize. We end up with the polybutadiene copolymerizing with the styrene monomer, to get a type of copolymer called a graft copolymer. This is a polymer with polymer chains growing out of it, and which are a different kind of polymer than the backbone chain. In this case, it's a polystyrene chain with chains of polybutadiene growing out of it.

These rubbery chains hanging off of the backbone chain do some good things for polystyrene. Polybutadiene and polystyrene homopolymers don't mix, mind you. So the polybutadiene branches try as best they can to phase separate, and form little globs, like you see in the picture below. But these little globs are always going to be tied to the polystyrene phase. So they have an effect on that polystyrene. They act to absorb energy when the polymer gets hit with something. They give the polymer a resilience that normal polystyrene doesn't have. This makes it stronger, not as brittle, and capable of taking harder impacts without breaking than regular polystyrene. This material is called *high-impact polystyrene*, or HIPS for short. Not all the chains in HIPS are branched like this. There are a lot chains of plain polystyrene and plain polybutadiene mixed in there, too. This makes HIPS something we call animmiscible blend of polystyrene and polybutadiene. But it is the grafted polystyrene-polybutadiene molecules that make the

High-impact polystyrene (HIPS)

whole system work by binding the two phases (the polystyrene phase and the polybutadiene phase) together. HIPS can be blended with a polymer called poly(phenylene oxide), or PPO. This blend of HIPS and PPO is made by GE and sold as Noryl^{TM}.

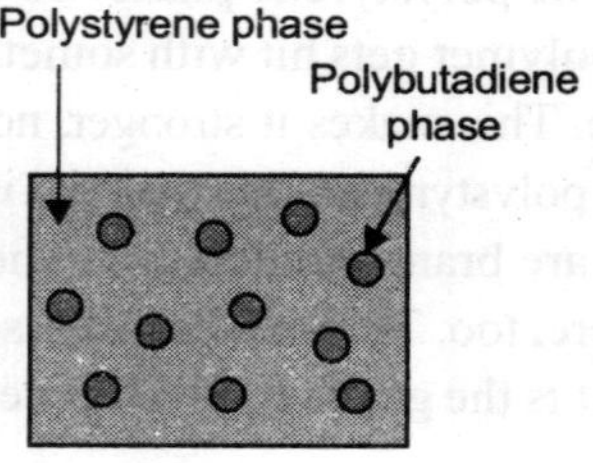

Polydicyclopentadiene

Polydicyclopentadiene is a polymer used to make really, really big things in one piece. With polydicyclopentadiene you can make a whole tractor cab in one piece, or a whole satellite dish antenna. That's not good enough for you? Then how about a 1500 gallon storage tank for dangerous chemicals? With polydicyclopentadiene that's no problem. But the very first use for it was the cowlings of snowmobiles, again molded in one piece. This was because it has very good impact resistance at low temperatures, where a lot of other polymers become brittle. Polydicyclopentadiene is made by a nifty reaction called ring-opening metathesis polymerization (ROMP) from the monomer *endo*-dicyclopentadiene.

ROMP

$-\!\!\!\![\;\;$—CH=CH$\;]_n$

endo-dicyclopentadiene poly(*endo*-dicyclopentadiene)

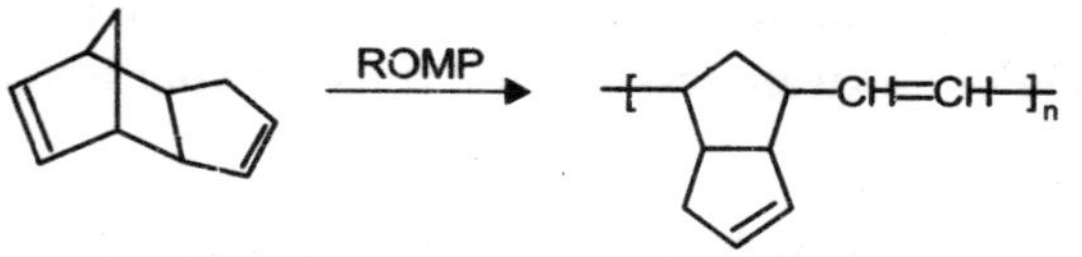

But it's not done yet! There's a double bond left in the bottom ring, as you can see in the picture of polydicyclopentadiene. These can undergo vinyl polymerization, to give us a crosslinked thermoset material.

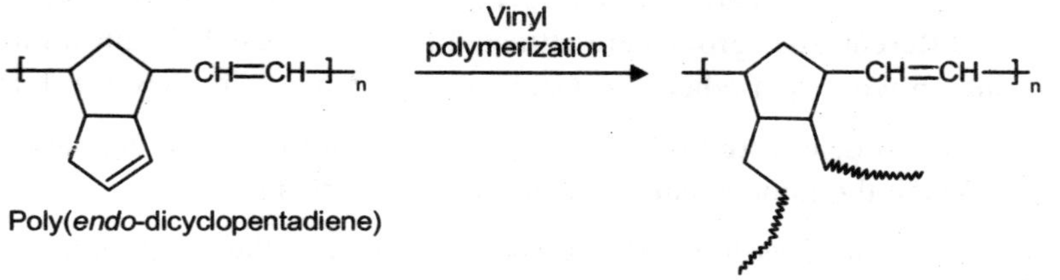

Vinyl polymerization

Poly(*endo*-dicyclopentadiene)

Crosslinked polymer

This thermoset is good stuff, but you can't mold a thermoset. So how do we make anything from it? The answer is to make it in chunks that are already shaped like we want them. The fancy name for this is called *reaction injection molding* or RIM for short. Put simply, we fill a mold full of the monomer, and polymerize it in the mold. That's how can make products from thermosets.

Polyesters

Polyesters are the polymers, in the form of fibres, that were used back in the seventies to make all that wonderful disco clothing, the kind you see being modeled on the right. But since then, the nations of the world have striven to develop more tasteful uses for polyesters, like those nifty shatterproof plastic bottles that hold your favorite refreshing beverages, like the blue bottle in the picture below. So you see, polyesters can be both plastics and fibres. Another place you find polyester is in balloons. Not the cheap ones that you use for water balloons, those are made of natural rubber. I'm talking about the fancy ones you get when you're in the hospital. These are made of a polyester film made by DuPont called Mylar. The balloons are made of a sandwich, composed of Mylar and aluminum foil. Materials like this, made of two kinds of material, are called composites.

A special family of polyesters are polycarbonates.

Polyesters have hydrocarbon backbones which contain ester linkages, hence the name.

$$\left[O-\overset{O}{\overset{\|}{C}}-C_6H_4-\overset{O}{\overset{\|}{C}}-O-CH_2-CH_2\right]_n$$

Ester group

The structure in the picture is called poly(ethylene terephthalate), or PET for short, because it is made up of ethylene groups and terephthalate groups (duh!). I realize that *terephthalate* is not the kind of word most English-speaking mouths are used to saying, but with practice you should be able to say it with only a slight feeling of awkwardness when it rolls off your tongue.

$$\left[O-\overset{O}{\overset{\|}{C}}-C_6H_4-\overset{O}{\overset{\|}{C}}-O-CH_2-CH_2\right]_n$$

Terephathalate group   Ethylene group

The ester groups in the polyester chain are polar, with the carbonyl oxygen atom having a somewhat negative charge and the carbonyl carbon atom having a somewhat positive charge. The positive and negative charges of different ester groups are attracted to each other. This allows the ester groups of nearby chains to line up with each other in crystal form, which is why they can form strong fibres.

The inventor who first discovered how to make bottles from PET was Nathaniel Wyeth. He's the brother of Andrew Wyeth the famous painter. But others had tried before.

The PET has too low a glass transition temperature, that is the temperature at which the PET becomes soft. Now reusing a soft drink bottle requires that the bottle be sterilized before it is used again. This means washing it at really high temperatures, temperatures too high for PET. Filling a jar with jelly is also carried out at high temperatures. Down at your local jelly factory, the stuff is shot into the jars hot, at temperatures which would cause PET to become soft. So PET is no good for jelly jars. There is a new kind of polyester that is just the thing needed for jelly jars and returnable bottles. It is poly(ethylene naphthalate), or PEN.

$$\left[O-\overset{O}{\overset{\|}{C}}-C_{10}H_6-\overset{O}{\overset{\|}{C}}-O-CH_2-CH_2\right]_n$$

Naphthalate group   Ethylene group

Poly(ethylene nephthalate), the polymer that bestows upon us the plastic jelly jar

The PEN has a higher glass transition temperature than PET. That's the temperature at which a polymer gets soft. The glass transition temperature of PEN is high enough so that it can withstand the heat of both sterilizing bottle washing and hot strawberry jelly. The PEN is so good at standing the heat that you don't even have to make the bottle entirely out of it. Just mixing some PEN in with the old PET gives a bottle that can take the heat a lot better than plain old PET. In the big plants where they make polyester, its normal to start off with a compound called dimethyl terephthalate. This is reacted with ethylene glycol is a reaction called *transesterification*. The result is bis-(2-hydroxyethyl)terephthalate and methanol. But if we heat the reaction to around 210 °C the methanol will boil away and we don't have to worry about it anymore.

$H_3C{-}O{-}\overset{O}{\overset{\|}{C}}{-}C_6H_4{-}\overset{O}{\overset{\|}{C}}{-}O{-}CH_3 + HO{-}CH_2{-}CH_2{-}OH$

Dimethyl terephthalate　　Ethylene glycol

$\longrightarrow HO{-}CH_2{-}CH_2{-}O{-}\overset{O}{\overset{\|}{C}}{-}C_6H_4{-}\overset{O}{\overset{\|}{C}}{-}O{-}CH_2{-}CH_2{-}OH + 2CH_3OH$

his-(2-hydroxyethyl)terephthalate　　Methanol

Then the bis-(2-hydroxyethyl)terephthalate is heated up to a balmy 270 °C, and it reacts to give the poly(ethylene terephtalate) and, oddly, ethylene glycol as a by product. Funny, we started off with ethylene glycol.

$HO{-}CH_2{-}CH_2{-}O{-}\overset{O}{\overset{\|}{C}}{-}C_6H_4{-}\overset{O}{\overset{\|}{C}}{-}O{-}CH_2{-}CH_2{-}OH \longrightarrow$

$\left[O{-}\overset{O}{\overset{\|}{C}}{-}C_6H_4{-}\overset{O}{\overset{\|}{C}}{-}O{-}CH_2{-}CH_2\right]_n + 2\ HO{-}CH_2{-}CH_2{-}OH$

Poly(ethylene terepthalate)　　Ethylene glycol

But in the laboratory, PET is made by other reactions. Terephthalic acid and ethylene glycol can polymerize to make PET when you heat them with an acid catalyst. It's possible to make PET from terephthoyl chloride and ethylene glycol. This reaction is easier, but terephthoyl chloride is more expensive than terephthalic acid, and it's a lot more dangerous.

$HO{-}\overset{O}{\overset{\|}{C}}{-}C_6H_4{-}\overset{O}{\overset{\|}{C}}{-}OH + HO{-}CH_2{-}CH_2{-}OH \xrightarrow[\text{Heat}]{\text{Acid}} PET$

Terephthalic acid　　Ethylene glycol

$Cl{-}\overset{O}{\overset{\|}{C}}{-}C_6H_4{-}\overset{O}{\overset{\|}{C}}{-}Cl + HO{-}CH_2{-}CH_2{-}OH \longrightarrow PET$

Terephthaloyl chloride　　Ethylene glycol

There are two more polyesters on the market that are related to PET. There is poly(butylene terephthalate) (PBT) and poly(trimethylene terephthalate). They are usually used for the same type of things as PET, but in some cases these perform better.

$$\left[-\overset{\displaystyle O}{\overset{\|}{C}}-C_6H_4-\overset{\displaystyle O}{\overset{\|}{C}}-O-CH_2-CH_2-CH_2-CH_2-O- \right]_n$$

Poly(butylene terephthalate)

$$\left[-\overset{\displaystyle O}{\overset{\|}{C}}-C_6H_4-\overset{\displaystyle O}{\overset{\|}{C}}-O-CH_2-CH_2-CH_2-O- \right]_n$$

Poly (trimethylene terephthalate)

Polyethylene

Polyethylene is probably the polymer you see most in daily life. Polyethylene is the most popular plastic in the world. This is the polymer that makes grocery bags, shampoo bottles, children's toys, and even bullet proof vests. For such a versatile material, it has a very simple structure, the simplest of all commercial polymers. A molecule of polyethylene is nothing more than a long chain of carbon atoms, with two hydrogen atoms attached to each carbon atom. That's what the picture at the top of the page shows, but it might be easier to draw it like the picture below, only with the chain of carbon atoms being many thousands of atoms long:

$$\sim\!\sim\!-\left(\overset{H}{\underset{H}{|}}\!\!\!\!\!\!\!\!\!\!C-\right)_{11}\!\sim\!\sim \qquad \text{i.e.}\ \sim\!\sim\!-(CH_2)_{11}-\!\sim\!\sim$$

Sometimes it's a little more complicated. Sometimes some of the carbons, instead of having hydrogens attached to them, will have long chains of polyethylene attached to them. This is called branched, or low-density polyethylene, or LDPE. When there is no branching, it is called linear polyethylene, or HDPE. Linear polyethylene is much stronger than branched polyethylene, but branched polyethylene is cheaper and easier to make.

A molecule of linear polyethylene, or HDPE

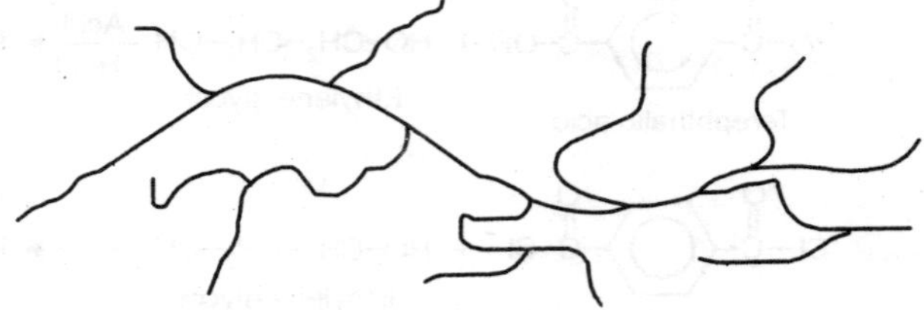

A molecule of branched polyethylene, or LDPE

Linear polyethylene is normally produced with molecular weights in the range of 200,000 to 500,000, but it can be made even higher. Polyethylene with molecular weights of three to six million is referred to as ultra-high molecular weight polyethylene, or UHMWPE. UHMWPE can be used to make fibres which are so strong they replaced Kevlar for use in bullet proof vests. Large sheets of it can be used instead of ice for skating rinks.

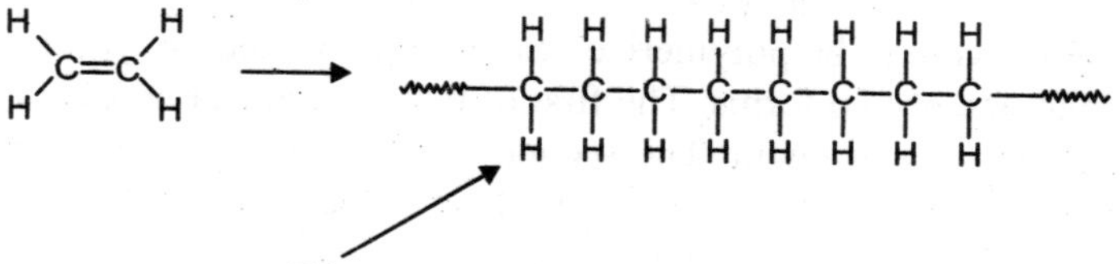

This can get tedious to draw, so wo often use shorthand like this.

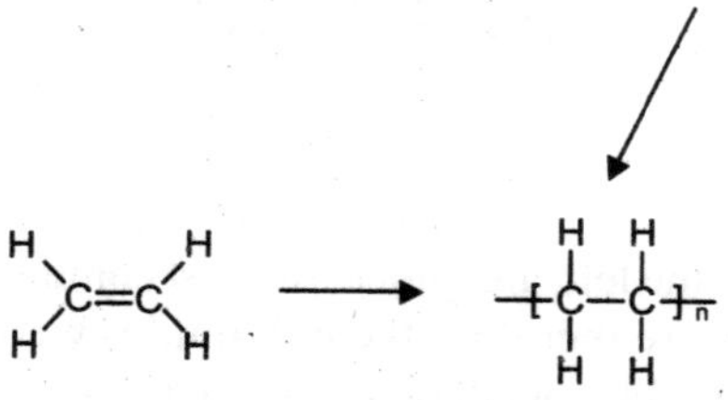

(Note: A line drawn between two atoms represents a pair of electrons shared by those atoms, which constitutes a chemical bond. Two lines represent two pairs of shared electrons, a double bond.)

And when we're feeling really lazy we just draw it like this:

$$CH_2 = CH_2 \longrightarrow [CH_2\text{-}CH_2]_n$$

Polyethylene is vinyl polymer, made from the monomer ethylene. Here's a model of the ethylene monomer. Branched polyethylene is often made by free radical vinyl polymerization. Linear polyethylene is made by a more complicated procedure called Ziegler-Natta polymerization. UHMWPE is made using metallocene catalysis polymerization. But Ziegler-Natta polymerization can be used to make LDPE, too. By copolymerizing ethylene monomer with a alkyl-branched comonomer such as one gets a copolymer which has short hydrocarbon branches. Copolymers like this are called *linear low-density polyethylene*, or LLDPE. The BP produces LLDPE using a comonomer with the catchy name 4-methyl-1-pentene, and sells it under the trade name Innovex™. LLDPE is often used to make things like plastic films.

Polyimides

Polyimides are a very interesting group of incredibly strong and astoundingly heat and chemical resistant polymers. Their strength and heat and chemical resistance are so great that these materials often replace glass and metals, such as steel, in many demanding industrial applications. Polyimides are even used in many everyday applications. Who knows, you may even have a few in your house. They are used for the struts and chassis in some cars as well as some parts under-the-hood because they can withstand the intense heat and corrosive lubricants, fuels, and coolants cars require. Do you have polyimides in your kitchen? They are also used in the construction of many appliances as well as microwave cookware and food packaging because of their thermal stability, resistance to oils, greases, and fats (YUM!), and their transparency to microwave radiation. They can also be used in circuit boards, insulation, fibres for protective clothing, composites, and adhesives.

By now you have probably guessed that a polyimide is a polymer that contains an imide group. *Yeah, I figured that out, but what is an imide?* I just knew you were going to ask that. An imide is a group in a molecule that has a general structure (drawn in red) which looks like this:

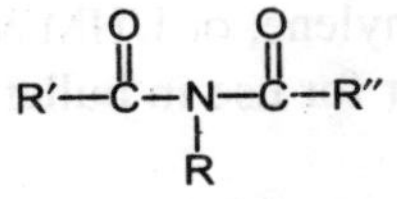

an imide

So if the molecule shown above were to be polymerized the product would be, you guessed it, a polyimide. Polyimides usually take one of two forms. The first of these is a linear structure where the atoms of the imide group are part of a linear chain. The second of these structures is a heterocyclic structure where the imide group is part of a cyclic unit in the polymer chain. Take a look.

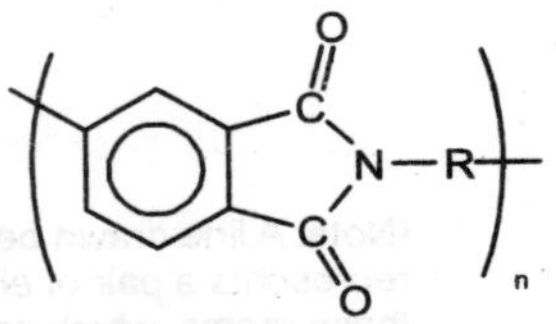

Aromatic heterocyclic polyimide 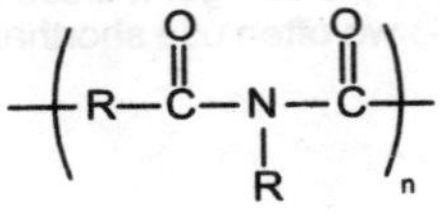   Linear polyimide

Aromatic heterocyclic polyimides, like the one on the left, are typical of most commercial polyimides, such as Ultem from G.E. and DuPont's Kapton, and are therefore the polyimides we will be spending most of our time on. These polymers have such incredible mechanical and thermal properties that they are used in place of metals and glass in many high performance applications in the electronics, automotive, and even the aerospace industries. These properties come from strong intermolecular forces between the polymer chains.

A polymer which contains a charge transfer complex consists of two different types of monomers, a donor and an acceptor. The donor is like a rich man with more money than he knows what to do with. It has plenty of electrons to go around because of its nitrogen groups. The acceptor, then, is like a mooching houseguest. Its carbonyl groups, like our houseguest's many vices of gambling, drinking and such, sucks away its electron density. The donor doesn't mind supporting the acceptor, in fact, with the acceptor around, the donor looks better. Charity looks good in some social circles (and it's good for a tax break too!). So the donor lends some of its electrons to the acceptor, holding them tightly together.

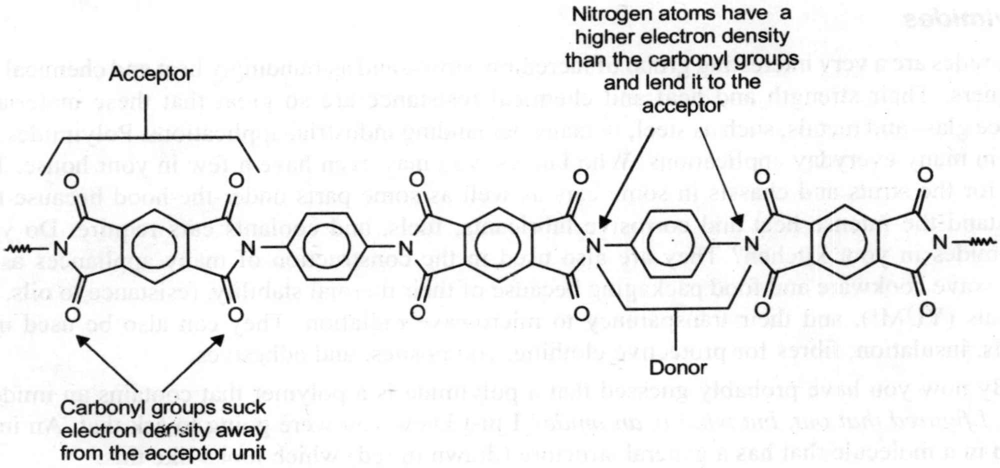

Acceptor

Nitrogen atoms have a higher electron density than the carbonyl groups and lend it to the acceptor

Donor

Carbonyl groups suck electron density away from the acceptor unit

The charge transfer complex works not only between adjacent units in the polymer chain, but also between chains. The chains will stack together like strips of paper, with donors and acceptors paired up.

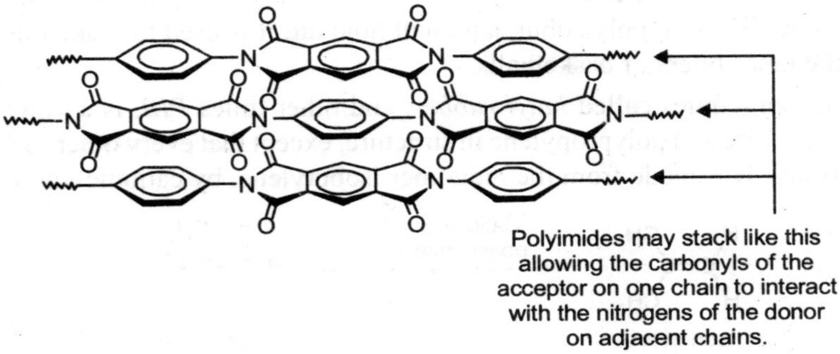

Polymides may stack like this allowing the carbonyls of the acceptor on one chain to interact with the nitrogens of the donor on adjacent chains.

This charge transfer complex holds the chains together very tightly, not allowing them to move around very much. When things can't move around on the molecular level, they can't move around in the whole material. This is why polyimides are so strong.

The charge transfer complex is so strong, it sometimes becomes necessary to make the polymer a little softer so it is able to be processed. This is accomplished with the bisphenol-A derived linkage shown in the polymer segment below.

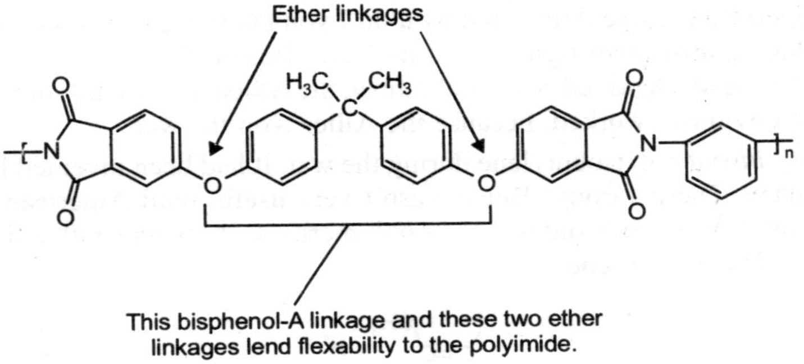

Ether linkages

This bisphenol-A linkage and these two ether linkages lend flexability to the polyimide.

Another interesting property of polyimides which makes them excellent for use in construction and transportation industries is they burn.

Why would you want to build a building or a car out of something that burns?

Good question, and I can answer it too. It is not the burning which industry likes, but polyimide's self extinguishing property which builders like. Self extinguishing? Yes, when an aromatic polyimide catches on fire, which by the way is difficult to begin with, a surface char develops which smothers the flame, blocking it from the fuel to burn. Then you just wipe it off, and it's just like the fire never happened.

Polyisobutylene

Polyisobutylene is a synthetic rubber, or elastomer. It is special because it is that only rubber that is gas impermeable, that is, it is the only rubber which can hold air for long periods of time. You may have noticed that balloons will go flat after a few days. This is because they are made of polyisoprene, which is not gas impermeable. Because polyisobutylene will hold air, it is used to make things like the inner liner of tires, and the inner liners of basketballs.

Polyisobutylene, sometimes called *butyl rubber*, and other times *PIB*, is a vinyl polymer, and is very similar to polyethylene and polypropylene in structure, except that every other carbon is substituted with *two* methyl groups. It is made from the monomer isobutylene, by cationic vinyl polymerization.

$$(H)_2C{=}C(CH_3)_2 \xrightarrow{\text{Cationic vinyl polymerization}} \left[\,CH_2{-}C(CH_3)_2\,\right]_n$$

Isobutylene → Polyacrylonitrile

And this is that monomer isobutylene:

Usually, a small amount of isoprene is added to the isobutylene. The polymerization is carried out at a right frosty -100 °C (or -148 °F) for you Americans out there. This is because the reaction is so fast we can't control it unless we freeze it colder than a brass toilet seat in the Yukon.

Polyisobutylene was first developed during the early 1940s. At that time, the most widely used rubber was natural rubber, polyisoprene. Polyisoprene was an excellent elastomer, and easy to isolate from the sap of the hevea tree. Huge plantations thrived in Malaysia which grew hevea trees to supply the world's rubber needs. There was only one slight problem, and that was that Malaysia had just been conquered by the Imperial Japanese Army, and wouldn't you know we just so happened to be fighting the Second World War against them right at that moment. Before the war was over more than sixty million people would be dead. Deprived of natural rubber, the Allied nations did some quick thinking and came up with PIB. It obviously worked, because the Allies won the war.

We didn't actually *invent* polyisobutylene during the war. It had been invented long before the war by chemists in Germany. There's irony! But it wasn't very useful until American chemists came up with a way to crosslink it. What they did was to copolymerize isobutylene with a little bit, say, around one percent, isoprene. This is isoprene:

$$H_2C{=}C(CH_3){-}CH{=}CH_2$$

Isoprene

When isoprene is polymerized with the isobutylene we get a polymer that looks like this:

$$\sim H_2C{-}C(CH_3)_2{-}CH_2{-}C(CH_3)_2{-}CH_2{-}C(CH_3)_2{-}CH_2{-}C(CH_3){=}CH{-}CH_2{-}C(CH_3)_2{-}CH_2{-}C(CH_3)_2{-}CH_2{-}C(CH_3)_2{-}CH_2\sim$$

Isoprene linkage

About one or two out of every hundred repeat units is an isoprene unit. These have double bonds, which means the polymer can be crosslinked by *vulcanization* just like natural rubber. What is this vulcanization?

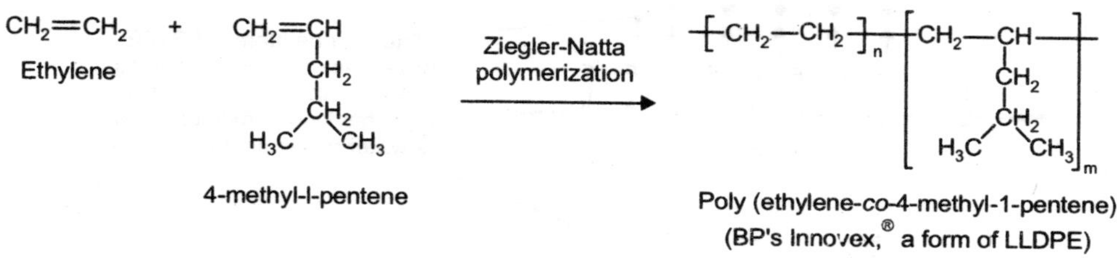

$CH_2{=}CH_2$ + $CH_2{=}CH{-}CH_2{-}CH(CH_3)_2$ $\xrightarrow{\text{Ziegler-Natta polymerization}}$ $-[CH_2-CH_2]_n-[CH_2-CH(CH_2CH(CH_3)_2)]_m-$

Ethylene

4-methyl-l-pentene

Poly (ethylene-*co*-4-methyl-1-pentene)
(BP's Innovex,® a form of LLDPE)

Polytetrafluoroethylene

Polytetrafluoroethylene is better known by the trade name Teflon®. It is used to make non-stick cooking pans, and anything else that needs to be slippery or non-stick. PTFE is also used to treat carpets and fabrics to make them stain resistant. What's more, it's also very useful in medical applications. Because human bodies rarely reject it, it can be used for making artificial body parts. Polytetrafluoroethylene, or PTFE, is made of a carbon backbone chain, and each carbon has two fluorine atoms attached to it. It may be easier to think of it as it's drawn in the picture below, with the chain of carbon atoms being thousands of atoms long.

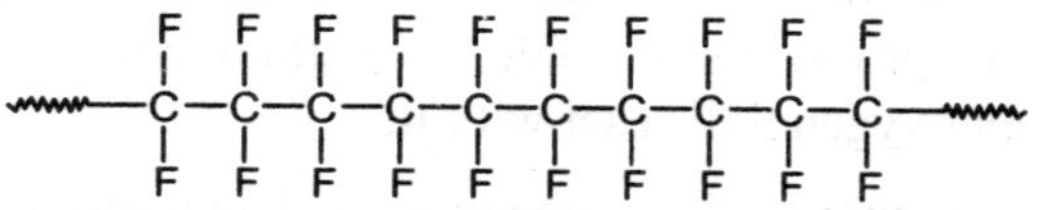

$\sim\sim -CF_2-CF_2-CF_2-CF_2-CF_2-CF_2-CF_2-CF_2-CF_2-CF_2- \sim\sim$

The PTFE is a vinyl polymer, and its structure, if not its behaviour, is similar to polyethylene. Polytetrafluoroethylene is made from the monomer tetrafluoroethylene by free radical vinyl polymerization.

$F_2C{=}CF_2 \xrightarrow{\text{Free radical vinyl polymerization}} -[CF_2-CF_2]_n-$

Tetrafluoroethylene

Polytetrafluoroethylene

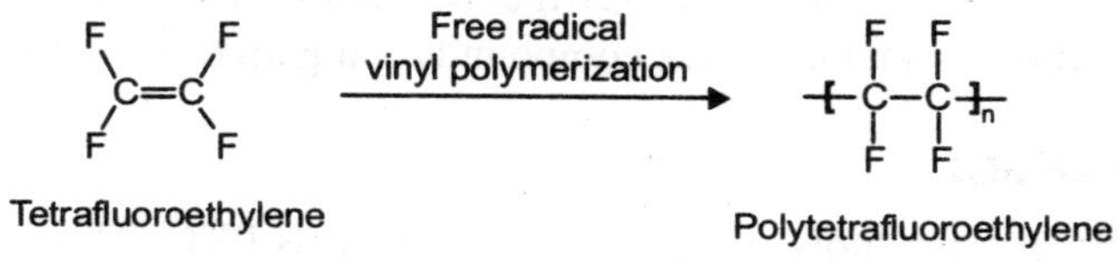

Fluorine is a very strange element. When it is part of a molecule, it doesn't like to be around other molecules, even the fluorine atoms on other molecules. But it likes other kinds of molecules even less. So a molecule of PTFE, being just chock full of fluorine atoms as it is, would like to be as far away from other molecules as it can get. For this reason, the molecules at the surface of a piece of PTFE will repel the molecules of just about anything that tries to come close to it. This is why nothing sticks to PTFE.

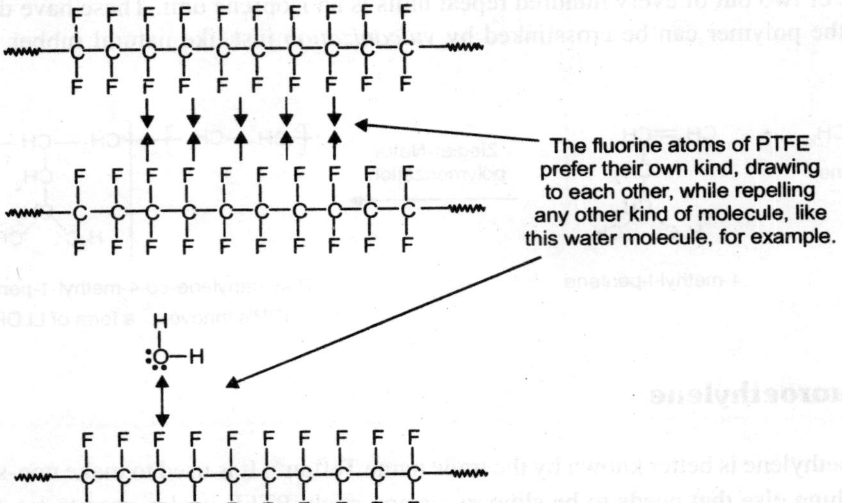

The fluorine atoms of PTFE prefer their own kind, drawing to each other, while repelling any other kind of molecule, like this water molecule, for example.

Because it's non-stick, PTFE means you can fry things without grease or butter. This means less fat and cholesterol, for a healthier heart.

The question remains then: If nothing will stick to this stuff, how does it stick to the surface of the frying pan? DuPont knows, but they're not telling!

There is the fact that the bond between the fluorine atom and the carbon atom is just really, really strong. The bond is almost bullet proof! It's so stable that nothing will react with it. Even when it gets as hot as a frying pan, not even oxygen will react with it!

Polytetraflouroethylene is another of those amazing accidental discoveries of science. In the late 1930s, when PTFE was discovered in DuPont's laboratories, DuPont was not at all concerned with non-stick frying pans or artificial heart valves. What they were really interested in was refrigeration. At the time, refrigerators used things like ammonia and sulfur dioxide as refrigerants. These are pretty nasty things to have leaking out of your refrigerator and into your kitchen. The quest was on, then, to make a non-toxic refrigerant. One of the compounds being investigated was tetrafluoroethylene.

Poly(vinyl chloride) (PVC)

Poly(vinyl chloride) is the plastic known at the hardware store as PVC. This is the PVC from which pipes are made, and PVC pipe is everywhere. The plumbing in your house is probably PVC pipe, unless it's an older house. The PVC pipe is what rural high schools with small budgets use to make goal posts for their football fields. But there's more to PVC than just pipe. The "vinyl" siding used on houses is made of poly(vinyl chloride). Inside the house, PVC is used to make linoleum for the floor. In the seventies, PVC was often used to make vinyl car tops.

The PVC is useful because it resists two things that hate each other: fire and water. Because of its water resistance it is used to make raincoats and shower curtains, and of course, water pipes. It has flame resistance, too, because it contains chlorine. When you try to burn PVC, chlorine atoms are released, and chlorine atoms inhibit combustion.

Structurally, PVC is a vinyl polymer. (well, duh!) It is similar to polyethylene, but on every other carbon in the backbone chain, one of the hydrogen atoms is replaced with a chlorine atom. It is produced by the free radical polymerization of vinyl chloride.

$$H_2C{=}CHCl \xrightarrow[\text{vinyl polymerization}]{\text{free radical}} {-}\!\!\left[CH_2{-}CHCl\right]_n\!\!{-}$$

Vinyl chloride — Poly(vinyl chloride)

The PVC was one of those odd discoveries that actually had to be made twice. It seems around a hundred years ago, a few German entrepreneurs decided they were going to make loads of cash lighting people's homes with lamps fueled by acetylene gas. Wouldn't you know it, right about the time they had produced tons of acetylene to sell to everyone who was going to buy their lamps, new efficient electric generators were developed which made the price of electric lighting drop so low that the acetylene lamp business was finished. That left a lot of acetylene laying around.

Leftover acetylene from the acetylene lamp fiasco

$$H{-}C{\equiv}C{-}H + HCl \longrightarrow H_2C{=}CHCl \longrightarrow {-}\!\!\left[CH_2{-}CHCl\right]_n\!\!{-}$$

Vinyl chloride — Poly(vinyl chloride)

So in 1912 one German chemist, Fritz Klatte decided to try to do something with it, and reacted some acetylene with hydrochloric acid (HCl). Now this reaction will produce vinyl chloride, but at that time no one knew what to do with it, so he put it on the shelf, where it polymerized over time. Not knowing what to do with the PVC he had just invented, he told his bosses at his company, Greisheim Electron, who had the material patented in Germany. They never figured out a use for PVC, and in 1925 their patent expired.

Wouldn't you know it, in 1926 the very next year, and American chemist, Waldo Semon was working at B.F. Goodrich when he independently invented PVC. But unlike the earlier chemists, it dawned on him that this new material would make a perfect shower curtain. He and his bosses at B.F. Goodrich patented PVC in the United States (Klatte's bosses apparently never filed for a patent outside Germany). Tons of new uses for this wonderful waterproof material followed, and PVC was a smash hit the second time around.

Poly(vinylidene fluoride)

Poly(vinylidene fluoride) of PVDF has a few things going for it. It has very high electrical resistance, and it has good flame resistance. Put these two bits of information together and it might dawn on you that this would make a good material for insulating electrical wires, especially ones that get hot during use. So you'll find it insulating the wires in the computer you're using right now. Electrical cables on airplanes are also insulated with PVDF, where it's important that practically everything on board be fireproof. It's also chemically resistant, so you'll find it used in the chemical industry to make pipes and bottles and such that hold chemicals. What about ultraviolet radiation? PVDF resists that, too. PVDF is often blended with poly(methyl methacrylate) (PMMA) to make it more resistant to UV light. PMMA degrades when exposed to UV radiation, so if we want to make PMMA windows for use outside we have to blend it with PVDF.

Another nifty thing about PVDF is that it is a *piezoelectric* material. What does that mean? That means that when it is placed in an electric field it will change its shape. *Why?*

Because fluorine is so much more electronegative than carbon, the fluorine atoms will suck electrons away from the carbon atoms to which they are attached. This means the -CF_2- groups in the chain will be very polar, with a partial negative charge on the fluorine atoms and a partial positive charge on the carbon atoms. So when they are placed in an electrical field, they will align. This causes the polymer sample to deform, all those -CF_2- groups trying to align.

The PVDF is made by free radical vinyl polymerization of the monomer vinylidene fluoride, like this:

$$H_2C{=}CF_2 \longrightarrow -\!\!\left[CH_2-CF_2\right]_n\!\!-$$

Vinylidene chloride   Poly(vinylidene fluoride)

Silicones

Silicones are used for a lot of things. They can be elastomers and lubricating oils. The caulking in your bathroom is probably made of a silicone. Silicones are also used to make the heat resistant tiles on the bottom of the space shuttle. Take a look at the picture on the right and you'll see how good silicones can be at deflecting heat. Back on earth, silicones are used to make hair conditioners that don't cause buildup.

$$\sim\!\!\sim -SiR_2-O-SiR_2-O-SiR_2-O-SiR_2-O-SiR_2-O-SiR_2-O- \sim\!\!\sim$$

Silicones are inorganic polymers, that is, there are no carbon atoms in the backbone chain. The backbone is a chain of alternating silicon and oxygen atoms. Each silicone has two groups attached to it, and these can be any organic groups. This polymer is called polydimethylsiloxane. It is the most common silicone. Want to see some others? Polymethylphenylsiloxane and polydiphenylsiloxane are also popular with the kids these days.

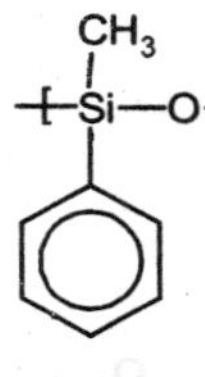

Polymethylphenylsiloxane 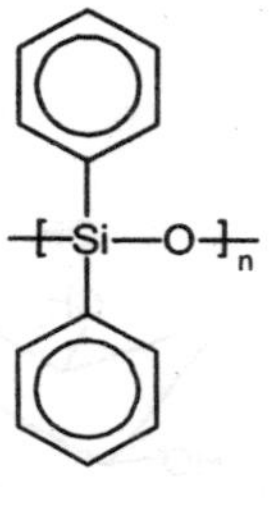 Polydiphenylsiloxane

"Polysiloxane" is the proper name for silicones. But when they were discovered it was thought that they had "silicone" groups in the backbone chain. When the real structure was discovered, it was too late, and the name stuck.

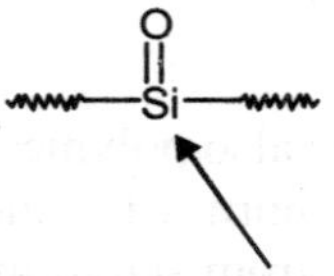

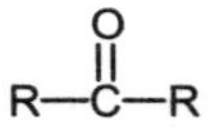

a ketone

Scientists once thought this was the structure of polyiloxanes, so they named them "silicones", by comparison carbon-containing ketones, shown on the right. Of course, this is not the structure.

Silicones make good elastomers because the backbone chain is very flexible. The bonds between a silicon atom and the two oxygen atoms attached to it are very flexible. The angle formed by these bonds can open and close like a scissors without much trouble. This makes the whole backbone chain flexible.

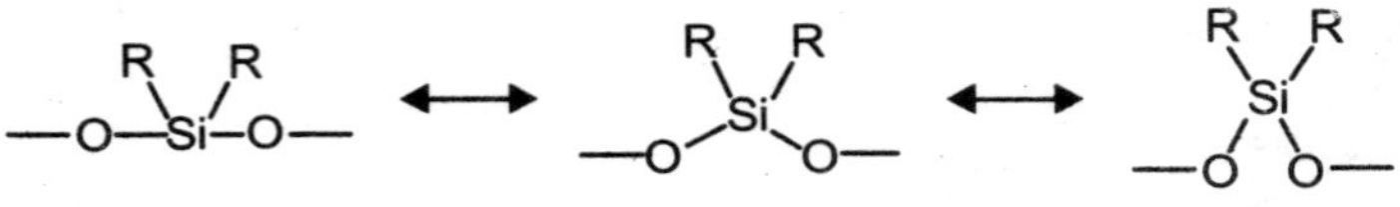

Polydimethylsiloxane does something really strange when you mix it with boric acid, or $B(OH)_3$. The mixture is soft and pliable, you can mold it into any shape easily with your fingers. But it is also very bouncy. What's more, push it gently and it gives way, but hit it hard with a hammer and it cracks! Strangely, if you spread it over newspaper, and pull it away, it gets printed with a mirror image of the newspaper text. No industrial use was ever found for his wonder material, but tons of it have been sold as toy called Silly Putty.

Starch

Starch is important because we eat it! Starch is found in potatoes, and in grains such as corn and wheat. Starch is made up of glucose repeat units.

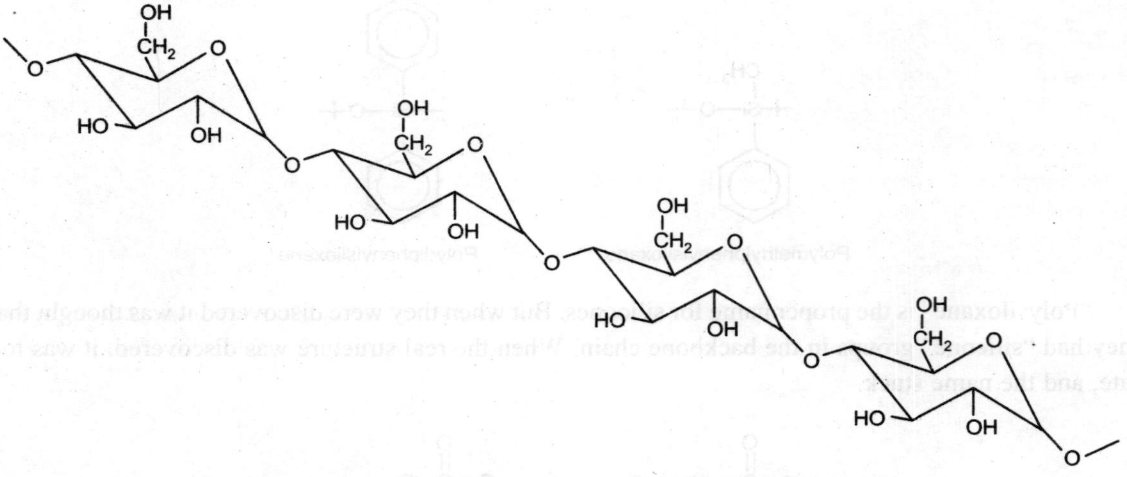

In your body, special proteins called enzymes (which are also polymers, by the way) break starch down into glucose, so your body can burn it for energy. If you're eating a healthy diet, you get most of your energy from starch in this way. Because it is made of sugar molecules it is called a polysaccharide. It is very similar to cellulose. To see just how the two are different, click here.

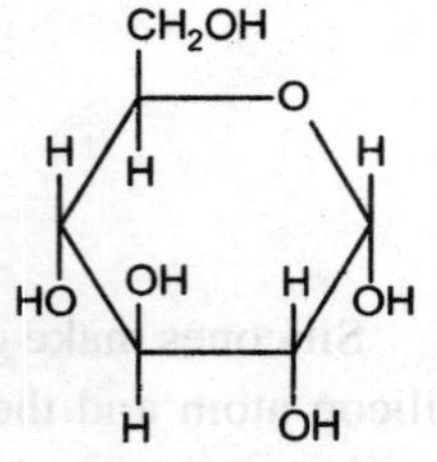

Starch has a few other uses other than food It's used in pressing clothes to keep them from wrinkling. It's also used to make a foam packing. Starch is biodegradable, so starch foam packing is an environmentally-friendly alternative to styrofoam packing.

Glucose

Cellulose

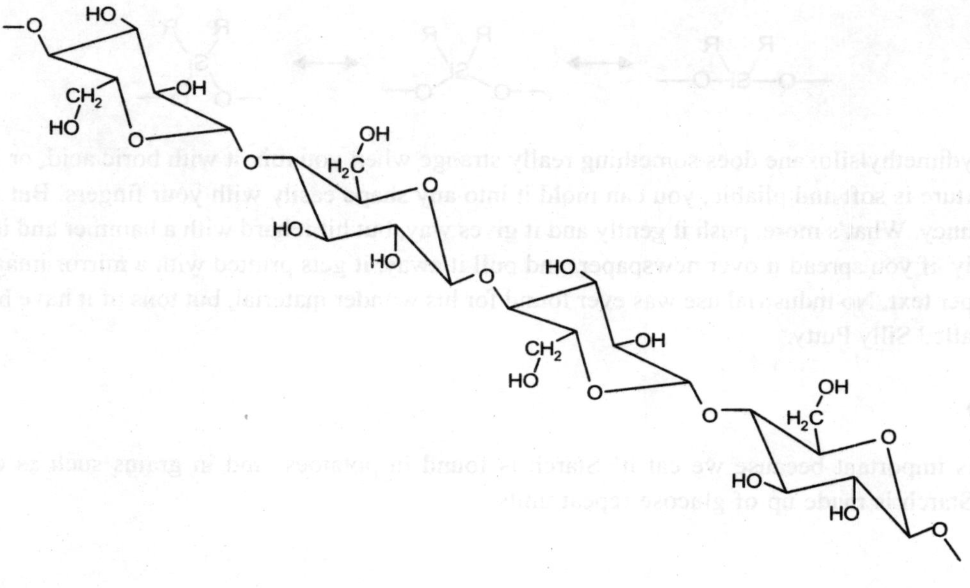

Cellulose is one of many polymers found in nature. Wood, paper, and cotton all contain cellulose. Cellulose is an excellent fibre. Wood, cotton, and hemp rope are all made of fibrous cellulose. Cellulose is made of repeat units of the monomer glucose. This is the same glucose which your body metabolizes in order to live, but you can't digest it in the form of cellulose. Because cellulose is built out of a sugar monomer, it is called a polysaccharide. Celluose has an important place in the story of polymers becuase it was used to make some of the first synthetic polymers, like cellulose nitrate, cellulose acetate, and rayon.

Glucose

Another cellulose derivative is hydroxyethylcellulose. It differs from plain ol' regular cellulose in that some or all of the hydroxyl groups of the glucose repeat unit have been replaced with hydroxyethyl ether groups.

Cellulose repeat unit

Hydroxyethylcellulose repeat unit

These hydroxyethyl groups get in the way when the polymer tries to crystallize. Because it can't crystallize, hydroxyethylcellulose is soluble in water. In addition to being a great laxative, it's used to thicken shampoos as well. It also make the soap in the shampoo less foamy, and it helps the shampoo clean better by forming *colloids* around dirt particles.

The hydroxyethylcellulose chain wraps around a particle of dirt, so it stays in suspension in water

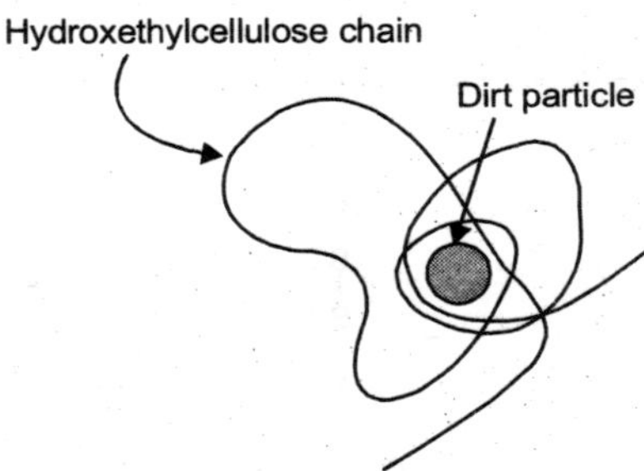

Normally, particles of dirt are insoluble in water. But a chain of hydroxyethylcellulose can wrap itself around a dirt particle. This mass can be though of as a snack cake, with the polymer chain as the cake and the dirt as the creamy filling. This snack cake is soluble in water, so by wrapping around the dirt like this, the hydroxyethylcellulose tricks the water into accepting the dirt. In this way, the dirt gets washed away instead of being deposited back onto your hair.

Co-polymer

When a polymer is made by linking only one type of small molecule, or monomer, together, it is called a *homopolymer*. When two different types of monomers are joined in the same polymer chain, the polymer is called a copolymer. Let's imagine now two monomers, which we'll cali *A* and *B*. *A* and *B* can be made into a copolymer in many different ways.

When the two monomers are arranged in an alternating fashion, the polymer is called, of course, an *alternating copolymer*:

—A—B—A—B—A—B—A—B—A—B—A—B—A—B—

alternating copolymer

In a *random copolymer*, the two monomers may follow in any order:

—A—A—B—A—B—B—A—B—A—A—B—B—B—A

random copolymer

In a *block copolymer*, all of one type of monomer are grouped together, and all of the other are grouped together. A block copolymer can be thought of as two homopolymers joined together at the ends:

—A—A—A—A—A—A—A—B—B—B—B—B—B—B—

block copolymer

A block copolymer you know very well, that is if you wear shoes, is SBS rubber. It's used for the soles of shoes and for tire treads, too.

When chains of a polymer made of monomer *B* are grafted onto a polymer chain of monomer *A* we have a *graft copolymer*:

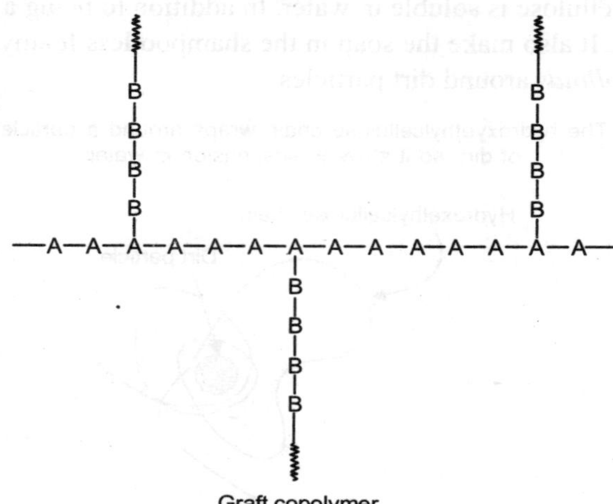

Graft copolymer

One kind of graft copolymer is high-impact polystyrene, or HIPS for short. It is a polystyrene backbone with chains of polybutadiene grafted onto the backbone. The polystyrene gives the material strength, but the rubbery polybutadiene chains give it resilience to make it less brittle.

Proteins

Proteins are one of many types of natural polymers, and they are the most versatile by far. You name it, proteins do it. What can proteins do? What can't they do! They can be catalysts. Some proteins, called enzymes, make certain chemical reactions in your body take place up to a million times faster than they would without enzymes. One protein called *hemoglobin* is found in your blood, and carries oxygen from your lungs to your cells. Another protein called *collagen* is a strong and tough material that makes up your skin, hair, and fingernails.

A protein is a naturally occurring polyamide. That is a polymer containing an *amide group* in the backbone chain. Want to see an amide group so you know just what it is I'm rambling about? Here's an amide group, and next to it a polyamide:

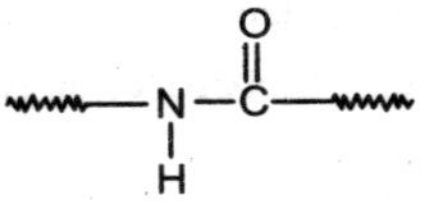

$$\sim\sim-\underset{\substack{|\\H}}{N}-\overset{\substack{O\\\|}}{C}-\sim\sim$$

An amide group

$$\sim\sim-R-\underset{\substack{|\\H}}{N}-\overset{\substack{O\\\|}}{C}-R-\underset{\substack{|\\H}}{N}-\overset{\substack{O\\\|}}{C}-R-\underset{\substack{|\\H}}{N}-\overset{\substack{O\\\|}}{C}-R-\underset{\substack{|\\H}}{N}-\overset{\substack{O\\\|}}{C}-\sim\sim$$

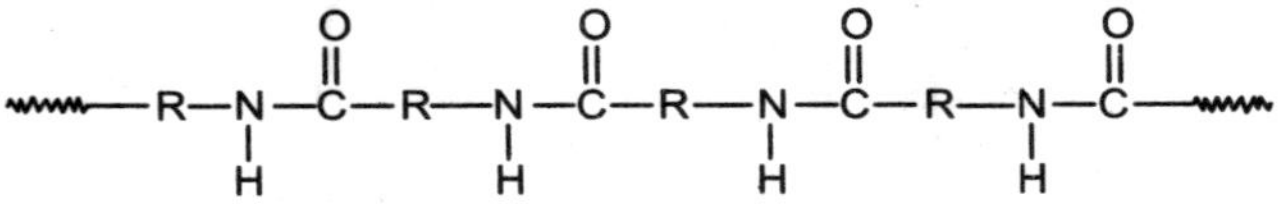

A polyamide

That *R* you see in betwixt the amide groups can be anything at all, but in proteins, the *R* is a single carbon atom, with two pendant groups attached. One pendant is a hydrogen atom, always, and the other is a wild card. It can be a lot of different things. In this picture we just represent it as *R'*.

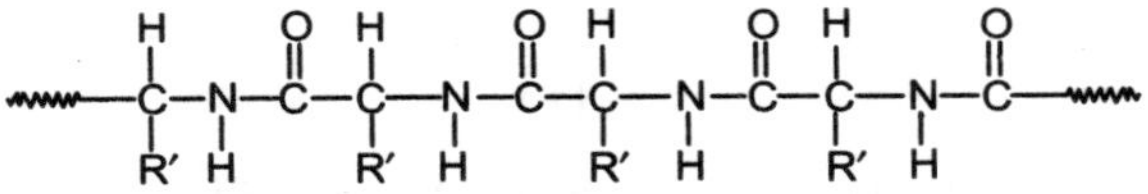

$$\sim\sim-\underset{R'}{\overset{H}{C}}-\underset{H}{N}-\overset{O}{C}-\underset{R'}{\overset{H}{C}}-\underset{H}{N}-\overset{O}{C}-\underset{R'}{\overset{H}{C}}-\underset{H}{N}-\overset{O}{C}-\underset{R'}{\overset{H}{C}}-\underset{H}{N}-\overset{O}{C}-\sim\sim$$

A protein

In your body, these proteins are made from monomers called *amino acids*, like this:

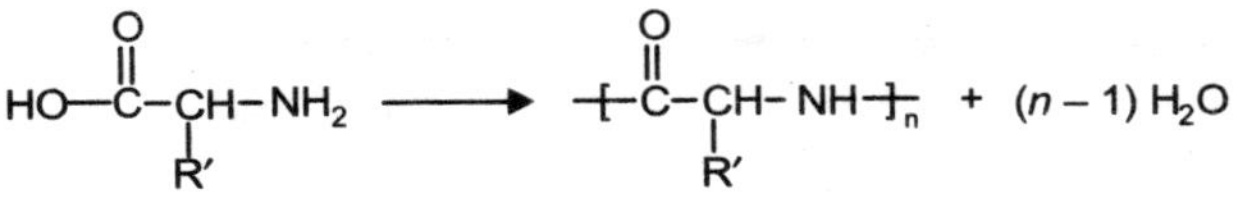

$$HO-\overset{\substack{O\\\|}}{C}-\underset{\substack{|\\R'}}{CH}-NH_2 \longrightarrow \left[-\overset{\substack{O\\\|}}{C}-\underset{\substack{|\\R'}}{CH}-NH-\right]_n + (n-1)\,H_2O$$

Amino acid Protein

There are a twenty different amino acids. Each one has a different *R'* group. Also, each protein has a specific sequence of the different amino acids in it. So in each protein, there is a different sequence of *R'* groups hanging off of the backbone chain. This sequence determines the properties of the protein. Proteins also make up silk. Silk is such nifty stuff that scientists tried to make synthetic silk. They tried to make synthetic polyamides, and what do you know, they did it! The artificial polyamides are called 'nylons'.

Structure of Ethylene

The molecular formula of ethylene is C_2H_4 and the possible structural formulæ are:

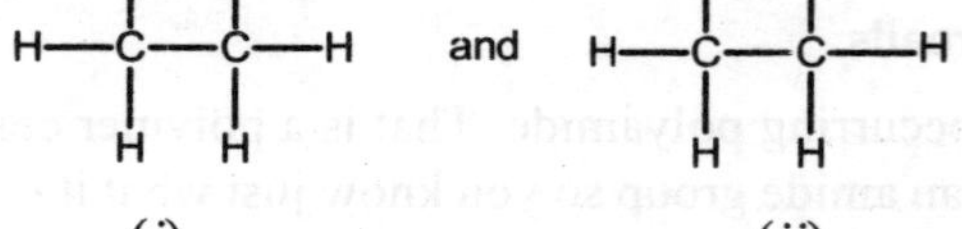

$$H-\overset{H}{\underset{H}{C}}-\overset{H}{\underset{H}{C}}-H \quad \text{and} \quad H-\overset{H}{\underset{H}{C}}-\overset{H}{\underset{H}{C}}-H$$

(*i*) (*ii*)

When ethylene combines with bromine to form ethylene di-bromide $C_2H_4Br_2$ the bromine adds on to the free valences. Ethylene di-bromide can therefore, be represented by one of the following formulæ

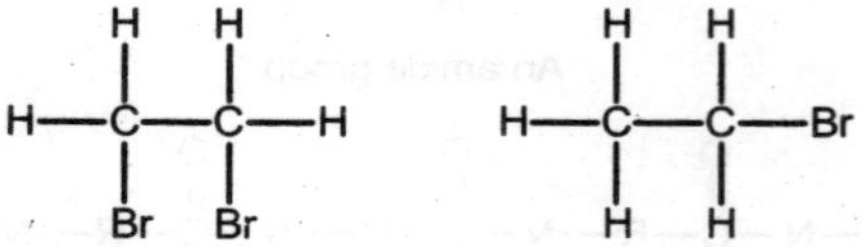

$$H-\overset{H}{\underset{Br}{C}}-\overset{H}{\underset{Br}{C}}-H \qquad H-\overset{H}{\underset{H}{C}}-\overset{H}{\underset{H}{C}}-Br$$

(*iii*) (*iv*)

The structure of acetaldehyde is known to be

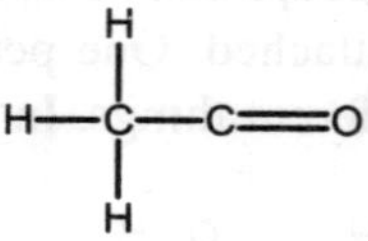

$$H-\overset{H}{\underset{H}{C}}-C=O$$

(*v*)

When acetaldehyde reads with phosphorous pentachloride, PBr_5, ethylene di-bromide $C_2H_4Br_2$ is produced. This is quite different form ethylene and has therefore a different structures

$$H-\overset{H}{\underset{H}{C}}-C=O + PBr_5 \longrightarrow H-\overset{H}{\underset{H}{C}}-\overset{H}{\underset{Br}{C}}-Br + POBr_2$$

Acetaldehyde + Phosphorous Pentabromide ⟶ Ethylene Dibromide + Phosphorous Oxybromide

Formulae (*iv*) and (*v*) are identical and represent ethylidene dibromide, therefore formula (*iii*) should represent ethylene dibromide. Which is isomeric with ethylene dibromide.

As ethyiene dibromide is obtained form the ethylene by the direct addition of Br_2, ethylene should have the constitution shown by the formula (*i*)

```
     H    H              H    H
     |    |              |    |
H—C—C—H      H—C═C—H
     |    |
     |    |
    (i)                 (ii)
```

The electronic formula is

```
   H   H
   ••  ••
H••C ∷ C••H
```

> The two free or residual valences of carbon atoms as shown in the above formula cannot exist therefore they are shared by the carbon atoms, giving rise to the bond, ethylene linkage or ethenial.

Bayer's Strain Theory

Union of two carbon atoms by a double bond should not be misunderstood to mean a firm structure. A double bond is a sign of weakness and reactivity rather than of strength and stability. Compounds containing a double bond easily add up two movements atoms (H_2, Cl_2, HCl) etc. to form addition compounds.

Bayer explained the nature of a double bond by this strain theory. According to Le Bell and Vant Hoffs views about the nature of carbon atom it is believed that four valances of a carbon atom are directed towards the four corners of a regular tetrahedron, at the centre of which lies the C atom. Angle between any two valences is 109°28′.

When a carbon atom is joined by single bonds, the direction of the four valences remain unaltered. In case of a double bond each valency has to bond half way

$$\left(\frac{109°28'}{2}\right).$$

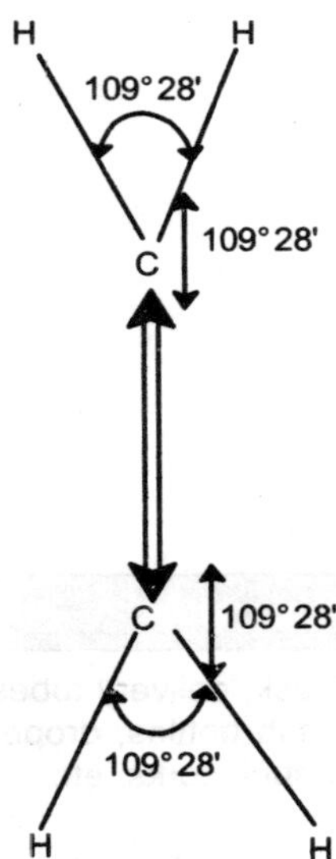

This bending of valences causes strain within the molecule and leads to its instability or reactivity. This is illustrated by the diagram in the right.

ACETYLENES OR ALKYNES

The acetylene hydrocarbons form a homologous series of the general formula C_nH_{2n+2}. They are more unsaturated than the olefines and characterized by the presence of a triple bond. They ate highly reactive and readily forms the additive compounds.

The most important member of this family is acetylene (Ethyne)

ACETYLENE

Occurrence

Acetylene is found in the coal gas and wood gas, It is formed during the incomplete combustion of many carbon compounds. The gas which escape when a Bunsen burner (strikes back) when contains 1% acetylene

Laboratory Preparation of Acetylene

Theory: Acetylene is prepared in the lab by the action of water on calcium carbide

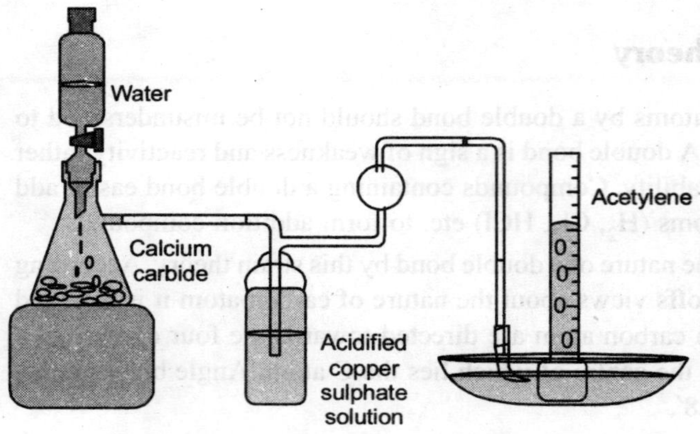

Fig. 7.3. Showing the lab preparation of Acetylene

Apparatus	Chemicals	Catalyst	Temperature
Flask, delivery tubes, wash bottles, dropping funnel, corks, etc.	Calcium Carbide, Water, copper sulphate	No	Room temperature

Reaction Equation:

$$\text{Ca}\begin{matrix}\diagup \text{C} \\ \diagdown \text{C}\end{matrix}\!\!\Big|\!\Big|\!\Big| + \text{H—O—H} \longrightarrow \text{Ca}\begin{matrix}\diagup \text{OH} \\ \diagdown \text{OH}\end{matrix} + \begin{matrix}\text{CH} \\ ||| \\ \text{CH}\end{matrix}$$

Calcium Carbide Water ⟶ Calcium Hydroxide acetylene

Precaution: *The gas produced forms an explosive mixture with air so no flame should be brought near the apparatus set up, and the calcium carbide should be much more than water.*

This process can also be used for the preparation of commercial acetylene. The calcium carbide is prepared commercially by heating quick lime and coke in an electric arc furnace.

$$CaO+3C \xrightarrow{2000^\circ C} \underset{\text{Calcium Carbide}}{CaC_2} + CO$$

Purification

Commercial calcium carbide contains traces of phosphides, nitrides and sulphur and the gas prepared in above method is therefore, impure and contains traces of PH_3 NH_3 and H_2S formed by the action of water on phosphides nitrites may be removed by passing gas through a wash-bottle containing acidified copper sulphate solution.

Other Methods of Acetylene Preparation

(1) By heating ethylene dibromide with alcoholic caustic potash

$$\begin{matrix} H_2C-Br \\ | \\ H_2C-Br \end{matrix} + 2KOH \longrightarrow \begin{matrix} CH \\ ||| \\ CH \end{matrix} + 2KBr + 2H_2O$$

+ Potassium Hydroxide → acetylene + Potassium Bromide + Water

These methods can also be used for preparation acetylene in the lab.

(2) From unsaturated Dicarbooxylic acid, by the electrolysis of sodium or potassium salt of fumaric acid

$$KOOC-CH=CH-COOK \xrightarrow[+H_2O]{\text{Electrolysis}} {}^{-}OOC-CH=CH-COO^{-} + 2K^{+}$$

Potassium Fumerate Acetate Ion

Anode$^+$: $H-C\equiv C-H + 2CO_2$

cathode-: $2K + 2H_2O \longrightarrow 2KOH + H_2$

(3) From iodoform by the action of silver

$$CHI_3 + 6Ag \longrightarrow \begin{matrix} H-C \\ ||| \\ H-C \end{matrix} + 6AgI$$

iodoform + Silver → acetylene + Silver Iodide

(4) With sulphuric acid: - Like the previous reaction addition takes place in the stages.

$$HC \equiv CH + H_2SO_4 \longrightarrow H-CH=CH-HSO_4 \xrightarrow{H_2SO_4} CH_3-CH(HSO_4)_2$$

acetylene sulfuric acid Ethylidine Sulphate Ethylidine Hydrogen Sulphate

Ethylidine hydrogen sulphate when distilled with water under goes hydrolysis and forms acetaldehyde

$$CH_3-CH(HSO_4)_2 \xrightarrow{2H_2O} CH_3-CH(OH)_2 \xrightarrow{\text{Breaks}} CH_3-CH=O$$

Ethyledine Hydrogen Sulphate Ethane-1,1-diol (unstable) acetaldehyde

The above reaction is represented by

$$HC \equiv CH + H_2O \xrightarrow[H_2SO_4 + HgSO_4]{\text{Warm dilute}} H_3C-CH=O$$

acetylene Water Acetaldehyde

Acetylene gas is absorbed in conc. Sulphuric acid and solution warmed with water and mercury sulphate. The conversion of acetylene into acetylene is very important in synthesis of various organic compounds.

(5) ***With Nitrogen***: Hydrocyanic acid is produced

$$C_2H_2 + N_2 \rightarrow 2HCN$$

Extremely poisonous reaction

Industrial Preparation of Acetylene

The main procedure is the same with the electrical arcing of the carbon and hydrogen in the reactor, the main difference is the supply of hydrogen and carbon, which is obtained from the methane gas. The natural gas is introduced in the first furnace, and the carbon is extracted, the hydrogen is obtained from the introduction of water into the reactor. The processed gas is send to the next reactor for converting the acetylene into fuel gas, with the introduction of oxygen. The acetylene comes out from the first reactor (Fig. ...).

Properties

Physical Properties

Acetylene is a colourless gas. The commercial gas possesses an unpleasant smell of garlic (phosphene) and is poisonous. It is slightly soluble in water more, so in alcohol and very soluble in acetone. The gas is generally stored in acetone. It is also adsorbed by colloidal or finely divided palladium.

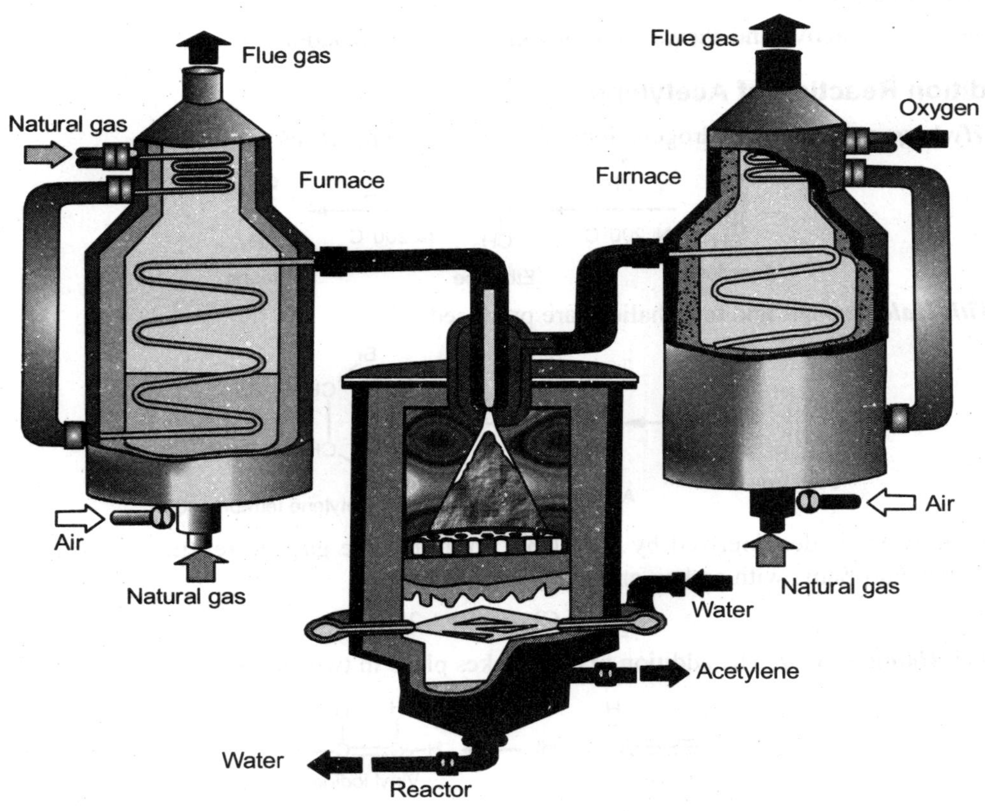

Fig. 7.4. The Industrial Preparation of Acetylene from Methane

Endpoint	Acetylene	Methylacetylene
Molecular weight (grams/mol)	26.04	40.07
Melting point	80.8°C[a]	101.5°C[a]
Boiling point (at 1016 hPa)	84°C[a]	23.2 °C[a]
Relative density (at 82 °C)	0.6208[a]	0.7062[a]
Vapour pressure (hPa).C)5,	6,969.2[b] (at 25°C)	155[a] (at 20°C)
Partition coefficient (Log Pow or Kow)	0.37[b]	0.94[b]
Water solubility (mg/l at 25°C)	1,230[a]	3,640[a]

a. Measured value; b. Estimated value.

Chemical Properties

Acetylene is very reactive and shows the following important reactions: -

(A) Addition Reaction of Acetylene

1. With Hydrogen: Catalytic hydrogenation, the olefines and paraffins are formed

$$\underset{\text{Acetylene}}{CH\equiv CH} + 2H \xrightarrow{\text{Ni } 200^{\circ}C} \underset{\text{Ethylene}}{CH_2=CH_2} \xrightarrow[\text{Ni } 200^{\circ}C]{+2H} \underset{\text{Ethane}}{CH_3-CH_3}$$

2. With Halogen: Di- and tetra-halides are produced

$$\underset{\text{acetylene}}{CH\equiv CH} \xrightarrow{Br} \underset{\text{Acetylene Dibromide}}{BrHC=CHBr} \xrightarrow{Br_2} \underset{\text{Acetylene tetrabromide}}{Br_2CH-CHBr_2}$$

Bromine water is decolourized by acetylene, with chlorine gas the reaction is violent and the mixture bursts into flame with carbon and hydrogen chloride

$$C_2H_2 + Cl_2 \rightarrow 2C + 2HCl$$

3. With Halogen Acid: The addition reaction takes place in two stages

$$\underset{\text{Acetylene}}{H-C\equiv C-H} + HI \longrightarrow \underset{\text{Vinyl iodide}}{H-C(H)=C(H)-I}$$

Vinyl iodide is still unsaturated and can therefore; react with another molecule of the acid, this reaction takes place accordingly with Markowinkoff's rule. According to this rule, the halogen or the electronegative group usually attaches itself to that carbon atom which is unified with the smaller number of hydrogen atoms.

$$\underset{\text{vinyl iodide}}{H-C(H)=C(H)-I} + \underset{\text{Hydrogen Iodide}}{HI} \longrightarrow \underset{\text{1,1-diiodoethane}}{H-C(H)(H)-C(H)(I)-I}$$

B. Formation of Metallic Derivatives

1. ***With silver and cuprous chloride:*** Acetylene gives precipitate with ammonical solution of silver nitrate and cuprous chloride

(i) $\underset{\text{acetylene}}{H-C\equiv C-H} + 2AgNO_3 \rightarrow \underset{\text{Silver acetylide}}{Ag-C\equiv C-Ag} + 2HNO_3$

$2HNO_3 + 2NH_4OH \rightarrow 2NH_4NO_3 + 2H_2O$

(ii) $H - C \equiv C - H + CuCl_2 \rightarrow Cu - C \equiv C - Cu + 2HCl$
acetylene cupric acetylide

$2HCl + 2NH_4OH \rightarrow 2NH_4Cl + 2H_2O$

The acetylide are explosive in the dry state. They regenerate acetylene with dilute acid. Hence the formation of acetylide is used as a distinctive test for acetylene and also serves to separate it from gaseous mixture.

2. ***With Sodium***: Sodium Acetylide is produced

$$H - C \equiv C - H \xrightarrow{Na(180°C)} H - C{=}C\text{-}Na \xrightarrow{Na(210°C)} Na - CC - Na$$

Acetylene Monosodium Acetylide Disodium Acetylide

3. ***Polymerization***: Acetylene polymerizes on passing through a red hot tube forming benzene

$$\underset{\text{Acetylene}}{3C_2H_2} \rightarrow \underset{\text{Benzene}}{C_6H_6}$$

4. ***With Bayer's Reagent***: Acetylene decolourizes Bayer's reagent – alkaline $KMnO_4$

In this reaction acetaldehyde is oxidised to oxalic acid

$$C_2H_2 + 4O \rightarrow (COOH)_2$$

5. ***Combustibility:*** Acetylene burns in air or oxygen producing a white smoky flame and very hot flame forming carbon dioxide and water.

$$2C_2H_2 + 5O_2 \rightarrow 4CO_2 + 2H_2O$$

Oxyacetylene flame, produced by burning a mixture oxygen and acetylene in a special torch.

Uses of Acetylene

Acetylene has been used since 1906 for welding and cutting metals; acetylene has earned an enviable reputation for being the hottest and most efficient of all the fuel gases.

Stored in maroon colour cylinders with a left hand cylinder valve connection acetylene has a low ignition temperature of 325°C and requires very little energy to ignite it.

Acetylene is the only fuel gas that can be used for welding steel due to its high heat distribution from the inner cone and its high flame temperature. In cutting, acetylene gives the fastest pre-heating and piercing times of any of the other fuel gas combinations. Burned with oxygen, acetylene produces a flame temperature of about 3160°C, the hottest of all the fuel gases.

Care must be taken when using the gas, as acetylene is an asphyxiant. Acetylene has a distinct garlic smell - recognisable above 2% concentrations in air.

Acetylene is a flammable gas referred to as fuel gas. When the acetylene gas is mixed with an oxidant and provided with an ignition source it will burn.

Cylinder Dimensions and Weight

The table below identifies key guiding principles on the product range of acetylene.

For more detailed information on cylinder sizes please click on the link below:

Cylinder Size	Approx. Dimensions Including valve and guard where supplied (mm)	Gross Weight (kg)	Volume (M^3)	Manifold Cylinder Pallet (MCP)	Approx. Dimensions Including Cylinders (mm)	Gross Weight (Kg)	Volume (M^3)
K	340 × 172	8	0.57	TF (12 × F)	1280 × 1500 × 850	1300	74
M	570 × 172	15	1.13	TG (12 × G)	1280 x 1500 × 850	1300	90
J	705 × 204	31	2.35	TN (12 × N)	1280 × 1500 × 850	1343	100.8
D	1050 × 213	59	3.33	T NB*	1280 × 1650 × 830	1343	100.8
H	1120 × 290	83	8.73	ZH*	1300 × 1260 × 8300	1800	104.76
G	1200 × 264	97	7.5				
F	1200 × 264	96	6.16				

Withdrawal

Before using acetylene, work out how much you will need. It may be that you should consider using a manifolded cylinder pallet (MCP) if you require significant amounts or high flow rates.

The amount of acetylene that can be taken from a cylinder is limited by the rate at which the solvent will give up the gas and the temperature at which the acetylene cylinder is being stored.

The table below shows acetylene withdrawal against storage temperature for cylinders between 6 and 8m^3 capacity (m^3/hr). These figures demonstrate that if you only use the acetylene occasionally, not continuously, you can withdraw around 50% more at any time. That is due to the fact that the monolithic mass and acetone can only release a certain amount of acetylene in a given time.

Temperature	Continuous	Intermittent
15° C	0.7	1.1
0 °C	0.4	0.6
-10 °C	0.3	0.5

(These figures are illustrative and must not be used to size high flow acetylene manifold installations, as these require special considerations.)

The above table only refers to full, large size cylinders. As the cylinder empties, the maximum withdrawal rates will fall regardless of your useage pattern. Should the flow limit be exceeded then acetone will be withdrawn along with the acetylene. This is characterised by the flame taking on a yellow or green tinge. As the stored temperature falls then it is more difficult for the acetone to give up the acetylene and the withdrawal rate reduces even further. It is recommended that acetylene should be

Structure of Acetylene

No.	Test	Paraffins C_nH_{2n+2} Ethane	Olefines (C_nH_{2n}) C_2H_4 Ethylene)	Acetylene C_nH_{2n-2} C_2H_2 Acetylene
1.	General nature	Saturated having single bond	Unsaturated having Doube bond (C = C)	Unsaturated having triple bond (C ≡ C)
2.	Bromine water	Not decolourized in dark in sunlight substitution	Decolourized even in dark forming products $CH_2{=}CH_2 \xrightarrow{Br_2} H_2C(Br){-}H_2C(Br)$	Decolourized even in dark forming addition product $CH{\equiv}CH \xrightarrow{Br_2} HC(Br){=}HC(Br) \xrightarrow{Br_2} HC(Br)_2{-}HC(Br)_2$ Acetylene Dibromide; Acetylene Tetrabromide
3.	Bayer's Reagent	No action	Decolorized Glycol is formed $CH_2{=}CH_2 \xrightarrow{H_2O + O} HC(OH){=}HC(OH)$	Decolourized various products like formate, oxalate are formed
4.	H_2SO_4	No action	Absorbed alkyl sulphate is formed $CH_2{=}CH_2 \xrightarrow{H_2SO_4} CH_2{=}CH_2\,H_2SO_4$ Ethyl Hydorgen Sulphate	Absorbed acetaledehyde is produced $CH{\equiv}CH \xrightarrow[H_2SO_4,\ HgSO_4]{+H_2O} H_3C{-}CH{=}O$ Acetylene; Acetaldehyde
5.	Ammonia $AgNO_3$	No action	No Action	White precipitate for silver acetylide $C(H){\equiv}C(H) \xrightarrow[2NH_4OH]{2\,AgNO_3} C(Ag){\equiv}C(Ag) + 2NH_4NO_3 + 2H_2O$ (White ppt)
6.	Ammonical $CuCl_2$	No action	No action	Red ppt of copper acetylide is produced $CH{\equiv}CH \xrightarrow[2NH_4OH]{CuCl_2} C(Cu){\equiv}C(Cu) + 2NH_4Cl + 2H_2O$ red

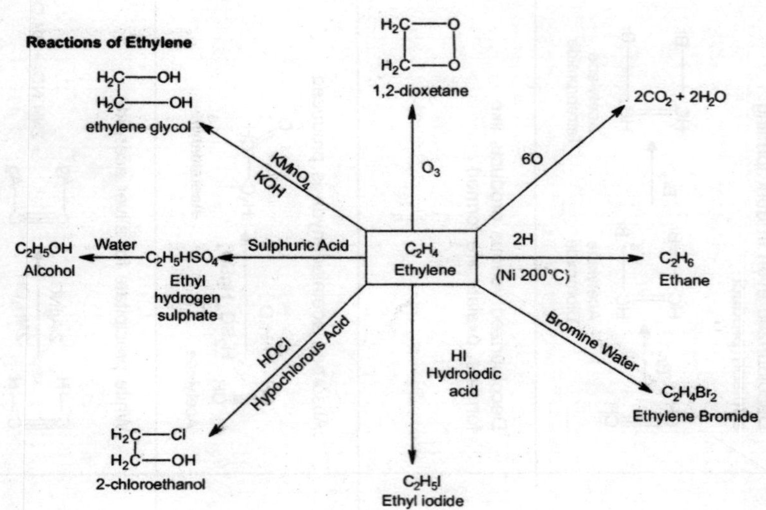
Reactions of Ethylene
H2C—O
H2C—O
1,2-dioxetane
H2C—OH
H2C—OH
ethylene glycol
KMnO4
KOH
O3
6O
2CO2 + 2H2O
C2H5OH
Alcohol
Water
C2H5HSO4
Ethyl hydrogen sulphate
Sulphuric Acid
C2H4
Ethylene
2H
(Ni 200°C)
C2H6
Ethane
HOCl
Hypochlorous Acid
HI
Hydroiodic acid
Bromine Water
C2H4Br2
Ethylene Bromide
H2C—Cl
H2C—OH
2-chloroethanol
C2H5I
Ethyl iodide

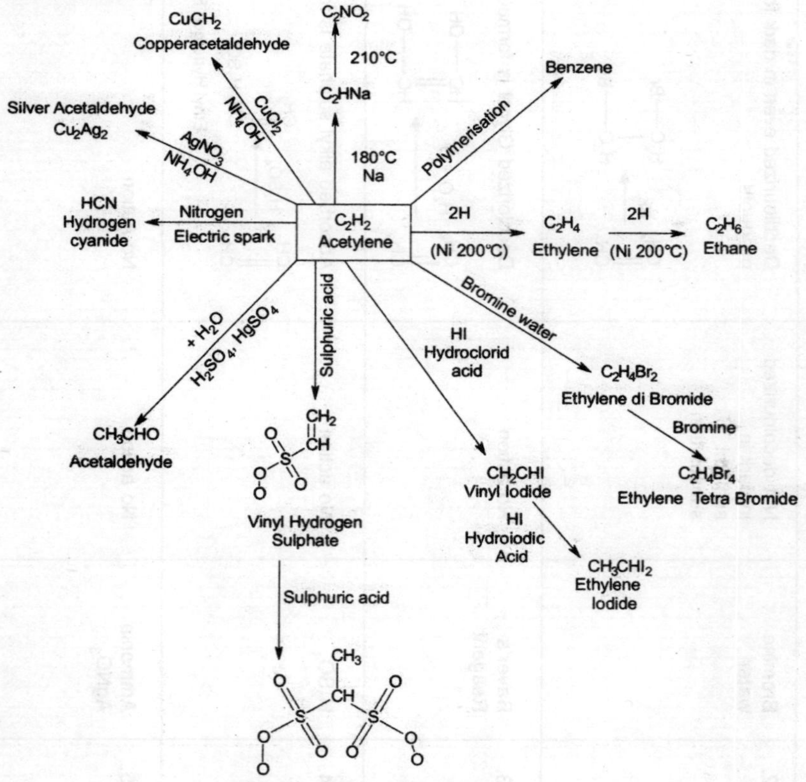
Reactions of Acetylene
CuCH2
Copperacetaldehyde
C2NO2
210°C
C2HNa
180°C
Na
Benzene
Polymerisation
Silver Acetaldehyde
Cu2Ag2
CuCl2
NH4OH
AgNO3
NH4OH
HCN
Hydrogen cyanide
Nitrogen
Electric spark
C2H2
Acetylene
2H
(Ni 200°C)
C2H4
Ethylene
2H
(Ni 200°C)
C2H6
Ethane
+ H2O
H2SO4, HgSO4
Sulphuric acid
Bromine water
HI
Hydroclorid acid
C2H4Br2
Ethylene di Bromide
Bromine
C2H4Br4
Ethylene Tetra Bromide
CH3CHO
Acetaldehyde
CH2
CH
Vinyl Hydrogen Sulphate
CH2CHI
Vinyl Iodide
HI
Hydroiodic Acid
CH3CHI2
Ethylene Iodide
Sulphuric acid
CH3
CH
Ethylidene Hydrogen Sulphate